NATURAL LANGUAGE PROCESSING AND ARTIFICIAL INTELLIGENCE

A Perspective Toward Current Trends, Challenges, and Applications

NATURAL LANGUAGE PROCESSING AND ARTIFICIAL INTELLIGENCE

A Perspective Toward Current Trends, Challenges, and Applications

Vikas Khullar, PhD
Aryan Chaudhary, PhD
Sunil Kumar, PhD
Editors

AAP | APPLE ACADEMIC PRESS

First edition published 2026

Apple Academic Press Inc.
1265 Goldenrod Circle, NE,
Palm Bay, FL 32905 USA

760 Laurentian Drive, Unit 19,
Burlington, ON L7N 0A4, Canada

CRC Press
2385 NW Executive Center Drive,
Suite 320, Boca Raton FL 33431

4 Park Square, Milton Park,
Abingdon, Oxon, OX14 4RN, UK

Library and Archives Canada Cataloguing in Publication

Title: Natural language processing and artificial intelligence : a perspective toward current trends, challenges, and applications / Vikas Khullar, PhD, Aryan Chaudhary, PhD, Sunil Kumar, PhD, editors.
Names: Khullar, Vikas, 1983- editor | Chaudhary, Aryan, editor. | Kumar, Sunil (Lecturer in information technology), editor.
Description: First edition. | Includes bibliographical references and index.
Identifiers: Canadiana (print) 2025021833X | Canadiana (ebook) 20250218437 | ISBN 9781998511266 (hardcover) | ISBN 9781998511273 (ebook)
Subjects: LCSH: Natural language processing (Computer science) | LCSH: Artificial intelligence.
Classification: LCC QA76.9.N38 N38 2026 | DDC 006.3/5—dc23

Library of Congress Cataloging-in-Publication Data

CIP data on file with US Library of Congress

ISBN: 978-1-998511-26-6 (hbk)
ISBN: 978-1-998511-27-3 (ebk)
DOI: 10.1201/9781998511273

About the Editors

Vikas Khullar, PhD
Associate Professor, Department of Computer Science and Engineering, Chitkara University, Punjab, India

Vikas Khullar, PhD, is currently working as an Associate Professor at the Department of Computer Science and Engineering, Chitkara University, Punjab, India. He has 15 years of experience in research and academia. During his doctorate dissertation, he developed new and unique assistive technologies for neurological disorders with the use of computer vision, machine learning, and deep learning along with IoT and embedded hardware. He has published more than 80 publications, and his major findings are published in reputed and indexed journals with significant impact factors. He has filed 50 design and utility patents to tackle several social and commercial problems. Dr. Khullar has worked on two innovative project grants, one on cyber security and another on vehicular security. He has authored and edited six books at various levels. At present, he is working in the domain of natural language processing, intelligent Internet of Things in society 5.0, assistive technologies for neurological disorders, federated learning, and trusted artificial intelligence. Dr. Khullar completed his PhD in Computer Science and Engineering.

Aryan Chaudhary, PhD
Chief Scientific Advisor, BioTech Sphere Research, India

Aryan Chaudhary, PhD, is an eminent figure in the healthcare and technology landscape, renowned for his groundbreaking contributions and innovative leadership. Formerly, the Research Head at Nijji HealthCare Pvt. Ltd., he has demonstrated unparalleled expertise in harnessing revolutionary technologies, including artificial intelligence (AI), deep learning, the Internet of Things (IoT), cognitive technology, and blockchain, to reshape the healthcare industry. With a relentless commitment to excellence, he has established himself as a thought leader, making a significant impact in the field. His extensive body of work includes influential academic papers on public health and digital health, published in prestigious international journals. A key focus of his research is the integration of IoT and sensor technology for efficient data collection through both one-time and ambulatory monitoring. A distinguished keynote speaker at numerous international and national conferences, he serves as a series editor for a *CRC* book series and as the editor of several books on biomedical science. His dedication to advancing scientific knowledge is further reflected in his role as a guest editor for special issues in renowned journals. Notably, he is currently serving as a guest editor for a special issue in the esteemed journal *EAI Endorsed Transactions on AI and Robotics* and has joined the editorial board of *Biomedical Science and Clinical Research (BSCR).* Recognized for his exceptional contributions, he has received prestigious accolades, including the *Most Inspiring Young Leader in HealthTech Space 2022* award by *Business Connect.* His influence extends to senior memberships in various international science associations, underscoring his impact in the field. Recently, he was appointed Chair of an Association for Computing Machinery (ACM) professional chapter and nominated as a member of the ACM U.S. Technology Policy Committee (USTPC) in the AI & Algorithms division. Further adding to his distinguished portfolio, he has been appointed as a Technical Committee Member for the 2024–2025 tenure with the Institute of Electrical and Electronics Engineers (IEEE) Robotics and Automation Society. Additionally, he has been elected as a Regular Fellow Member of the International Organization for Academic and Scientific Development (IOASD) and admitted as a Member of the International Association for Carbon Capture (MIACC). His influence also extends to serving as a Technical Committee Member of the IEEE Consumer Technology Society (CTSoc) for Healthcare and Wellbeing (CSH).

Sunil Kumar, PhD
Associate Professor, School of Information Technology,
AURO University, Surat, Gujarat, India

Sunil Kumar, PhD, is an Associate Professor in the School of Information Technology, AURO University, Surat, Gujarat India. He is an academician and collaborative researcher with over 16 years of teaching and research experience in computer science and engineering, having several international publications from IEEE Xplore, ACM, Springer Nature, Elsevier, and IGI Global, USA. He is a certified IBM trainer for Python, machine learning, deep learning, and IT network security. He is also a certified Oracle Java Programmer, Ethical Hacker, and AWS Certified Cloud Solution Architect Associate. Dr. Kumar has delivered and conducted many expert talks, guest lectures, and workshops in the field of cyber security, machine learning, and data analytics. He has received best paper awards at many international conferences. He has published two textbooks on artificial intelligence and one on cyber security. He has also guided several undergraduate, postgraduate, and PhD research works. He was granted five national and international patents. He is a reviewer for many international SCI/Scopus-indexed journals and is an active member of Computer Society of India (CSI) and the Institute of Electrical and Electronics Engineers (IEEE),and a lifetime member of the Institution of Engineers (India).

Contents

Contributors

Abhineet Anand
Apex Institute of Technology–CSE, Chandigarh University, Mohali, Punjab, India

Mohit Angurala
Apex Institute of Technology–CSE, Chandigarh University, Mohali, Punjab, India

Sangeetha Annam
Chitkara University Institute of Engineering and Technology, Chitkara University, Punjab, India

Anushka
Chitkara University Institute of Engineering and Technology, Chitkara University, Punjab, India

Ruqqaiya Aziz
NED University, Karachi, Pakistan

Ekta Bhaggi
GNA University, Phagwara, Punjab, India

H. G. Charitra
Assistant Professor, School of Liberal Arts and Sciences, Presidency University, Bengaluru, India

Sachin Chaudhary
School of Computer Science and Applications, IIMT University, Meerut, Uttar Pradesh, India

Mamta Devi
CT Group of Institutions, Punjab, India

Shikha Gupta
Apex Institute of Technology–CSE, Chandigarh University, Mohali, Punjab, India

V. G. Hemant
School of Computer Science Engineering, Lovely Professional University, Punjab, India

Noman Islam
Iqra University, Karachi, Pakistan

Christian Kaunert
International Security, Dublin City University, Ireland

Gagandeep Kaur
Chitkara University Institute of Engineering and Technology, Chitkara University, Punjab, India

Prabhjot Kaur
Chitkara University Institute of Engineering and Technology, Chitkara University, Punjab, India

Veerpal Kaur
Department of Computer Science and Engineering, Lovely Professional University, Punjab, India

Vikas Khullar
Chitkara University Institute of Engineering and Technology, Chitkara University, Punjab, India

Komal
School of Computer Science and Engineering, Lovely Professional University, Punjab, India

Bhupendra Kumar
School of Computer Science and Applications, IIMT University, Meerut, Uttar Pradesh, India

Sunil Kumar
School of Information Technology, Auro University, Gujarat, India

Meghna
School of Civil Engineering, IIMT University, Meerut, Uttar Pradesh, India

Anand Muni Mishra
Department of Computer Science and Engineering, Chandigarh University, Mohali, Punjab, India

Mafaza Mohi
NED University, Karachi, Pakistan

Laiba Nadeem
NED University, Karachi, Pakistan

G. Ranganatham
Associate Professor, Department of Commerce, School of Business Studies, Central University of Karnataka, India

Anshu Sharma
School of Computer Science and Engineering, Lovely Professional University, Punjab, India

Anurag Sharma
Department of Computer Science and Engineering, GNA University, Phagwara, Punjab, India

Kewal Krishan Sharma
School of Computer Science and Applications, IIMT University, Meerut, Uttar Pradesh, India

Neharika Sharma
Department of Computer Science and Engineering, GNA University, Phagwara, Punjab, India

Vikas Sharma
Department of Computer Applications, SRM Institute of Science and Technology,
Delhi NCR Campus, Ghaziabad, U.P., India

Ghazala Shafi Sheikh
Iqra University, Karachi, Pakistan

Bhupinder Singh
Sharda School of Law, Sharda University, Greater Noida, Uttar Pradesh, India

Davinder Singh
Department of Computer Science and Engineering, GNA University, Phagwara, Punjab, India

K. Umashankar
Head of Learning and Development, Presidency University, Bengaluru, India

Tarun Kumar Vashishth
School of Computer Science and Applications, IIMT University, Meerut, Uttar Pradesh, India

Varsha
School of Computer Science and Engineering, Lovely Professional University, Punjab, India

Parag Verma
Datafoundry Private Limited, Bangalore, Karnataka, India; AICG, Algihaz Holding, Riyadh, Saudi Arabia

Yagyata
Chitkara University Institute of Engineering and Technology, Chitkara University, Punjab, India

Preface

Artificial intelligence (AI) and natural language processing (NLP) are among the few fields within the vast landscape of technological advancements that have experienced rapid growth and possess the potential to drive transformative change. The convergence of these two disciplines has opened up previously unexplored areas in human-computer interaction, fundamentally reshaping the way we express thoughts, process data, and make decisions. This book takes readers on an in-depth journey through the intricate subfields of AI-driven NLP, exploring its methodologies, emerging trends, challenges, and applications across diverse domains.

In the first chapter, we explore the foundational principles of AI for NLP, laying the groundwork for the rest of the book. From rule-based systems to modern deep learning techniques, this chapter traces the evolution of AI-driven approaches to understanding and generating human language. Chapter 2 delves into the dynamic landscape of AI-based NLP, analyzing prevailing methodologies, emerging trends, and ongoing challenges. Given the continuous advancements in this field, staying updated is essential for both professionals and enthusiasts. The symbiotic relationship between machine learning and NLP is the focus of Chapter 3. This chapter provides an in-depth examination of the interdependencies that drive progress in both fields, shaping the future of intelligent language processing systems. Successfully navigating the AI landscape requires a profound understanding of NLP-specific structures and algorithms, a journey that begins in Chapter 4. This chapter serves as a roadmap for building robust NLP systems, covering everything from traditional models to cutting-edge deep learning architectures.

Chapter 5 offers a comprehensive evaluation of techniques that integrate linguistic principles with modern AI frameworks, bridging the gap between language structures and deep learning models through a detailed comparative analysis. Chapter 6 dissects AI architectures and algorithms within NLP ecosystems, shedding light on the intricate relationships between technology, data, and real-world applications. In an era of information overload, distinguishing between authentic and misleading news is crucial. Chapter 7 explores how machine learning and deep learning techniques can be leveraged for accurate classification, safeguarding the integrity of information ecosystems. Chapter 8 harnesses the power of big data for information retrieval and text mining, enabling the extraction of valuable insights from vast repositories of textual data.

In Chapter 9, a conceptual model is introduced to enhance organizational success by integrating emotional intelligence with AI. Chapter 10 shifts the focus to the intersection of future mobility and intelligent virtual assistants, showcasing the transformative impact of AI-driven chatbots on digital consumer experiences. Chapter 11 provides a deep dive into Federated Learning in NLP, offering insights into secure and efficient strategies for collaborative model training across decentralized data sources.

Chapter 12 examines large language model (LLM)-based human-like content generation, demonstrating AI's ability to produce articles, social media posts, and marketing content with remarkable fluency and coherence. Chapter 13 explores the capabilities of LLM-powered chatbots and virtual assistants, which enable interactive and human-like conversations, revolutionizing user experiences. Chapter 14 highlights the potential of transformer-based Optical Character Recognition (OCR) in streamlining the extraction of handwritten medical information, paving the way for advancements in medical data management. Finally, Chapter 15 delves into the fascinating domain of speech emotion detection using deep learning, uncovering the subtle patterns that underlie human emotional expression.

As we embark on this journey through the multidimensional world of AI-driven NLP, we encourage readers to delve deeper, challenge assumptions, and explore new frontiers in this rapidly evolving field. Let us work together to unlock the mysteries of language, cognition, and computing, shaping a future where intelligence knows no boundaries.

CHAPTER 1

Natural Language Processing and Artificial Intelligence: A Perspective Toward Current Trends, Challenges, and Applications

MOHIT ANGURALA[1,*], SHIKHA GUPTA[1], ABHINEET ANAND[1], and VIKAS KHULLAR[2]

[1]*Apex Institute of Technology–CSE, Chandigarh University, Mohali, Punjab, India*

[2]*Chitkara University Institute of Engineering and Technology, Chitkara University, Punjab, India*

**Corresponding author.*

ABSTRACT

This chapter digs into the dynamic and transformational convergence of natural language processing (NLP) and artificial intelligence (AI), offering a thorough examination of current trends, problems, and applications in this emerging subject. We investigate the impact of deep learning (DL) and neural networks, highlighting their critical significance in tasks ranging from language translation to sentiment analysis. Transfer learning (TL) appears as an important trend, allowing pre-trained models to be fine-tuned for specific NLP applications. The integration of multimodal AI, which combines text with visuals and audio, is being investigated for its influence on a variety of applications. However, difficulties exist, such as linguistic ambiguity, ethical problems, and the interpretability of complicated models. Despite these challenges, the applications of NLP and AI are extensive, including chatbots, sentiment analysis, and language translation.

Natural Language Processing and Artificial Intelligence: A Perspective Toward Current Trends, Challenges, and Applications. Vikas Khullar, Aryan Chaudhary, & Sunil Kumar (Eds.)

DOI: 10.1201/9781998511273-1

1.1 INTRODUCTION

Natural language processing (NLP) and artificial intelligence (AI) have emerged as revolutionary forces in technology, altering how people connect with and comprehend machines. The convergence of these fields has fueled innovation, opening up new avenues and paving the way for intelligent systems capable of interpreting, creating, and responding to human language. This chapter takes a detailed look at the current trends, difficulties, and applications in the dynamic world of NLP and AI. Transfer learning (TL), a cornerstone of today's NLP and AI scene, has brought a fresh way to model creation. Pre-trained models, such as BERT and GPT, serve as fundamental knowledge bases for certain tasks, reducing the requirement for substantial training on task-specific datasets. This method not only speeds up the creation process, but it also makes use of the richness of information included in these models, pushing the frontiers of what is possible in language-related activities [1, 2].

Another important trend is the incorporation of multimodal AI, which recognizes the necessity of mixing text with various modalities such as visuals and audio. This comprehensive approach improves AI systems' capacity to perceive and respond to different inputs, matching the complexities of human communication. Image captioning, speech-to-text, and visual question-answering are examples of how multimodal AI might expand the reach of intelligent systems beyond standard text-based interfaces. Despite this tremendous advancement, obstacles in the development and implementation of NLP and AI systems remain. Ambiguity and context awareness continue to be severe barriers, particularly in jobs requiring a nuanced grasp of linguistic subtleties. Resolving these issues is critical for the future progress of AI systems, ensuring that they can efficiently navigate the complexity of human communication.

Ethical concerns are on the horizon, necessitating a rigorous study of bias in NLP models. The potential for biased language processing to promote social stereotypes raises ethical concerns that must be addressed in order to support responsible AI research. Furthermore, the lack of interpretability in complicated deep learning (DL) models raises questions about accountability and transparency, particularly in real-world applications like healthcare and finance [3].

As we traverse the complexities of current trends and face obstacles, the applications of NLP and AI are expanding across several fields.

The impact is profound and multifaceted, ranging from the proliferation of chatbots and virtual assistants, which improve customer service and automate routine tasks, to sentiment analysis, which provides businesses with insights into public opinion, and language translation, which breaks down global communication barriers. Finally, the combination of NLP and AI is an innovation frontier with the potential to reshape human-machine interactions. As we investigate cutting-edge trends, confront obstacles, and discover applications, we must take a responsible and ethical approach to ensure that the revolutionary power of NLP and AI serves society. This chapter functions as a route map.

The NLP and AI environment is constantly changing, driven by continuing research, technical improvements, and an increasing awareness of the complexity inherent in language processing. We see a convergence of advances that not only improve the capacities of intelligent systems but also change the possibilities of human-machine connection as we dive into the current trends influencing this dynamic subject.

1.2 NEURAL NETWORKS AND DEEP LEARNING

The widespread effect of DL, particularly via the lens of neural networks, is at the forefront of present NLP and AI advances. These complex models, inspired by the structure and function of the human brain, have proven useful in interpreting linguistic subtleties. RNNs and transformer architectures, in particular, have emerged as cornerstones of language comprehension, allowing machines to grasp contextual information and relationships within text. DL models provide exceptional accuracy in tasks such as language translation, sentiment analysis, and text summarization [4]. The capacity of neural networks to learn hierarchical representations of language helps to design systems that can not only interpret but also generate human-like writing, allowing for more natural and contextually aware interactions between humans and machines.

1.2.1 TRANSFER LEARNING

The growth of TL in NLP has ushered in a paradigm change in model creation. Pre-trained models, such as bidirectional encoder representations from transformers (BERT) and generative pre-trained transformers (GPT), act as knowledge libraries that have been trained on large datasets and cover a wide range of linguistic patterns and structures. Because of this richness of prior information, these models may be fine-tuned for specific tasks with smaller datasets, minimizing the need for lengthy training.

Transfer learning is especially useful when labeled data is limited or training from scratch is computationally demanding. This trend not only speeds up the construction of NLP models but also allows knowledge to be transferred across domains, resulting in more robust and adaptive language processing systems.

1.2.2 MULTIMODAL ARTIFICIAL INTELLIGENCE (AI) INTEGRATION

The incorporation of multimodal AI has gained popularity in the quest for a more thorough understanding of human communication [5]. This method goes beyond typical text-based analysis by bringing other modalities into the language processing paradigm, such as pictures, audio, and video. For example, the combination of text and images has resulted in applications such as picture captioning and visual question answering, in which AI systems can comprehend and create written descriptions based on visual inputs.

Multimodal AI not only deepens language understanding but also broadens the range of applications. Speech-to-text and gesture recognition are two examples of how combining various modalities increases AI systems' adaptability, allowing them to accommodate a greater range of user inputs and interactions.

1.2.3 CONTINUAL ADVANCEMENTS IN PRE-TRAINING MODELS

The environment of pre-training models is constantly evolving, with breakthroughs and adjustments. Text-to-text transfer transformer (T5) models, for example, stress a unifying framework for multiple NLP tasks, considering them as a text generation issue. This transition away from task-specific structures and toward more adaptable, all-encompassing frameworks demonstrates a shift away from task-specific architectures and toward more generic models that can adapt to varied activities [6]. The trend toward more advanced pre-training models indicates not only a broader understanding of language but also the potential for AI systems to develop a more generalized intelligence capable of adapting to a wide range of linguistic tasks without the need for task-specific architectures.

1.2.4 ATTENTION MECHANISMS AND EXPLAINABILITY

Another important development is the integration of attention processes in NLP models, which contributes to improved performance and interpretability. Models can use attention mechanisms to focus on certain regions of input sequences, simulating human cognitive processes. This not only increases the speed of language processing jobs, but it also provides some transparency, allowing users to understand how the model makes certain judgments.

Explainability is regarded as an important feature of AI research, particularly in applications with real-world implications. As models become more complicated, the capacity to comprehend and explain their judgments becomes increasingly important in establishing confidence and guaranteeing accountability.

1.2.5 ZERO-SHOT AND FEW-SHOT LEARNING

Emerging methods like zero-shot and few-shot learning address the difficulty of adapting models to new tasks with less labeled data. In zero-shot learning, models are trained to perform tasks without explicit examples, relying on generalization and prior knowledge from pre-training. Few-shot learning, on the other hand, entails training models with a limited number of instances, allowing them to be more adaptive to a wider range of tasks. These techniques show a shift toward more flexible and adaptive models that can quickly learn and adapt to new tasks or domains despite having minimal labeled data. Finally, contemporary NLP and AI advances demonstrate a concentrated attempt to push the bounds of language understanding and creation. These developments, which range from the complicated structures of neural networks to the flexibility of TL and the incorporation of multimodal AI, collectively contribute to the creation of more sophisticated, adaptable, and contextually aware intelligent systems [7]. As research and innovation continue to propel the field forward, the future offers even more game-changing breakthroughs, further reinforcing NLP and AI's position in determining how humans communicate with and through machines.

1.3 CHALLENGES IN NATURAL LANGUAGE PROCESSING (NLP) AND ARTIFICIAL INTELLIGENCE (AI)

Natural language processing (NLP) has various problems that impede its smooth incorporation into a variety of applications. This section dives into crucial factors such as ambiguity and context comprehension, ethical problems and biases, lack of interpretability, explainability, accountability, and the dynamic nature of language in investigating the terrain of these challenges.

1.3.1 AMBIGUITY AND CONTEXT UNDERSTANDING

Despite great advances in NLP, the difficulty of ambiguity and context comprehension remains a formidable impediment. Human language is intrinsically complicated, and the same words or phrases can have several meanings depending on the situation. NLP models frequently fail to grasp linguistic nuances, posing difficulties in effectively detecting user intent and delivering contextually relevant replies. The ambiguity introduced by the intricacy of human communication is strongly ingrained in language complexities, comedy, and cultural nuances. To resolve this uncertainty, models must go beyond surface-level patterns and integrate a deeper grasp of context, idioms, and colloquial terms. Attaining this degree of sophistication is a continuing research problem, especially in applications.

1.3.2 ETHICAL AND BIAS CONCERNS

With the increasing integration of AI into different aspects of society, the essential problem of ethical issues and biases in NLP models has come to the fore. Language models trained on large datasets unintentionally acquire data biases, mirroring and possibly increasing existing social prejudices. This raises issues regarding fair and unbiased language processing, especially when these systems are used in sensitive domains like employment procedures, judicial systems, and content regulation [8].

Addressing ethical problems entails not just recognizing the presence of prejudices but also actively attempting to mitigate and eliminate them. This demands a comprehensive strategy that includes varied and representative dataset curation, transparent model-building methods, and continuous monitoring to detect and correct biases as they appear. To set norms and standards for responsible AI deployment, engineers, ethicists, and legislators must collaborate to strike the correct balance between innovation and ethical responsibility.

1.3.3 LACK OF INTERPRETABILITY

The black-box nature of DL models presents a huge interpretability difficulty. Understanding how and why a model makes a choice is critical, especially when the conclusions have real-world implications. Lack of interpretability can undermine user confidence, inhibit regulatory compliance, and raise ethical problems in applications like as healthcare diagnostics, financial decision-making, and legal procedures.

Interpretability is a complicated problem with technological and ethical implications. Technical attempts are being undertaken to build explainable AI (XAI) approaches, such as attention mechanisms and model-agnostic methodologies, to shed light on complicated models' decision-making processes. Obtaining interpretability without losing model performance, on the other hand, remains a tricky balance.

1.3.4 EXPLAINABILITY AND ACCOUNTABILITY

The greater difficulty of explainability and accountability in AI systems is related to the lack of interpretability [9]. As AI applications reach vital industries such as healthcare and finance, transparency in decision-making becomes crucial. Users should be able to grasp the reasoning behind AI-driven judgments, whether they are medical specialists, financial analysts, or end users, in order to trust and rely on these systems.

To achieve explainability, it is necessary to not only make the decision-making process clear but also to hold AI systems accountable for their actions. It is critical to establish clear lines of accountability, especially in cases where AI impacts significant choices. This involves determining who is accountable for the results of AI-driven choices and providing processes for dealing with errors, biases, or unexpected effects.

1.3.5 ADAPTING TO DYNAMIC AND EVOLVING LANGUAGE

Human language is dynamic and continually changing because of societal upheavals, new slang, and new phrases. Adapting NLP models trained on static datasets to these dynamic language changes is difficult. Language's fast growth creates a temporal gap between training data and real-time language utilized by users, potentially causing mistakes in language interpretation and production [10].

Adapting to dynamic language involves systems for continual learning that can absorb new linguistic patterns and trends. For NLP models, research on dynamic adaptation, continuous learning, and lifetime learning is critical to ensuring that these systems stay relevant and successful in changing linguistic landscapes. Understanding the influence of cultural and geographical differences on language dynamics adds another degree of complication to this task. Finally, resolving these issues in NLP and AI necessitates a multifaceted and multidisciplinary approach. To guarantee that intelligent systems push the frontiers of language comprehension safely, ethically, and openly, technological breakthroughs, ethical considerations, and legal frameworks must all align [11]. Finding answers to these difficulties will be critical in unleashing the full potential of NLP and AI for the benefit of society as the field advances.

1.4 APPLICATIONS OF NATURAL LANGUAGE PROCESSING (NLP) AND ARTIFICIAL INTELLIGENCE (AI)

NLP and AI have moved beyond theoretical frameworks to become important components of a wide range of applications that have a significant influence on our daily lives. We will look at three significant applications that demonstrate the adaptability and revolutionary impact of NLP and AI in this exploration: chatbots and virtual assistants, sentiment analysis, and language translation.

1.4.1 CHATBOTS AND VIRTUAL ASSISTANTS

Chatbots and virtual assistants represent a paradigm change in human-computer interaction, leveraging the capabilities of NLP and AI to deliver smooth and intuitive communication interfaces. Chatbots, powered by advanced language models, engage users in natural language discussions, answering questions, providing help, and automating regular activities. Chatbots have become vital in customer support, providing rapid replies to questions, assisting customers through troubleshooting processes, and improving the overall user experience. Virtual assistants, such as Amazon's Alexa and Apple's Siri, use conversational AI to operate smart gadgets, play music, and perform a variety of activities, making the notion of a digital helper a realistic reality. The capacity of chatbots and virtual assistants to recognize user intent, assess context, and create contextually relevant replies is critical to their success. Continuous advances in NLP, particularly in the areas of intent detection and dialogue management, are refining these applications, making them more complex, adaptive, and capable of comprehending human subtleties.

1.4.2 SENTIMENT ANALYSIS

Sentiment analysis, often known as opinion mining, is a strong NLP tool that extracts and understands opinions conveyed in textual data. Sentiment analysis is used by businesses and organizations to acquire insights into public opinion, consumer feedback, and social media conversations [12–14]. The capacity to differentiate between positive, negative, and neutral attitudes enables data-driven decision-making, focused marketing tactics, and proactive consumer involvement. Sentiment methods in social media monitoring sift through massive volumes of user-generated material to evaluate public opinion regarding goods, businesses, or events. This real-time feedback loop enables firms to respond to client issues quickly, adjust marketing efforts, and create a favorable brand image. Furthermore, sentiment analysis is used in financial markets, where the capacity to examine news stories, social media posts, and financial data assists investors in determining market mood. This proactive research helps investors make better-informed investment decisions, resulting in a more dynamic and responsive financial landscape.

1.4.3 LANGUAGE TRANSLATION

Language translation exemplifies NLP and AI's worldwide influence, breaking down linguistic boundaries and enabling cross-cultural contact. Modern translation systems, powered by neural machine translation models, have considerably increased translation accuracy and fluency. DL algorithms are used by platforms such as Google Translate to grasp the context of words, capturing not just literal translations but also the subtleties and idiomatic expressions that are inherent in good communication. This has ramifications not only for personal communication but also for multinational corporations. Language translation in healthcare promotes communication between patients and healthcare professionals by removing language barriers and guaranteeing correct information transmission. Similarly, language translation solutions help organizations interact with varied markets, localize information, and grow their worldwide reach in international commerce.

1.4.4 TEXT SUMMARIZATION

Text summarization is a complex NLP technique that entails condensing enormous amounts of information into brief, logical summaries. The necessity for automatic summarization has become more obvious as the amount of textual material on the internet has grown exponentially [1, 15, 16]. Text summarization algorithms in news aggregation evaluate news stories and provide succinct summaries, giving readers a rapid overview of relevant events. This not only saves time but also assists in information digestion, which is especially important in today's fast-paced news environment. Text summarization technologies aid academics in navigating enormous amounts of material, extracting significant ideas from articles and research papers. This speeds up the literature review process, allowing researchers to keep up to date on advancements in their disciplines.

1.4.5 HEALTHCARE APPLICATIONS

NLP and AI have found extensive applications in the healthcare sector, altering how doctors analyze, interpret, and use textual data [3]. NLP is useful for transforming unstructured clinical notes into structured data, allowing for more effective clinical recordkeeping. This not only decreases the administrative load on healthcare workers, but it also improves patient data accessibility and usability for research and analysis. To aid in illness diagnosis, NLP models are used to examine medical records, pathology reports, and other textual data. These models help to provide more accurate and prompt diagnoses by recognizing trends and extracting pertinent information, potentially improving patient outcomes. NLP assists in the sorting of enormous databases of scholarly literature, patents, and clinical trial reports in pharmaceutical research. This speeds up the drug development process by finding prospective drug candidates, understanding molecular interactions, and staying up to date on the most recent advances in the area. As telemedicine has grown in popularity, virtual health assistants that use NLP to engage with patients have emerged. These virtual assistants may schedule appointments, remind people to take their medications, and provide basic medical information, improving the accessibility and efficiency of healthcare services.

1.4.6 LEGAL AND COMPLIANCE

NLP methods enable legal practitioners to comb through huge legal databases, case law, and legislation to retrieve essential information. This speeds up the legal research process, allowing attorneys to develop better arguments and provide more informed advice to clients. NLP is used to analyze legal documents and contracts. These technologies may extract essential clauses, detect potential hazards, and ensure legal compliance. This saves time while also lowering the danger of oversight in complicated legal agreements. NLP systems are used to monitor regulatory changes, ensuring that legal teams are always up to date on changing compliance requirements. This proactive compliance management strategy assists firms in mitigating risks and adapting to changing legal climates. NLP is used to analyze legal texts and contracts. These technologies may extract essential clauses, detect potential hazards, and ensure legal compliance. This saves time while also lowering the danger of oversight in complicated legal agreements.

1.4.7 EDUCATIONAL TECHNOLOGY

NLP and AI applications in education are transforming the learning experience for both students and instructors. NLP-driven intelligent tutoring systems assess students' interactions and reactions, adapting learning experiences to individual requirements. These systems can provide individualized feedback, adaptive learning routes, and real-time evaluations, thereby improving the efficacy of educational activities. NLP applications make automated grading of assignments and assessments possible. These systems can give fast and objective feedback to students by evaluating written responses, relieving educators from manual grading responsibilities and allowing them to focus on more customized aspects of instruction. Language learning apps use NLP to test language competency, provide targeted exercises, and deliver interactive courses. Language acquisition becomes more entertaining and successful when these apps adjust to individual learning methods.

1.4.8 FINANCIAL SERVICES

NNLP and AI technologies are transforming operations in the financial industry, such as data analysis, client interactions, and risk management. For algorithmic trading, NLP is used in financial analysis and news sentiment analysis. These systems feed algorithmic trading techniques by extracting insights from news articles, social media, and financial reports, thereby improving decision-making in turbulent markets. For customer service and engagement, financial institutions use NLP-powered chatbots. These virtual assistants can answer routine questions, provide account information, and assist with typical transactions, thus enhancing client experiences and operational efficiency. NLP systems help identify fraud by evaluating textual data from financial transactions. Unusual linguistic patterns, such as anomalies in transaction descriptions or account activity, may alert financial institutions to suspected fraudulent conduct, allowing them to take prompt preventive action.

1.4.9 FUTURE DIRECTIONS AND EMERGING APPLICATIONS

The applications described provide a view into the current environment of NLP and AI, but the area is still evolving, with new applications ready to redefine its scope and influence. NLP and AI applications are becoming more common in the field of human resources and recruitment. NLP algorithms can help with resume screening by extracting relevant information and recognizing essential credentials, lowering the risk of bias in the first screening phases. Furthermore, using NLP to analyze applicant replies and gain insights into communication skills and cultural fit, interview analysis enables recruiters to make more educated recruiting decisions. Another frontier where NLP and AI are being used is in environmental monitoring and climate change. The study of scientific literature assists researchers in finding patterns, analyzing intervention effects, and contributing to a worldwide knowledge of environmental concerns [17–20]. Policy and advocacy NLP applications aid in the monitoring and analysis of policy papers, allowing environmental groups and policymakers to stay informed about legislative changes and advocate for sustainable practices. In the field of Mental Health and Well-Being, NLP applications are being used to address issues, such as using sentiment analysis techniques to track mental health trends using textual data such as social media postings or journal entries [21, 22]. Additionally, NLP-driven therapeutic chatbots are being created to provide compassionate chats, coping skills, and resources for mental health development.

1.5 CONCLUSION

In conclusion, this chapter's investigation of NLP and AI provides a thorough grasp of the current trends, problems, and applications in this dynamic and transformational discipline. The integration of DL and neural networks, as demonstrated by models such as BERT and GPT, has had a substantial influence on language-related tasks such as translation and sentiment analysis. TL has emerged as a critical concept, allowing for the fine-tuning of pre-trained models for specific NLP applications, thereby speeding development and improving flexibility. The chapter also provides information on the developing field of multimodal AI, in which the combination of text with graphics and music expands AI systems' capabilities. Despite these advances, challenges such as ambiguity and context comprehension, ethical concerns about biases in NLP models, and the intrinsic lack of interpretability in sophisticated DL models remain. As we navigate the panorama of problems, NLP and AI applications span a wide range of areas. These applications demonstrate the pervasive impact of NLP and AI, from the transformative influence of chatbots and virtual assistants in reshaping human-computer interactions to the invaluable insights derived from sentiment analysis for businesses and language translation breaking down global communication barriers. The chapter delves into the complexities of NLP and AI challenges, highlighting the importance of addressing ambiguity, ethical considerations, and interpretability issues. Ambiguity and context knowledge continue to be key barriers to effective communication, necessitating better insights into linguistic nuances. Ethical issues, notably biases in language processing, require a proactive approach to ensuring fair and impartial outcomes, particularly in crucial sectors such as employment and finance. The lack of interpretability in DL models complicates accountability and transparency, especially in high-stakes applications such as healthcare.

Following that, the chapter delves into emergent applications, emphasizing the growing relevance of NLP and AI in a variety of industries. NLP and AI continue to redefine possibilities, from transforming human resources and recruiting procedures with better resume screening and interview analysis to addressing environmental concerns through scientific literature analysis and policy lobbying. The promise of these technologies in enhancing well-being is demonstrated by mental health applications that use sentiment analysis for monitoring and therapeutic chatbots for assistance.

KEYWORDS

- **artificial intelligence**
- **deep learning**
- **drug development**
- **molecular interactions**
- **multimodal AI**
- **natural language processing**
- **pharmaceutical research**
- **transfer learning**

REFERENCES

1. Andreev, N. D. (1967). The intermediary language as the focal point of machine translation. In A. D. Booth (Ed.), *Machine Translation* (pp. 3–27). North Holland Publishing Company.

2. Borst, F., et al. (1989). Analyse automatique de comptes rendus d'hospitalisation. In *Informatique Et Santé, Informatique Et Gestion Des Unités De Soins, Comptes Rendus Du Colloque AIM-IF* (pp. 246–256)
3. Chi, E., Lyman, M. S., Sager, N., Friedman, C., & Macleod, C. (1985). A database of computer-structured narrative: Methods of computing complex relations. In *Proceedings of the Annual Symposium on Computer Application in Medical Care* (p. 221).
4. Ahonen, H., Heinonen, O., Klemettinen, M., & Verkamo, A. I. (1998). Applying data mining techniques for descriptive phrase extraction in digital document collections. In *Research and Technology Advances in Digital Libraries, 1998. ADL 98. Proceedings. IEEE International Forum on* (pp. 2–11).
5. Bangalore, S., Rambow, O., & Whittaker, S. (2000). Evaluation metrics for generation. In *Proceedings of the First International Conference on Natural Language Generation-Volume 14* (pp. 1–8).
6. Baud, R. H., Rassinoux, A. M., & Scherrer, J. R. (1991). Knowledge representation of discharge summaries. In *AIME 91* (pp. 173–182).
7. Baud, R. H., Rassinoux, A. M., & Scherrer, J. R. (1992). Natural language processing and semantical representation of medical texts. *Methods of Information in Medicine, 31*(2), 117–125.
8. Briscoe, E. J., Grover, C., Boguraev, B., & Carroll, J. (1987). A formalism and environment for the development of a large grammar of English. In *IJCAI 87* (pp. 703–708).
9. Berger, A. L., Della Pietra, S. A., & Della Pietra, V. J. (1996). A maximum entropy approach to natural language processing. *Computational Linguistics, 22*(1), 39–71.
10. Bondale, N., Maloor, P., Vaidyanathan, A., Sengupta, S., & Rao, P. V. (1999). Extraction of information from open-ended questionnaires using natural language processing techniques. *Computer Science and Informatics, 29*(2), 15–22.
11. Benson, E., Haghighi, A., & Barzilay, R. (2011). Event discovery in social media feeds. In *Proceedings of the 49th Annual Meeting of the Association for Computational Linguistics: Human Language Technologies* (Vol. 1, pp. 389–398).
12. Bengio, Y., Ducharme, R., & Vincent, P. (2001). A neural probabilistic language model. In *Proceedings of NIPS.*
13. Cho, K., Van Merriënboer, B., Bahdanau, D., & Bengio, Y. (2014). *On the Properties of Neural Machine Translation: Encoder-Decoder Approaches.* arXiv preprint arXiv: 1409.1259.
14. Bahdanau, D., Cho, K., & Bengio, Y. (2015). Neural machine translation by jointly learning to align and translate. In *ICLR 2015.*
15. Androutsopoulos, I., Paliouras, G., Karkaletsis, V., Sakkis, G., Spyropoulos, C. D., & Stamatopoulos, P. (2000). *Learning to Filter Spam E-Mail: A Comparison of a Naive Bayesian and a Memory-Based Approach.* arXiv preprint cs/0009009.
16. Baclic, O., Tunis, M., Young, K., Doan, C., Swerdfeger, H., & Schonfeld, J. (2020). Artificial intelligence in public health: Challenges and opportunities for public health made possible by advances in natural language processing. *Canada Communicable Disease Report, 46*(6), 161.
17. Blanzieri, E., & Bryl, A. (2008). A survey of learning-based techniques of email spam filtering. *Artificial Intelligence Review, 29*(1), 63–92.
18. Chomsky, N. (1965). *Aspects of the Theory of Syntax*. MIT Press.

19. Carreras, X., & Marquez, L. (2001). *Boosting Trees for Anti-Spam Email Filtering. arXiv preprint cs/0109015.*
20. Choudhary, N. (2021). LDC-IL: The Indian repository of resources for language technology. *Language Resources & Evaluation, 55*, 855–867.
21. Chalkidis, I., Fergadiotis, M., Malakasiotis, P., Aletras, N., & Androutsopoulos, I. (2020). *LEGAL-BERT: The Muppets Straight Out of Law School. arXiv preprint arXiv:2010.02559.*
22. Chouikhi, H., Chniter, H., & Jarray, F. (2021). Arabic sentiment analysis using BERT model. In *International Conference on Computational Collective Intelligence* (pp. 621–632).

CHAPTER 2

Artificial Intelligence-Based NLP: Methods, Trends, and Challenges

PRABHJOT KAUR[1,*] and ANAND MUNI MISHRA[2]

[1]*Chitkara University Institute of Engineering and Technology, Chitkara University, Punjab, India*

[2]*Department of Computer Science and Engineering, Chandigarh University, Mohali, Punjab, India*

**Corresponding author.*

ABSTRACT

Unsupervised or semi-supervised learning approaches are receiving more and more attention in "natural language processing (NLP)" research. These learning methods can learn from data that hasn't been manually annotated with the essential solutions or by merging manually and automatically annotated data. The goal of NLP, which sits in the middle of computational linguistics and computer science, is to transform spoken and written natural human languages into structured, reusable data. A text's meaning can be ascertained using linguistic, statistical, and "artificial intelligence (AI)" techniques, and NLP can even be used to generate responses that resemble those of a human being. In recent years, NLP has drawn a lot of interest for its ability to computationally represent and analyze human language. It now has a wide range of uses, including machine translation, email spam detection, information extraction, and summarization, as well as medical applications and question-answering. "AI," one of these technologies, has demonstrated the significant value of boosting immersive experiences and enabling human-like intelligence in virtual agents. This chapter also makes

Natural Language Processing and Artificial Intelligence: A Perspective Toward Current Trends, Challenges, and Applications. Vikas Khullar, Aryan Chaudhary, & Sunil Kumar (Eds.)

DOI: 10.1201/9781998511273-2

a useful effort to study the function of AI, including deep learning (DL) architectures and machine learning (ML) algorithms, in collaboration with NLP. The thorough analysis of AI-based approaches in many technical areas (such as NLP, machine vision, blockchain, and networking) is the main contribution. Several major AI-assisted fields, such as healthcare, manufacturing, smart cities, and gaming, are also investigated as prospective NLP applications.

2.1 INTRODUCTION

A language is a structured system of rules, and a collection of symbols employed to articulate ideas and convey information. Natural language processing (NLP) caters to individuals who may lack the time or expertise to acquire new languages or achieve proficiency in their existing ones, particularly when they possess limited knowledge of machine-specific languages [1]. In essence, NLP represents a branch of linguistics and artificial intelligence (AI) with the central goal of enabling computers to understand and process human language, encompassing both spoken and written forms. Its development seeks to streamline user-computer interactions, allowing users to communicate with computers using natural language. Natural language comprehension and natural language generation (NLG) are two major subfields of NLP that both improve the ability to interpret and produce textual content. The majority of NLP research is done by computer scientists, but linguistics experts, psychologists [2], philosophers, and others have also expressed interest. The fact that NLP advances our understanding of human language is one of the most intriguing features of this field. Different theories and methods that address the issue of using natural language to communicate with computers are related to the topic of NLP [3].

With the creation of decision-making systems using cognitive intelligence similar to that of humans, "AI" is a rapidly expanding field. The potential of AI in education is vast, as it can revolutionize the field by analyzing extensive datasets, including both unstructured data like audio and videos and structured data like SQL databases [4, 5]. AI brings machine learning (ML) techniques into play through learning management systems (LMSs), incorporating methods like "deep learning (DL)" and "transfer learning (TL)" to apply previously acquired knowledge to address new, related challenges. Furthermore, it leverages "NLP" techniques to listen to student feedback, analyze it, and make predictions regarding their sentiments about the learning environment. AI has the potential to revolutionize various aspects of the education sector, such as LMSs, curriculum creation, online teaching, assistance with job transfers, teacher preparation, research training, and assessment procedures [6]. The research data from the institution are varied and include feedback from students in text format as well as recordings of classes in video and audio formats. "AI" is described as "the activity devoted to making machines intelligent, and intelligence is that quality that enables an entity to function appropriately and with foresight in its environment." AI has been widely incorporated into educational institutions' various service delivery models for students [7]. NLP is one of the AI techniques that are most frequently utilized for student opinion mining. It is crucial in interpreting end-user input or thoughts. The majority of organizations devote time and money to studying end-user opinions.

The goal of the NLP sector of AI is to create algorithms and construct models that can use language like that of humans. It is frequently used in Google searches and translations, as well as in virtual assistants like "Alexa" and "Siri." NLP offers the ability to execute sentiment analysis and text summarization, automate question-answering, and analyze and extract information from unstructured sources. NLP holds the key to releasing the full potential of AI [8] in the field of biomedical sciences because the primary means of information gathering and dissemination in public health and medicine is natural language (communication). The majority of current NLP platforms are constructed using models improved by ML methods. ML methods consist of a model, data, a loss function that measures the extent to which the prototype represents the data, and an approach to train (develop) the model [9]. DL, a branch of ML, has been used to fuel significantly enhanced NLP models as a result of recent advances in these fields.

Artificial intelligence approaches include ML and DL. A set of algorithms, collectively referred to as ML, has the capability to analyze data, learn from it, and apply the acquired knowledge. A multi-layer neural network with processing layers is used in DL techniques to train on new concepts and create connections with previously learned information [10]. With ideas like the continuous bag-of-words and skip-gram models, DL improves NLP. Different DL methods are utilized for text categorization, including convolutional neural networks (CNNs), recurrent neural networks (RNNs), their superior cases of gated recurrent units (GRUs), and long short-term memory (LSTM).

2.1.1 COMPONENTS OF NATURAL LANGUAGE PROCESSING

NLP can be divided into two categories: "natural language understanding (NLU)" and "NLG." These categories involve the work of understanding and producing text. The general categorization of NLP is shown in Figure 2.1. This section's goal is to discuss NLG and NLU (linguistic).

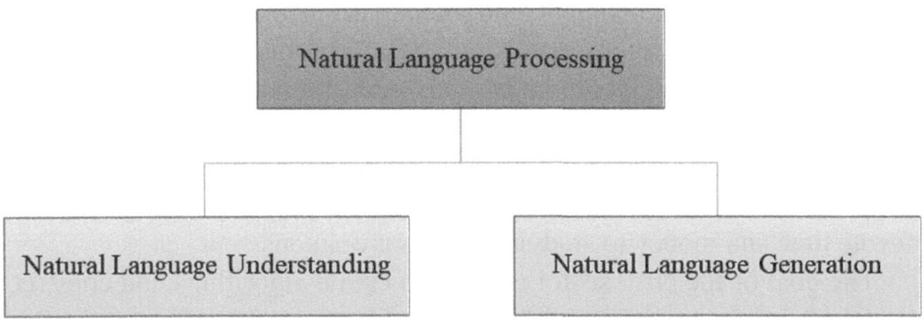

FIGURE 2.1 Types of natural language processing.

2.1.1.1 NATURAL LANGUAGE UNDERSTANDING (NLU)

Machines can achieve an understanding of natural language through NLU. NLU enables machines to analyze and extract various elements from language, such as concepts, entities, emotions, keywords, and more. It finds application in customer care systems to comprehend and address the issues reported by clients in their verbal or written communications [11]. NLU includes research on language context, meaning, and many linguistic forms, all of which are fundamental aspects explored in the field of linguistics. Therefore, it is crucial to comprehend the numerous key NLP terms as well as the various NLP levels. Next, we go over a few of the terms that are frequently used at various NLP levels. The branch of linguistics known as phonology, shown in Figure 2.2, deals with the orderly pattern of sound. The word "phonology" comes from the Ancient Greek terms "phono" (which means sound or voice) and the suffix "-logy" (which refers to speech or word). Phonology, within the domain of linguistics, broadly encompasses the study of the sounds of language, concentrating on the arrangement and behavior of these sounds. Nikolai Trubetzkoy defined phonology as "the study of sound within the framework of a language's system." Phonology is the study of how sounds are used to convey meaning in any human language [12]. Morphemes, which are the smallest units of meaning, are essential components of words. Morphology, in turn, deals with the structure and nature of words. For instance, "pre-cancellation" can be divided into three separate morphemes: "pre" (a prefix), "cancella" (the root), and "tion" (a suffix). All morphemes have consistent meanings, allowing humans to deconstruct unfamiliar words into morphemes to comprehend their meanings.

Both humans and NLP systems have the ability to comprehend the meaning of specific words at the lexical level. This word-level understanding is improved through various processing techniques, one of which involves assigning a part-of-speech (POS) tag to each word. In this process, the most likely POS tag is assigned to words that can function depending on the context in which they appear, in various sections of speech. Semantic representations can replace words with multiple meanings at the lexical level. The specific nature of this representation in NLP systems varies depending on the semantic theory being employed [13]. Consequently, the study of word structure at the lexical level takes into account both lexical meaning and POS categorization. In order to better understand the text, paragraphs, phrases, and individual words have been broken down in this study.

After the PoS tagging at the lexical level, words are further organized into phrases, phrases into clauses, and clauses into sentences at the syntactic level. This analysis of the language's grammatical structure emphasizes how sentences should be constructed correctly. This level's outcome is a phrase that showcases the systematic relationships among words, and this process is often referred to as parsing because it identifies sentences that convey more meaning than individual words alone. At the syntactic level, aspects such as morphology, PoS, and word order that are not fully considered at the lexical level are examined [14]. Word dependencies can change as the order of words in sentences is altered, and this can impact how sentences are understood. The most crucial semantic task involves determining the correct meaning of a sentence. Machines cannot utilize the same intuitive techniques that humans use to determine the meaning of a sentence; instead, humans rely on their comprehension of language and the notions contained within it. Semantic processing uses the logical structure of the sentence to find the most pertinent keywords in order to understand the connections among terms and ideas in a sentence.

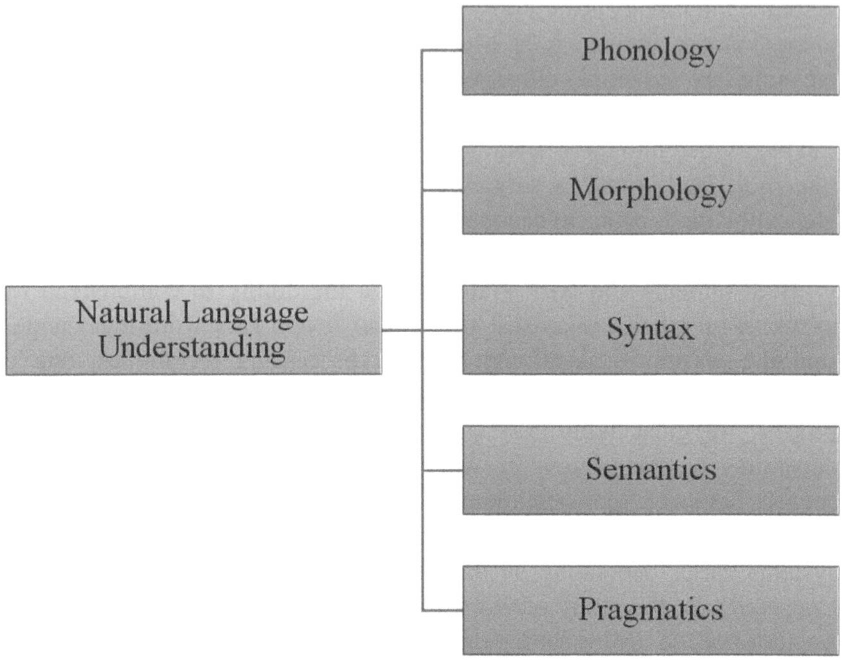

FIGURE 2.2 Classification of natural language understanding (NLU).

The discourse level of NLP works with longer conversations, whereas the structure and definition levels only work with sentences. By creating connections between phrases and words that maintain consistency, it analyzes logical structure [15]. By analyzing the relationships between sentences and extracting linguistic structures from texts at various levels, it emphasizes the text's elements that convey significance. Coreference resolution and anaphora resolution are the two most prevalent levels. Anaphora resolution involves identifying the thing the anaphor refers to in order to resolve the references in the text that have the same sense. Pragmatic-level analysis is concerned with the knowledge or information that comes from sources external to the content of a document. It revolves around what the speaker and the listener deduce or infer from each other during communication [16]. It delves into the unspoken or implicit aspects of sentences. To fully grasp what is being discussed in a text, it is essential to possess expertise in the relevant field. An in-depth understanding of the context is key to obtaining a meaningful interpretation of the text. When a sentence lacks specificity and the context is silent, pragmatic ambiguity arises. Pragmatic ambiguity is the term used to describe when individuals interpret similar material distinctly depending on its context.

NLP has the goal of accommodating one or more algorithmic or systemic specializations. It can merge both language comprehension and language generation through NLP assessment measures within an algorithmic system. This is useful not only for language understanding but also for language creation. NLP is applied in various domains, including multilingual event detection. An innovative modular approach for event extraction in texts written in Dutch, Italian, and English has been developed by an author. The greatest multilingual NLP tools are used by this system in a modular approach. In addition to handling more complex tasks like time-based standardization, semantic role labeling, and cross-language named object connection, the pipeline integrates modules for fundamental NLP execution. Ambiguity is one of the primary problems with natural language, which happens when a single sentence can have multiple meanings [17]. This typically occurs at the lexical, semantic, and syntactic levels. One statement can be broken down into different syntactical forms when there is syntactic-level uncertainty. The possibility of misinterpretation of a word's meaning is known as semantic ambiguity. Lexical level ambiguity is when a single word has the potential to mean several different things. Each of these levels has the potential to raise questions that can be answered by understanding the entire statement. Numerous approaches, including minimizing ambiguity, preserving ambiguity, interactive disambiguation, and weighting ambiguity, can be used to resolve the issue.

2.1.1.2 NATURAL LANGUAGE GENERATION (NLG)

"Natural language generation (NLG)" is the method of constructing coherent words, phrases, and sentences from an internal model. NLG is a component of NLP and typically involves four stages, as depicted in Figure 2.3: goal identification, goal planning, utilizing context and available communication sources, goal realization, and transforming the goal into text.

Text must be generated by a speaker or application, as well as a generator or program that transforms the application's objectives into a flowing phrase suitable for the situation. The following tasks are intertwined during the language generation process [18]. Content choice: Selected information should be added to the set. Depending on how this data is broken down into representational units, certain components may need to be eliminated while others may automatically be included.

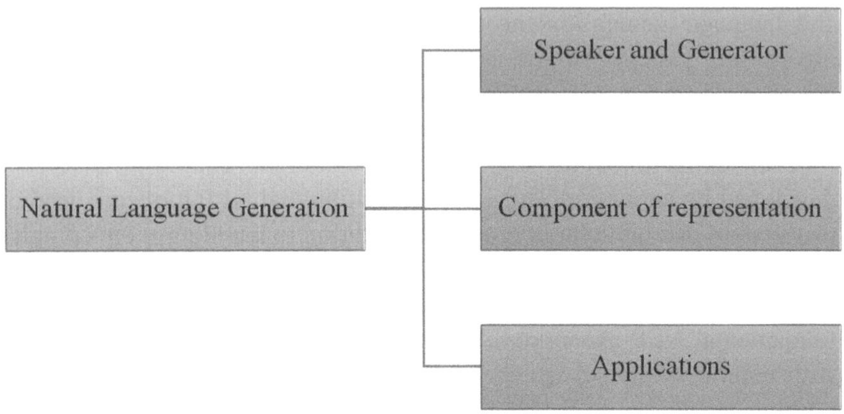

FIGURE 2.3 Components of natural language generation (NLG).

2.2 MODELS OF ARTIFICIAL INTELLIGENCE (AI)-BASED NATURAL LANGUAGE PROCESSING (NLP)

Feature extraction and feature selection are necessary data pre-processing steps to transform text data into quantitative vector representations before submitting the student response data to standard ML algorithms or ML methodologies like topic modeling. This section will discuss current methods for feature extraction, feature selection, and subject modeling.

2.2.1 FEATURE EXTRACTION

In order to process student response data and prepare that data for ML modeling, feature extraction techniques are commonly employed [19]. These techniques are used to derive relevant information from the text. Some of the techniques used for feature extraction from text data include Bag of words (BoW), term frequency-inverse document frequency (TF-IDF), and Word Embedding. These methods help convert textual data into a format that ML models can effectively work with. A lexicon of well-known terms, as well as an estimate of the words' presence, are two components of the BoW feature extraction technique. Only known words in a document are taken into consideration by the BoW [20]. It won't take the word choice or sentence structure of a document into account if the context of the words is disregarded. Based on their weights, TF-IDF determines the significance of every phrase or term within a document. The IDF of a term indicates how prevalent or uncommon it is in a corpus. A term's value approaches zero the more frequently it occurs in a corpus. TF-IDF is the product of multiplying TF and IDF. Evaluations of Word2Vec and GloVe's word embedding strategies were conducted using DL techniques. CNNs and RNNs, as well as ensembles made up of CNN and LSTM networks, were examined in a study. There were eight different DL algorithm combinations used. The GloVe method outperformed Word2Vec by roughly 5–7%, according to their comparison findings. Sangeetha et al.'s innovative method for analyzing and locating students' emotions in feedback was put forth. For semantic and grammatical analysis, feature extraction algorithms commonly convert student response data into tokens of words [21]. Word embedding algorithms such as GloVe, Cove, Word2Vec, and BERT are designed to restrict a word's dimensionality to circumstances involving groups based on neural networks. The frequency of phrases in a document is provided by BoW. DL and ML approaches are frequently utilized to compare the performances of feature extraction algorithms.

2.2.2 FEATURE SELECTION

Feature selection refers to the process of reducing the complexity of information in the context of features. The performance of the ML approach would be maintained or enhanced as a result. The goal of reduction criteria is to decrease the complexity of a model while preserving its accuracy over time. In a dataset with n features, there are 2^n possible feature subsets, and a high number of features can make modeling impractical. Considerations including the size of the feature subsets, the elimination of non-contributing features, and the grouping of related features are taken into account when assessing the stability and robustness of feature subsets [22]. There are three primary categories of feature subset assessment techniques: filters, wrappers, and embedded methods. These techniques help in selecting the most relevant and informative features while eliminating those that may not contribute significantly to the model's performance.

2.2.2.1 FILTER METHODS

Filter methods use a threshold to prioritize the important traits and choose the best candidates. Before real modeling, filter methods rank the characteristics and select them, as seen in Figure 2.4. Before training a model, this method filters out the low-relevance features. Two metrics are evaluated while ranking the features using the feature significance technique. The first step is to evaluate each feature's ability to predict the desired variable or variables. These are referred to as dependence or correlation measures. Mutual information, the chi-squared statistic, Markov blankets, and minimal-redundancy-maximum-relevancy processes are used to ascertain a feature's association with a target variable. Redundancy, which evaluates the features with redundant information, is the second metric in the feature significance procedure [23]. Comparing pertinent metrics among the independent variables determines the duplicate features, as reported in the redundant feature analysis in an article. To estimate the redundancy in other characteristics, it first identifies the features that are most useful in predicting the target variables.

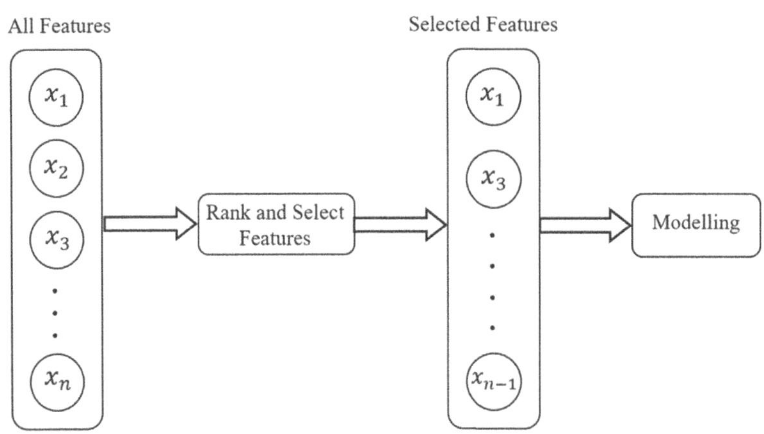

FIGURE 2.4 Filter rank method.

2.2.2.2 WRAPPER METHODS

Wrapper approaches use a predetermined classifier to find a subset of features, and then predefined classifiers are used to assess the performance of the subset of features. A ML technique is employed in wrapper approaches to improve the performance of feature selection. Figure 2.5 illustrates how a classifier is trained using a subset of the features that have been chosen. The classifier's performance is then assessed [24]. A wrapper technique for selecting features sequentially is known as sequential forward selection (SFS). Based on how well the classifier performs, an optimal subset of features is iteratively extracted using a greedy search technique. Features are picked one by one iteratively from a pool of all accessible features.

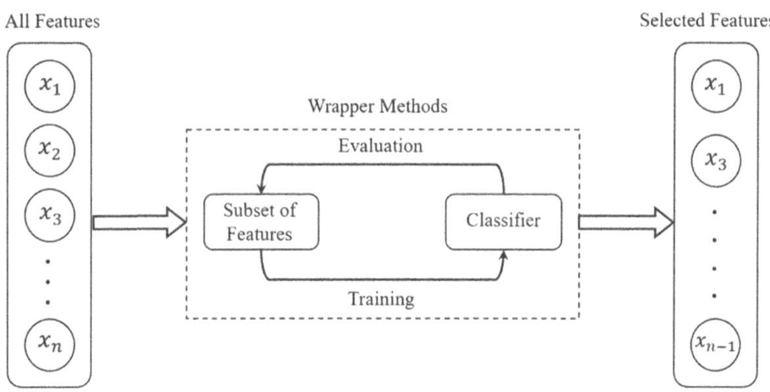

FIGURE 2.5 Wrapper method architecture.

2.2.2.3 *EMBEDDED METHODS*

Filter and wrapper methods are typically combined in embedded methods. They overcome problems with wrapper methods' poor computation speed and low precision in filter techniques. The most accurate features are examined by embedded algorithms, which improve the classifier's performance. Figure 2.6 illustrates how embedded approaches calculate the performance of each feature subset. Regularization is one of the most widely used embedding techniques, and it involves penalizing a model's complexity for L1 regularization [25] approaches to lower the level of the model's variance or overfitting.

For feature selection, the authors suggested the query expansion ranking (QER) method. They used classifiers including Naive Bayes multinomial, a support vector machine (SVM), maximum entropy modeling, and a decision tree to compare it to other feature selection techniques like information gain, Chi-square, document frequency difference, and optimal orthogonal centroid [26]. The performance of the model was evaluated by the authors using Turkish and English review databases. In the Naive Bayes multinomial classifier, the suggested feature selection method performed better than the other feature selection methods. Similar methods were used in the teaching assessment system to analyze student input. The number of attributes in the data was decreased to 18 in the study using the filter approach, which was chosen for feature selection. For classifying student comments, three ML algorithms—ID3, J48, and Naive Bayes—were employed [27]. Their effectiveness was compared to that of vote ensemble learning. With an accuracy of 87.16%, the voting ensemble learning combined with Chi-square feature selection surpassed conventional ML methods.

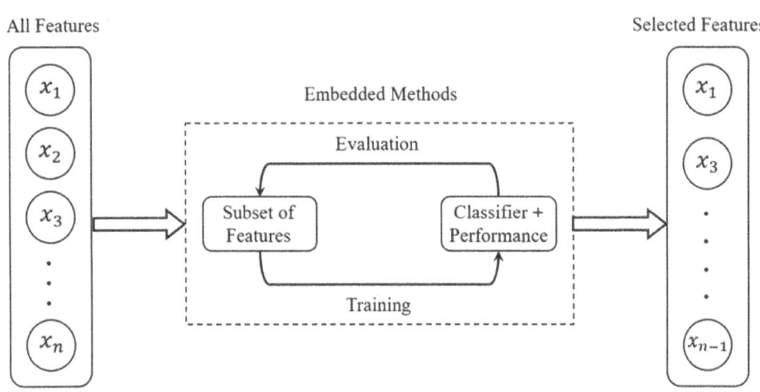

FIGURE 2.6 Embedded method approaches.

Based on an investigation of student comments, several writers developed a social mining model architecture to improve learning and e-learning. The goal of this strategy was to improve teaching methods and suggest courses for teachers in higher education. The random forest importance measure method, which computes each word's weights and filters them based on higher weights, was utilized as part of the study's feature selection procedure [28, 29]. SVM with kernels for linear, radial, poly, and random forest classifiers received the chosen features. K-fold cross-validation was used to train the ML classifiers, and the SVM model with a radial kernel outperformed the competition with an accuracy of 85.17%. Similarly, an improved sentiment analysis strategy uses an information gain filter method to choose the most pertinent aspects of student comments. In the study, SVM, random forest, and Naive Bayes classifiers were utilized, and their performances before and after feature selection were compared. Before the selection, random forest outperformed the other two classifiers with an accuracy of 81.6%, while SVM outperformed with an accuracy of 85.9%.

2.3 APPLICATIONS OF NATURAL LANGUAGE PROCESSING (NLP)-BASED ARTIFICIAL INTELLIGENCE (AI)

Next, we examine some of the topics along with the pertinent work that has been done in those directions. NLP may be used in numerous areas, including machine translation, email spam detection, information extraction, summarization, and question answering, among others.

2.3.1 MACHINE TRANSLATION

Ensuring data accessibility and availability to a global audience poses a challenge, particularly because the majority of the world's population is now online. One significant obstacle in this regard is the diversity of languages. There are numerous languages, each with its own unique syntax and sentence structure [30]. Machine translation, which often employs statistical engines like Google Translate, is used to convert words from one language to another. However, the challenge with machine translation technology lies not in translating individual words but in preserving the meaning, grammar, and tense of complete sentences. Statistical ML plays a key role in this process by collecting as much parallel data as possible from both languages (Language A and Language B) to determine the likelihood that elements in Language A correspond to elements in Language B. This helps improve the accuracy and effectiveness of translation systems by leveraging large datasets and statistical analysis to bridge language barriers.

2.3.2 TEXT CATEGORIZATION

Massive volumes of data are processed by categorization systems, which subsequently assign this data to predetermined categories or indexes.

These categories can include official papers, military casualty reports, market data, newswires, and more. A technology developed by The Carnegie Group has been employed to streamline the indexing process for Reuters' articles, significantly reducing the time required by human indexers or staff. Several businesses use categorization systems to classify issue tickets or complaint requests before directing them to the appropriate departments or individuals. Another notable application of text categorization is in email spam filters. These filters serve as the first line of defense against unsolicited emails and have become increasingly important. NLP technology plays a crucial role in addressing the core issues of false negatives and false positives in spam filtering. It enhances the ability to extract meaning from strings of text. Email filtering solutions utilize a set of protocols to identify incoming messages as either spam or legitimate [31]. This technology aids in distinguishing between unwanted and desired emails, contributing to more efficient email management. Spam filters come in many different varieties. Content filters examine the message's content to determine whether it contains spam or not. Header filters examine the email header for fictitious data. All emails from receivers who are on a general blacklist will be blocked. User-defined criteria are used in rules-based filters. For instance, stopping mail that contains a certain word or comes from a certain individual. Permission filters require the recipient's prior approval before anyone can send a message. Challenge-response filters need a code to be entered by anyone sending an email to send it.

2.3.3 SPAM FILTERING

Text categorization and anti-spam filtering utilize various ML techniques to classify text. Some of these techniques include rule learning, memory-based learning, SVMs, decision trees, maximum entropy models, hash forests, and a rule encoding method. These methods are preferred because they involve training classifiers on data rather than manual creation, which allows for more efficient and accurate categorization [32]. One of the simplest yet effective methods is the naive Bayes classifier, which is favored for its performance in text categorization and spam filtering. There are two main models applied in text categorization, both of which assume the presence of a predetermined vocabulary. In the first model, a document is generated by selecting a subset of the vocabulary and using the chosen words at least once, regardless of their order and frequency. This model is known as the Multivariate Bernoulli model and captures information on the presence or absence of words in a document. The second model generates a document by selecting a group of word occurrences and placing them in any order. This model, also known as the Multinomial model, records data on how frequently a word is used in a document, regardless of the specific placement or order of words.

2.3.4 INFORMATION EXTRACTION

Information extraction aims to identify and extract relevant terms from text data. This process involves extracting entities such as names, places, events, dates, times, and prices, which is a valuable technique for summarizing information that is essential to a user's specific needs across various applications. For instance, in the context of a domain-specific search engine, automatic recognition of significant data can enhance the precision and efficiency of directed searches. Hidden Markov models (HMMs) [33] are commonly used to identify pertinent information in research publications. These text extracts play a crucial role in enabling searches within specific fields, improving the presentation of search results, and establishing connections between citations and research papers. In text categorization, two models are typically applied, both relying on a predetermined vocabulary. The first model creates a document by selecting a subset of vocabulary words and utilizing them at least once, irrespective of their order or frequency. This model is known as the Multivariate Bernoulli model and captures information about the presence or absence of words in a document. The second model generates a document by selecting a group of word occurrences and placing them in any order. This model, also referred to as the Multinomial model, records data on how frequently each word is used within a document, regardless of its specific order or placement.

2.3.5 MEDICINE

NLP is used in the industry as well. One of the expansive NLP efforts in the medical arena is the Linguistic String Project-Medical Language Processor. To identify potential side effects of any medication while highlighting or flagging data items, the LSP-MLP assists clinicians in extracting and summarizing information on any signs or symptoms, drug dosage, and response data. The Specialist System is being developed by the national library of medicine [34]. Its intended use is as an information extraction tool for Medline abstracts and other biomedical knowledge bases. Medical Subject Headings (MeSH), *Dorland's Illustrated Medical Dictionary*, and standard English dictionaries were utilized to develop the lexicon. The Hôpital Cantonal de Genève's Centre d'Informatique Hospitalière is currently developing an electronic archiving environment with NLP functionality. The initial stage involved the archiving of patient records. A proper NLP system named RECIT has been created utilizing a technique called proximity processing at a later stage after the LSP-MLP was modified for French. The objective of the Linguistic String Project-Medical Language Processor and similar initiatives is to create a robust, multilingual system that can understand and analyze medical sentences effectively [35]. These systems aim to convert free text medical data into a language-independent knowledge representation, allowing for comprehensive and standardized analysis of medical information regardless of the language in which it is originally recorded.

2.4 NATURAL LANGUAGE PROCESSING (NLP) MODELS WORK WITH ARTIFICIAL INTELLIGENCE (AI)

2.4.1 BI-DIRECTIONAL ENCODED REPRESENTATION OF TRANSFORMERS (BERT)

Bi-directional encoded representation of transformers, which stands for bidirectional encoder representations from transformers, was created by Google in 2018 and pre-trained on more than 2.5 billion internet terms and 800 million book corpus words. BERT, an open-source ML framework, reads the text simultaneously in both directions and analyzes the language using a sizable database of specialized labeled training data, as the name implies. The trained data originates from a pre-trained RNN utilizing an unlabeled plain text corpus, which qualifies as unsupervised ML since the expected output of a sentence or paragraph is not labeled by a human expert [36]. Additionally, the pre-trained model acts as a foundational body of knowledge upon which further training can be based and adapted to the unique user's requirements. TL is the name for this procedure. Each Google user account's BERT NLP model is altered as a result.

The two essential steps of BERT are pre-training and fine-tuning. Challenges involving the production of language from sequences include (Figure 2.7):

- Answering inquiries;
- Summary of an abstract;
- Sentence forecast;
- Creating conversational responses.

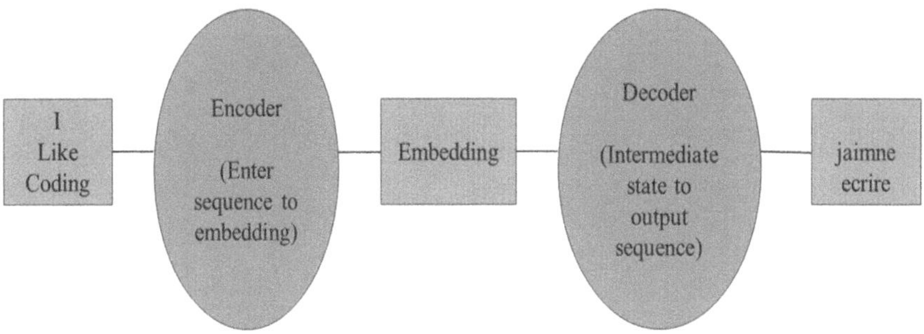

FIGURE 2.7 BERT architecture working.

2.4.2 GENERATIVE PRE-TRAINER-3 (GPT-3)

Generative Pre-Trainer, or GPT. While some of the most sophisticated NLP models are constructed on top of BERT, OpenAI's GPT-3 model, along with the GPT series, is the most recent model. GPT and GPT-2 are earlier GPT models. GPT-3, which is trained on 175 billion parameters, is a large-scale transformer-based language model similar to BERT. BERT-like models were surpassed by the GPT-3 model's sheer size by a factor of about 470. In actuality, it was trained using practically all of the Internet's data. In contrast to BERT, which involves a complex fine-tuning procedure where the model collects user-specific input data for TL, GPT-3 has a text-in and text-out API that enables users to access and reprogram the model using instructions [37]. Like BERT, it starts with unsupervised learning as a foundation, but tailoring the model to a particular user's needs avoids the requirement for a significant degree of TL. In some specific tasks, all that was required to outperform any rival was a pre-trained model integrating a few instances (better than BERT). Additionally, since the model is made available by OpenAI via an API, tuning is not an option. Tasks that can be performed by BERT and its variants, as well as BERT-like models, can be accomplished using GPT-3, which is widely regarded as being more accurate than BERT and its variants (Figure 2.8).

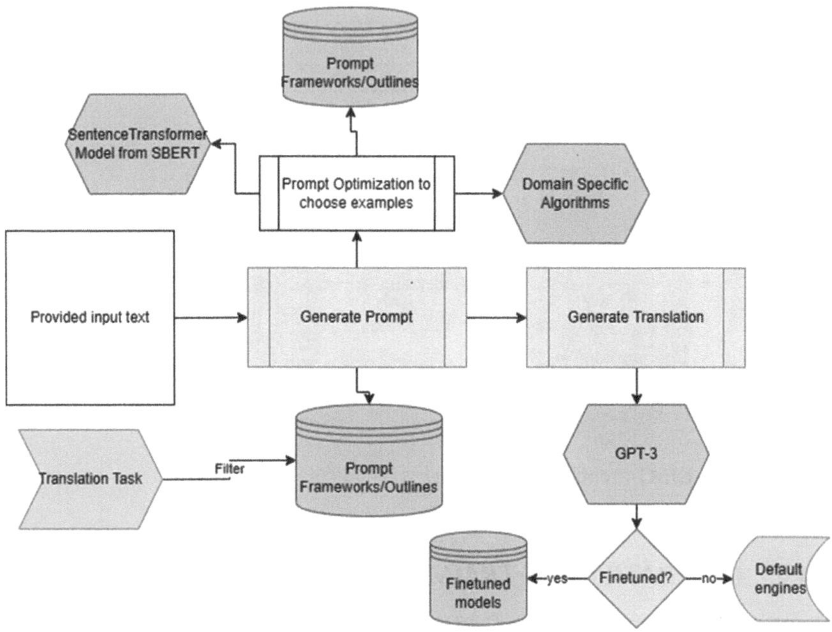

FIGURE 2.8 GPT-3 architecture working.

2.4.3 XLNET

Instead of being completely distinct from BERT, XLNet is a highly promising model. It is an NLP pre-training technique using generalized autoregressive (AR) [38]. The goal of XLNet is to continue using forward and backward traversal with the AR model. To address this issue, it focuses on the use of permutation and particularly suggests a new goal called permutation language modeling (Figure 2.9).

2.4.4 ALBERT

ALBERT is a Google-proposed NLP model that is based on the BERT architecture, as was the case with BERT, as previously mentioned.

For self-supervised learning of language representations, the acronym ALBERT stands for A Lite BERT. The three primary alterations that set it apart from the fundamental model are detailed in the next section. The input layer and hidden layer embedding in the BERT model and XLNet are the same size, while for ALBERT, the model divides the two embedding matrices, resulting in two smaller sizes [39]. When compared to BERT, this separation leads to an 80% reduction in parameters with barely any impact on model performance. Sharing of parameters between layers implies that to increase effectiveness and reduce redundancy, parameters from several model layers may be shared. Overall, parameters are reduced by 70% as a result (Figure 2.10).

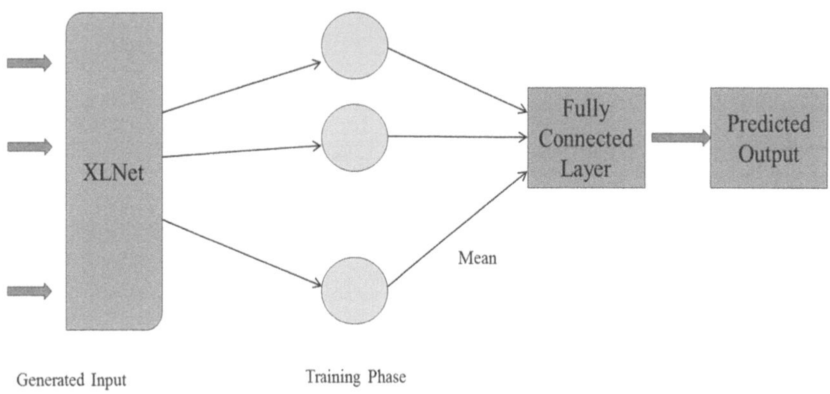

FIGURE 2.9 XLNet architecture working.

Several stages are suggested by the author:

- Sharing only the feed-forward network parameter;
- Share only attention-based parameters;
- Share all settings. Unless otherwise noted, authors utilize the default configuration.

2.5 OPPORTUNITIES

To attain the best possible health outcomes within and across populations, public health primarily develops and puts into action interventions that target modifiable causes of ill health. The ability to accurately estimate

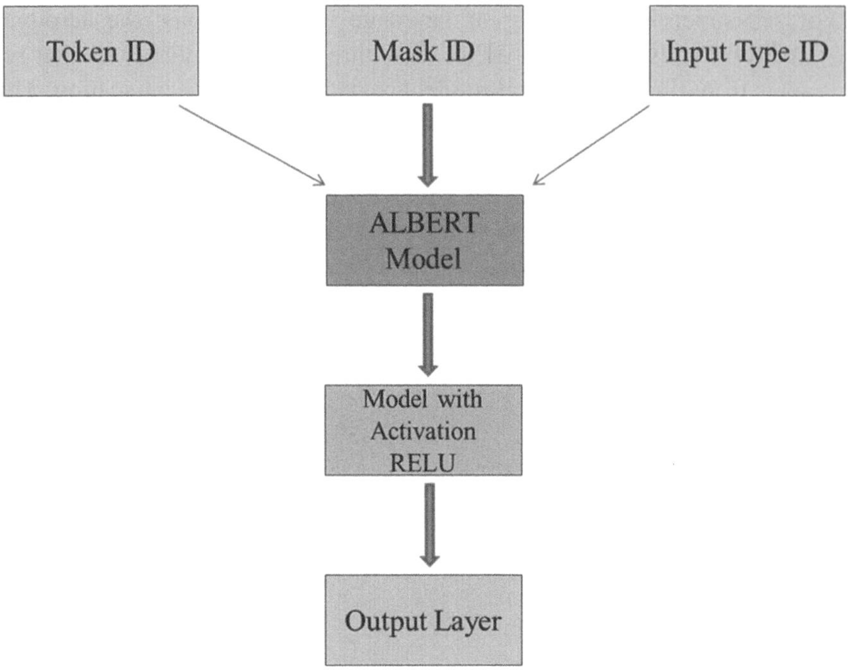

FIGURE 2.10 ALBERT model representation.

the burden of disease or disease risk factors in the community, identify populations that are disproportionately affected or at risk, identify the best approaches, and monitor outcomes are necessary for success. The PICO concept (patient/problem, intervention/exposure, comparison, outcome) is the best way to describe this evidence-informed approach to decision-making. PICO offers the best method for identifying knowledge that can be used to formulate and respond to particular clinical or public health inquiries [8, 35]. The PICO framework's elements-based comprehensive and systematic examination and synthesis of data serve as the standard foundation for evidence-informed decision-making. Information is created and released at an unparalleled rate nowadays, including scientific literature, technical reports, health records, social media, surveys, registries, and other documents. NLP has greatly expanded the possibilities for text-based research and evidence-based decision-making by enabling the quick analysis of enormous amounts of unstructured or semi-structured material.

To enable the quick identification of people, interventions, and outcomes of interest needed for disease surveillance, disease prevention, and health promotion, NLP is emerging as a potentially formidable tool.

One recent study showed that NLP techniques can foretell the presence of depression before it manifests in the medical record [35]. When prompt guidance is required, such as during an outbreak, the ability to perform real-time text mining of scientific publications for a specific PICO concept gives decision-makers the chance to quickly offer recommendations on disease prevention or management that are supported by the most recent body of evidence. By involving people and offering individualized help or guidance, NLP-powered chatbots and question-answering platforms have the potential to enhance health promotion initiatives.

2.6 CHALLENGES

In this section, challenges in implementing NLP techniques in the education domain are discussed.

2.6.1 DOMAIN-SPECIFIC LANGUAGE

In the context of applying NLP to education, it is essential to understand the fundamental elements of the instructional situation. Categorizing academic datasets or student feedback relies on this understanding, and it is one of the challenges faced in this field. Educational feedback, generated from a variety of sources such as surveys, questionnaires, and LMSs, can be substantial. NLP techniques alone may not be able to unveil the hidden semantic meaning in such texts without a background understanding or specialized training in the relevant educational domains. Interpreting and categorizing this data effectively often require domain-specific knowledge and context, which can be a complex aspect of educational NLP applications [40]. For students, teachers, and universities studying computer science or information technology in higher education, a domain-specific NLP approach is necessary. Using "named entity recognition (NER)," the authors were able to extract tech-related abilities and create a customized multi-level course recommendation system (RS). This is a domain-specific NER built to scrape information from various websites, such as job ads, course descriptions, and massive open online courses (MOOCs), and to improve the system with annotated corpora from Stack Overflow and GitHub. The embedding, training, testing, and validation datasets were created from the annotated stack overflow data. GitHub data with the Stack Overflow test dataset was utilized to evaluate the models. The training and testing data were provided to a bidirectional long short-term memory (Bi-LSTM) model and the recommended computer science and information technology named entity recognition (CSIT-NER) models for the purpose of training and evaluation. To create a corpus, the scraped data were embedded and extracted entities [4, 10]. Using Cloud API and Google Translate API, the authors suggested a way to extract entities and their relations. The authors used automatic text generation from the subject area or teacher-created topics to determine relevance for calculating grades. Creating corpora for a specific application domain can be facilitated by utilizing data scraping techniques to extract entities or concepts from a large database. After extracting this data, manual annotation can be employed to refine and categorize the entities or concepts, which is crucial for building high-quality corpora that are tailored to the specific needs of the application or domain. This combined approach helps ensure that the corpora contain accurate and relevant information for training and testing NLP models in the chosen domain.

2.6.2 SARCASM

In NLP tasks such as sentiment annotation and opinion analysis, the ability to decode sarcasm is crucial. This is especially important for understanding how students feel about the way courses are taught and the overall educational system, as sarcasm can often be used to convey sentiments that are not directly expressed. Several studies have specifically focused on automatic sarcasm detection, and authors have identified a research gap in this area after reviewing the literature on traditional sarcasm detection methods [41]. To accurately predict sarcasm labels, which are often concealed attributes, it is necessary to consider both the context of the text and the phrases used both before and after a sarcastic statement. This broader context helps improve the accuracy of sarcasm detection, which is a valuable component of sentiment and opinion analysis. Short text, lengthy text, transcripts, conversations, and other categories were employed to separate the datasets used for sarcasm detection in that study. The authors described three distinct methods for detecting sarcasm, including rule-based, statistical, and DL methods.

Sarcasm can be recognized using a rule-based technique based on important sarcasm indications that are recorded as evidence. Punctuation, sentiment-lexicon-based features, unigrams, word embedding similarity, frequency of the rarest words, sentiment flips, and other features are crucial components of the statistical classifiers used to detect sarcasm [42]. The classification of sarcasm involves various approaches, including the use of conventional ML algorithms such as SVMs, logistic regression, decision trees, Naive Bayes, HMMs, and ensemble classification techniques. These algorithms are applied to automatically detect sarcasm in text. In addition to conventional ML, DL techniques are also used for sarcasm detection. RNN models and LSTM techniques can be applied individually or in combination with CNNs to improve the accuracy of sarcasm detection.

2.6.3 AMBIGUITY

Ambiguity is frequent in natural languages since it depends on the context and how the reader interprets the text. Ambiguity in ML language processing is more difficult due to challenges in context decoding. Ambiguity may result from a sentence's lexical, syntactic, or structural characteristics. A sentence with structural ambiguity has many syntactic structures. A grammatical construct error in a sentence's sub-part that results in grammatical uncertainty in the entire sentence is known as syntactic ambiguity. Lexical ambiguity occurs when two words with the same form also have two different meanings [43]. The ambiguity problem must be addressed to properly analyze feedback. A study addressed word meaning disambiguation by making modifications to the BERT language representation model. In this study, researchers selected the top context-gloss pairs from a collection of related pairs to help disambiguate word meanings. These context-gloss pairs were categorized into pairs with both positive and negative sentiment by the authors, and positive and negative gloss pairs were merged with sample phrases from WordNet 3.0. Adding annotations to the combo made it easier to produce more training samples. In terms of score, the proposed BERT model performed 77% better than other current cutting-edge models.

2.6.4 EMOTICONS AND SPECIAL CHARACTERS

In student feedback and other text data, the presence of unique symbols, emoticons, and unusual characters is important for opinion mining, as these elements often convey the sentiments and emotions of the text. However, processing emoticons and accurately assigning the proper emotion tags to them can be a challenging task for NLP [41]. In a study published in 2020, the scientists examined cross-cultural responses to the newly discovered coronavirus and used emoticons to validate the sentiment polarity and emotion inferred from the tweets of participants. Combining feature extraction techniques like GloVe and word embeddings with an LSTM-based DL model was used. Various emoticons with their corresponding Unicode were used to authenticate the six emotions of happiness, surprise, sadness, anger, fear, and disgust. According to some authors, there are three difficulties in processing emoticons in a big dataset, including real-world emojis. They were query-by-emoji, emoji anticipation, and emoji processing. The Bi-LSTM model for text-to-emoji baseline findings and the CNN model for image-to-emoji were both employed by the authors as DL models. The two algorithms were then integrated to provide a multimodal technique approach for processing emoticons. The technique can be applied to analyzing emoticons used in student comments and understanding their perspectives.

2.6.5 ASPECT-BASED SENTIMENT ANALYSIS

The author claims that a sentiment analysis tool that is frequently underused in education was unable to discover opinions on a variety of subjects. The majority of research focuses on applying lexicon-based or machine-learning techniques to categorize positive or negative sentiment in student comments or feedback at the document level [44]. The issues and challenges involved in extracting different aspects were examined. The three topics under examination were aspect extraction, aspect sentiment analysis, and sentiment evolution. Explicit aspect extraction, implicit aspect extraction, aspect-level sentiment analysis, entity-level sentiment analysis, multi-word sentiment analysis, recognition of components in sentiment evolution, and forecasting sentiment evolution over social data were all used to further categorize each topic.

2.6.6 DATA IMBALANCE

Data imbalance is a common and significant issue in the field of AI, occurring when one class of data has significantly more samples than another class. This challenge also affects NLP, which is a branch of AI [40]. Gathering a large volume of labeled data, particularly in the education sector, can be difficult as it often requires manual annotation by subject-matter experts. When DL algorithms are trained on such imbalanced data, it can lead to biased classification performance. One potential solution to address this problem is TL, where a DL model is trained on a substantial corpus of student feedback data to perform similar tasks on other data sources. This can help improve classification performance in NLP tasks. Other methods might include sampling strategies to under- or over-sample minority groupings, which could necessitate text.

2.7 CONCLUSION

Research on "NLP" is giving more and more consideration to unsupervised or semi-supervised learning methodologies. These learning techniques can combine manually and automatically labeled data to learn from data that has not been manually annotated with the key solutions. The aim of NLP, a field that straddles computational linguistics and computer science, is to convert spoken and written natural human languages into structured, reusable data. Using linguistic, statistical, and "AI" methods, the meaning of a text can be determined. NLP can even be used to produce responses that are like those of a human being. Due to its capacity to represent and evaluate human language computationally, NLP has attracted a lot of attention lately. Machine translation, email spam detection, information extraction and summarization, as well as medical and question-answering, are just a few of the many applications it presently has. One of these technologies, "AI," has proven to have a substantial impact on increasing immersion and enabling intelligent virtual agents. Additionally, in this chapter, we work with NLP to investigate the role of AI, including DL architectures and ML algorithms. The key contribution is the extensive examination of AI-based strategies in numerous technical fields (such as NLP, machine vision, blockchain, and networking). Prospective NLP applications are also being researched in a number of significant AI-assisted industries, including healthcare, manufacturing, smart cities, and gaming.

KEYWORDS

- **artificial intelligence**
- **deep learning**
- **machine learning**
- **natural language processing**
- **semi-supervised learning**
- **unsupervised learning**

REFERENCES

1. Bengio, Y., Ducharme, R., & Vincent, P. (2001). A neural probabilistic language model. In *Proceedings of NIPS*.
2. Benson, E., Haghighi, A., & Barzilay, R. (2011). Event discovery in social media feeds. In *Proceedings of the 49th Annual Meeting of the Association for Computational Linguistics: Human Language Technologies* (Vol. 1, pp. 389–398). Association for Computational Linguistics.
3. Berger, A. L., Della Pietra, S. A., & Della Pietra, V. J. (1996). A maximum entropy approach to natural language processing. *Computational Linguistics, 22*(1), 39–71.
4. Blanzieri, E., & Bryl, A. (2008). A survey of learning-based techniques of email spam filtering. *Artificial Intelligence Review, 29*(1), 63–92.
5. Bondale, N., Maloor, P., Vaidyanathan, A., Sengupta, S., & Rao, P. V. (1999). Extraction of information from open-ended questionnaires using natural language processing techniques. *Computer Science and Informatics, 29*(2), 15–22.
6. Ahonen, H., Heinonen, O., Klemettinen, M., & Verkamo, A. I. (1998). Applying data mining techniques for descriptive phrase extraction in digital document collections. In *Research and Technology Advances in Digital Libraries, 1998. ADL 98. Proceedings. IEEE International Forum on* (pp. 2–11). IEEE.
7. Alshawi, H. (1992). *The Core Language Engine*. MIT Press.
8. Andreev, N. D. (1967). The intermediary language as the focal point of machine translation. In A. D. Booth (Ed.), *Machine Translation* (pp. 3–27). North Holland Publishing Company.
9. Dhiman, P., Kaur, A., Iwendi, C., & Mohan, S. K. (2023). A scientometric analysis of deep learning approaches for detecting fake news. *Electronics, 12*(4), 948.
10. Androutsopoulos, I., Paliouras, G., Karkaletsis, V., Sakkis, G., Spyropoulos, C. D., & Stamatopoulos, P. (2000). *Learning to Filter Spam E-Mail: A Comparison of a Naive Bayesian and a Memory-Based Approach*. arXiv preprint cs/0009009.
11. Baclic, O., Tunis, M., Young, K., Doan, C., Swerdfeger, H., & Schonfeld, J. (2020). Artificial intelligence in public health: Challenges and opportunities for public health

made possible by advances in natural language processing. *Canada Communicable Disease Report, 46*(6), 161.
12. Borst, F., Sager, N., Nhàn, N. T., Su, Y., Lyman, M., Tick, L. J., & Scherrer, J. R. (1989). Analyse automatique de comptes rendus d'hospitalisation. In P. Degoulet, J. C. Stephan, A. Venot, & P. J. Yvon (Eds.), *Informatique Et Santé, Informatique Et Gestion Des Unités De Soins, Comptes Rendus Du Colloque AIM-IF, Paris* (pp. 246–256).
13. Briscoe, E. J., Grover, C., Boguraev, B., & Carroll, J. (1987). A formalism and environment for the development of a large grammar of English. In *IJCAI 87* (pp. 703–708).
14. Carreras, X., & Marquez, L. (2001). *Boosting Trees for Anti-Spam Email Filtering.* arXiv preprint cs/0109015.
15. Chalkidis, I., Fergadiotis, M., Malakasiotis, P., Aletras, N., & Androutsopoulos, I. (2020). *LEGAL-BERT: The Muppets Straight Out of Law School.* arXiv preprint arXiv:2010.02559.
16. Khullar, V., Singh, H. P., Miro, Y., Anand, D., Mohamed, H. G., Gupta, D., Kumar, N., & Goyal, N. (2022). IoT fog-enabled multi-node centralized ecosystem for real-time screening and monitoring of health information. *Applied Sciences, 12*(19), 9845.
17. Chi, E. C., Lyman, M. S., Sager, N., Friedman, C., & Macleod, C. (1985). A database of computer-structured narrative: Methods of computing complex relations. In *Proceedings of the Annual Symposium on Computer Application in Medical Care* (p. 221). American Medical Informatics Association.
18. Bahdanau, D., Cho, K., & Bengio, Y. (2015). Neural machine translation by jointly learning to align and translate. In *ICLR 2015.*
19. Bangalore, S., Rambow, O., & Whittaker, S. (2000). Evaluation metrics for generation. In *Proceedings of the First International Conference on Natural Language Generation* (Vol. 14, pp. 1–8). Association for Computational Linguistics.
20. Baud, R. H., Rassinoux, A. M., & Scherrer, J. R. (1991). Knowledge representation of discharge summaries. In *AIME 91* (pp. 173–182). Springer.
21. Baud, R. H., Rassinoux, A. M., & Scherrer, J. R. (1992). Natural language processing and semantical representation of medical texts. *Methods of Information in Medicine, 31*(2), 117–125.
22. Baud, R. H., Alpay, L., & Lovis, C. (1994). Let's meet the users with natural language understanding. *Knowledge and Decisions in Health Telematics: The Next Decade, 12*, 103.
23. Feldman, S. (1999). NLP meets the Jabberwocky: Natural language processing in information retrieval. *Online-Weston Then Wilton, 23*, 62–73.
24. Friedman, C., Cimino, J. J., & Johnson, S. B. (1993). A conceptual model for clinical radiology reports. In *Proceedings of the Annual Symposium on Computer Application in Medical Care* (p. 829). American Medical Informatics Association.
25. Gao, T., Dontcheva, M., Adar, E., Liu, Z., & Karahalios, K. (2015). DataTone: Managing ambiguity in natural language interfaces for data visualization. In *UIST '15: Proceedings of the 28th Annual ACM Symposium on User Interface Software & Technology* (pp. 489–500). https://doi.org/10.1145/2807442.2807478.
26. Ghosh, S., Vinyals, O., Strope, B., Roy, S., Dean, T., & Heck, L. (2016). *Contextual LSTM (CLSTM) Models for Large-Scale NLP Tasks. arXiv preprint arXiv:1602.06291.*

27. Glasgow, B., Mandell, A., Binney, D., Ghemri, L., & Fisher, D. (1998). MITA: An information-extraction approach to the analysis of free-form text in life insurance applications. *AI Magazine, 19*(1), 59.
28. Diab, M., Hacioglu, K., & Jurafsky, D. (2004). Automatic tagging of Arabic text: From raw text to base phrase chunks. In *Proceedings of HLT-NAACL 2004: Short Papers* (pp. 149–152). Association for Computational Linguistics.
29. Doddington, G. (2002). Automatic evaluation of machine translation quality using n-gram co-occurrence statistics. In *Proceedings of the Second International Conference on Human Language Technology Research* (pp. 138–145). Morgan Kaufmann Publishers Inc.
30. Drucker, H., Wu, D., & Vapnik, V. N. (1999). Support vector machines for spam categorization. *IEEE Transactions on Neural Networks, 10*(5), 1048–1054.
31. Dunlavy, D. M., O'Leary, D. P., Conroy, J. M., & Schlesinger, J. D. (2007). QCS: A system for querying, clustering, and summarizing documents. *Information Processing & Management, 43*(6), 1588–1605.
32. Elkan, C. (2008). *Log-Linear Models and Conditional Random Fields. Retrieved from:* http://cseweb.ucsd.edu/welkan/250B/cikmtutorial.pdf *(accessed on 13 March 2025).*
33. Emele, M. C., & Dorna, M. (1998). Ambiguity-preserving machine translation using packed representations. In *Proceedings of the 36th Annual Meeting of the Association for Computational Linguistics and 17th International Conference on Computational Linguistics,* (*Vol. 1,* pp. 365–371). Association for Computational Linguistics.
34. Koehn, P. (2005). Europarl: A parallel corpus for statistical machine translation. In *MT Summit 2005.*
35. Fan, Y., Tian, F., Xia, Y., Qin, T., Li, X. Y., & Liu, T. Y. (2020). Searching better architectures for neural machine translation. *IEEE/ACM Transactions on Audio, Speech, and Language Processing, 28*, 1574–1585.
36. Fang, H., Lu, W., Wu, F., Zhang, Y., Shang, X., Shao, J., & Zhuang, Y. (2015). Topic aspect-oriented summarization via group selection. *Neurocomputing, 149*, 1613–1619.
37. Fattah, M. A., & Ren, F. (2009). GA, MR, FFNN, PNN, and GMM-based models for automatic text summarization. *Computer Speech & Language, 23*(1), 126–144.
38. Kaur, R., Ginige, J. A., & Obst, O. (2023). AI-based ICD coding and classification approaches using discharge summaries: A systematic literature review. *Expert Systems with Applications, 213*, 118997.
39. Srivastava, V., Khaparde, A., Kothari, A., & Deshmukh, V. (2023). NLP-based AI-powered Sanskrit voice bot. *Artificial Intelligence Applications and Reconfigurable Architectures*, 95–124.
40. Kusal, S., Patil, S., Choudrie, J., Kotecha, K., Vora, D., & Pappas, I. (2023). A systematic review of applications of natural language processing and future challenges with special emphasis in text-based emotion detection. *Artificial Intelligence Review, 56*, 15129–15215.
41. Qiu, Y., & Wei, G. W. (2023). Artificial intelligence-aided protein engineering: From topological data analysis to deep protein language models. *Briefings in Bioinformatics, 24(5), bbad289.*
42. Castaneda, J., Calvet, L., Benito, S., Tondar, A., & Juan, A. A. (2023). Data science, analytics, and artificial intelligence in e-health: Trends, applications, and challenges. *SORT-Statistics and Operations Research Transactions*, *47*, 81–128.

43. Joudar, S. S., Albahri, A. S., Hamid, R. A., Zahid, I. A., Alqaysi, M. E., Albahri, O. S., & Alamoodi, A. H. (2023). Artificial intelligence-based approaches for improving the diagnosis, triage, and prioritization of autism spectrum disorder: A systematic review of current trends and open issues. *Artificial Intelligence Review, 56*, 1–65.
44. Lima, K. A., Md Hasib, K., Azam, S., Karim, A., Montaha, S., Noori, S. R., & Jonkman, M. (2023). A novel data- and model-centric artificial intelligence-based approach in developing high-performance named entity recognition for Bengali language. *PLOS ONE, 18*(9), e0287818.

CHAPTER 3

Exploring the Synergy of Machine Learning and Natural Language Processing

GAGANDEEP KAUR[1,*], VEERPAL KAUR[2], ANUSHKA[1], and YAGYATA[1]

[1]*Chitkara University Institute of Engineering and Technology, Chitkara University, Punjab, India*

[2]*Department of Computer Science and Engineering, Lovely Professional University, Punjab, India*

**Corresponding author.*

ABSTRACT

The fusion of machine learning (ML) and natural language processing (NLP) during the last few decades has resulted in an innovative epoch of revolutionary potential in the domain of language understanding and development. The present research focuses on the intricate relationship between ML and NLP, highlighting the significant achievements and real-world applications that have emerged from their integration within the discipline. The merging of ML and NLP provides computers with the ability to not just comprehend but also communicate text that progressively approaches information generated by individuals. This chapter delves into the fundamental concepts that underlie the mutual dependence of ML algorithms and NLP models.

The current study focuses on how ML algorithms have been developed and enhanced to address the complexities of language, with applications including language translation, sentiment analysis, chatbots, and text summarization. Intriguing case studies and actual instances from

Natural Language Processing and Artificial Intelligence: A Perspective Toward Current Trends, Challenges, and Applications. Vikas Khullar, Aryan Chaudhary, & Sunil Kumar (Eds.)

DOI: 10.1201/9781998511273-3

real-world settings augment the analysis for advanced intelligent systems capable of interpreting the context, sentiment, and intricacies inherent in human language. This work lays the framework for advanced intelligent systems capable of interpreting the context, sentiment, and intricacies inherent in human language, focusing on the significance of pre-trained models such as GPT-3 and BERT. Various consequences in numerous sectors such as healthcare, banking, customer service, and others can be seen as we move through the domain of interconnection. The integration of ML along with NLP is becoming more dominant. In a nutshell, this research sheds light on the noteworthy trend of ML and NLP, emphasizing the disruptive impact of their mutually beneficial collaboration in the arena of digital technology. It also provides an opportunity for in-depth inquiry, comprehending the complexities of language, and utilizing the boundless powers of this combination for the benefit of society, scientific pursuits, and other businesses.

3.1 INTRODUCTION

Machine learning (ML) has become an indispensable tool for pacesetters due to the rise of data across industries and the pervasiveness of technology in modern life. For automating processes, providing predictive assessments, and making well-informed judgments, ML algorithms are now requisite. An individual requires specialized skills in order to simplify procedures, extract critical information from big data sets, and then use that knowledge to make better decisions. To revolutionize a few businesses in the present interconnected and carefully changed world, including medical services and finance, AI is required. To procure the boundless benefits of this field, which is nowadays seeing a gigantic commotion in information creation, a top-to-bottom experience with ML is fundamental for all [1, 2].

To work with the explanation of human language, regular language handling (NLP) is vital to the human-machine connection. The combination of NLP and ML addresses a kind of cooperative energy that goes beyond the constraints of state-of-the-art technology. Because of the accessibility of information and algorithmic methodologies that are suitable for NLP, comparable results are attainable with ML. NLP is used in ML to improve its ability to comprehend and interact with the environment in a manner that is comparable to human behavior. A wide assortment of applications, including interpretation, opinion analysis, chatbots, and various interventions, has cultivated the area of ML and NLP. Various segments stand to benefit enormously from the integration, such as client development, content creation, diagnostic advancements, and authoritative record analysis. The integration of AI, ML and NLP is a cutting-edge major advantage. In the age of data-driven learning, there has been a major change in how individuals engage with technology, utilizing the extraordinary potential of human discourse [3–5].

3.2 LITERATURE REVIEW

The study gives a brief overview of various works relevant to the research field, which is discussed below:

Mahesh et al. [6] analyzed rapid advances in the field and the need for a comprehensive review of existing algorithms. There is a discussion of the types of algorithms that were considered, the inclusion/exclusion criteria, and the approaches used for evaluating and comparing them. Additionally, factors such as accuracy, scalability, interpretability, and applicability to different types of data are briefly discussed. Summarizing the key findings of the review and possibly offering recommendations for practitioners, researchers, and policymakers was included.

Osisanwo et al. [7] explored the importance of supervised ML in various applications. The article discusses rapid advancements in the field and the need to review classification algorithms comprehensively. It discusses different types of supervised ML algorithms. The various types have been further compared on the basis of aspects such as accuracy, precision, recall, and F1-score, among others. Comparative analyses were presented in tables. As a result, it can shed light on which algorithms are best suited to specific types of tasks, as well as highlight potential future research directions.

Khanum et al. [8] discussed the various applications of unsupervised ML, with a focus on automation, classification, and maintenance. The study explores and classifies several unsupervised ML techniques, with an emphasis on those that are pertinent to maintenance, automation, and classification. Examples and case studies illustrating the practical applications of unsupervised ML algorithms have also been discussed. The work includes the evaluation of several algorithms' performances in relation to particular parameters, such as computing efficiency, scalability, and adaptability to various datasets. The study highlights the benefits and drawbacks of every class of unsupervised ML techniques.

Elizabeth et al. [9] highlight that the computerized method of text analysis known as NLP is predicated on a number of theories and technologies. Moreover, as this is a field of study and improvement that is continuously developing, there is certainly not a solitary definition that has been settled upon that would fulfill everyone.

Nadkarni et al. [10] examine this article on normal language processing (NLP) means to give an outline and rules on the development of contemporary NLP frameworks. Features of clinical NLP drives, a summation of typical sub-issues in the field, and a verifiable development of NLP are all included in the work. It also incorporates a summation of contemporary NLP plans (including the Apache Foundation's Unstructured Data Management Architecture), a discussion of possible future directions for NLP, and a brief introduction to ML methods used in NLP.

Jones et al. [11] discuss that NLP expects to give a top-to-bottom assessment of the improvement of regular language processing (NLP) from the last part of the 1940s to the present, meaning to find successive examples that relate to various difficulties. The area of regular semantic handling has gone through different authentic stages, each recognized by massive changes in core relating to computerized reasoning, logico-syntactic style, machine interpretation, and the openness of broad phonetic information. Each period of the advancement of NLP frameworks and critical thinking highlighted a development of difficulties and approaches. An in-depth look at the development of NLP throughout its history is provided in this overview, with an emphasis on the significant advancements made at each stage.

Hristea et al. [12] highlight the importance of the convergence of ML and natural language processing (NLP), emphasizing both theoretical progress and real-world applications. The study provides a concise overview of the relevant subjects, highlighting the significant developments made in the fields of NLP and ML. The introductory section provides researchers with context by elucidating the objectives of the special issue, which comprise the progression of knowledge and the analysis of recent developments in these specific disciplines.

Glaz et al. [13] examine that NLP and ML are vital for contemporary medical research, presenting a potential paradigm shift. These models frequently confirm clinical views rather than yielding entirely novel evidence. One of the challenges is the dependence on vague cohorts, particularly social media users. Strengthening NLP performance requires investigating language-specific factors in a range of languages. NLP and ML offer insights into patient data that were previously unavailable, but before implementing these technologies in mental health care, ethical issues need to be resolved, highlighting the role that these technologies play as auxiliary tools in clinical practice.

Tusar et al. [14] examine that customer satisfaction is essential for business growth in the fiercely competitive business world of today. Businesses spend a lot of money trying to understand and meet the needs of their customers, but manual analysis is difficult. Sentiment analysis—a hybrid of NLP and ML—is suggested as a way to improve customer satisfaction by gleaning insights from public opinion. To effectively analyze sentiment, this paper presents machine learning algorithms (support vector machine, logistic regression, naive Bayes, random forest) and natural language processing techniques (bag-of-words and TF-IDF). Using SVM and Logistic Regression with the Bag-of-Words technique, the study achieves 77% accuracy on a large, imbalanced, multi-class dataset.

Chaitrali et al. [15] focus on intelligent chatbots, or virtual assistants, that can comprehend and reply to natural language queries. These chatbots are used in customer service centers and banks with the goal of increasing productivity and decreasing the time it takes to process customer requests. The main objective of the chatbot is to enable customers to communicate with it in simple English, mimicking the customer service experience by providing prompt and relevant responses. The suggested system describes the architecture, methodology, and a comparison of seven classification algorithms for input classification using a dataset from bank FAQs. By offering solutions for customer service centers, phone answering systems, and help desks, the chatbot expands its application to everyday life.

3.3 FUNDAMENTALS OF MACHINE LEARNING

The primary idea is around the application of training data in order to provide computers with the ability to enhance their overall performance on a certain activity over the course of time. Learning with supervision, learning on one's own without supervision, and learning through reinforcement are three essential components, each of which is tailored to a specific learning environment [16].

3.3.1 MACHINE LEARNING BASICS

ML is a branch of artificial intelligence (AI) that focuses on the development of algorithms and statistical models. These algorithms empower computers to learn and make predictions or decisions without requiring explicit, rule-based programming for specific tasks. The fundamentals of ML include:

1. **Learning from Data:** ML is primarily about learning from data. ML algorithms are intended to identify patterns and relationships in data rather than depending on established programming rules.
2. **Learning Categories:**
   i. **Supervised Learning:** The algorithm in supervised learning is trained on a labeled dataset. Each example in the dataset is composed of input data (features) and the desired output (target). Because the algorithm learns to map inputs to outputs, it is well-suited for tasks that involve classification and regression.
   ii. **Unsupervised Learning:** Unsupervised learning requires interacting with unlabeled data. The objective of the algorithm is to identify patterns, clusters, or hidden structures in the data. Clustering and dimensionality reduction are two common approaches.
   iii. **Reinforcement Learning (RL):** When an agent interacts with its environment in order to attain a goal, RL is applied. The agent learns by receiving input in the form of rewards or penalties and strives to maximize aggregate benefits over an extended period.

ML is a diverse and rapidly expanding field. These basic ideas constitute the foundation for understanding its fundamental principles. Achieving competence in ML entails digging into specific algorithms, approaches, and tools, along with keeping updated on the domain's latest advancements [17].

3.3.2 TYPES OF MACHINE LEARNING ALGORITHMS

Machine learning algorithms can be categorized into several types, each suited for different tasks and data. Here are some common types of ML algorithms:

1. **Supervised Learning Algorithms:**
   i. **Regression:** Used for foreseeing continuous numerical values. Examples include linear regression and polynomial regression.
   ii. **Classification:** Data is categorized into specified classes or labels using this method. Logistic regression, decision trees, random forests, and SVM are a few examples.
2. **Unsupervised Learning Algorithms:**
   i. **Clustering:** These algorithms cluster data points according to similarity or closeness. K-Means and hierarchical clustering are prominent strategies.
   ii. **Dimensionality Reduction:** These algorithms aim to reduce the number of features (dimensions) in the data while conserving its key properties. Illustrations involve principal component analysis (PCA) and t-distributed stochastic neighbor embedding (t-SNE).
   iii. **Meta-Learning Algorithms:** Meta-learning is concerned with training models for learning how to learn. These algorithms try to improve the learning process and make it easier to adapt to new activities.

The choice of algorithm depends on the specific problem you are trying to solve, the nature of your data, and the available resources. In practice, it is often necessary to experiment with different algorithms to determine which one performs best for a given task.

3.3.3 SUPERVISED LEARNING

Over the decade, bank frauds have been increasing day by day. In order to detect and reduce fraud, supervised ML techniques can be used. With supervised ML algorithms, these patterns can be separated from genuine transactions. Supervised ML algorithms can identify fraudulent activity much quicker and with more accuracy than traditional-based systems. It is based on supervision. This algorithm involves training the machine with a labeled dataset and then predicting the output based on the training. The mapping of input variables (x) and output variables (y) is the goal of supervised learning.

Some practical applications of supervised ML, as shown in Figure 3.1, include speech recognition, fraud detection, spam filtering, image segmentation, and medical diagnosis [18].

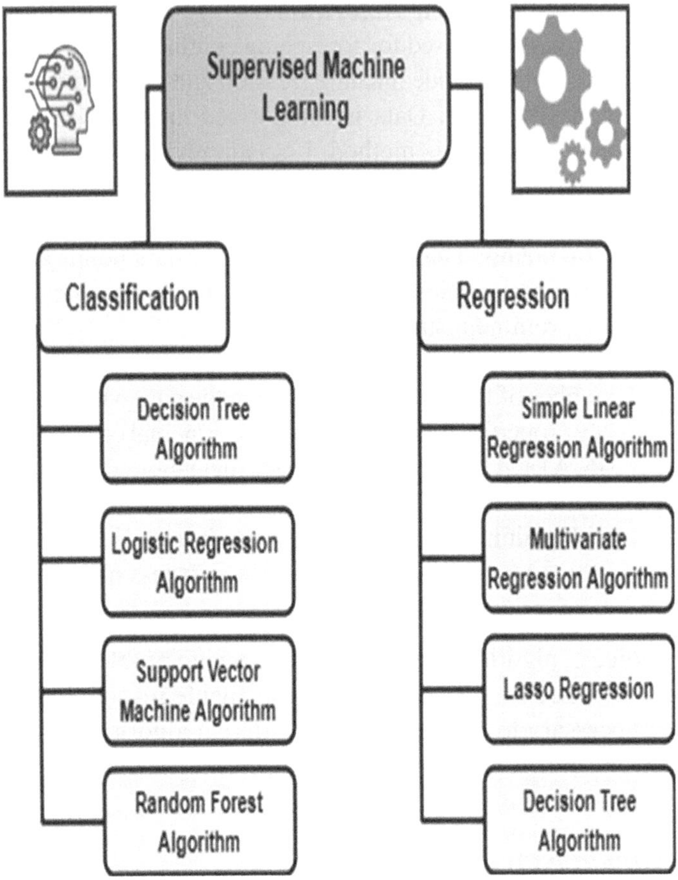

FIGURE 3.1 Types of supervised learning.

3.3.4 UNSUPERVISED LEARNING

Unsupervised learning is a subset of ML in which the algorithm discovers patterns and connections in the data on its own, without any help from the user or any prior knowledge. In other words, it is a method of finding structure or patterns in data that are not labeled, classified, or categorized. In unsupervised learning, as shown in Figure 3.2, an algorithm is given a set of input data and asked to find patterns, clusters, or similarities in the data without being given a predetermined result to follow. Finding useful patterns or insights in the data is the aim.

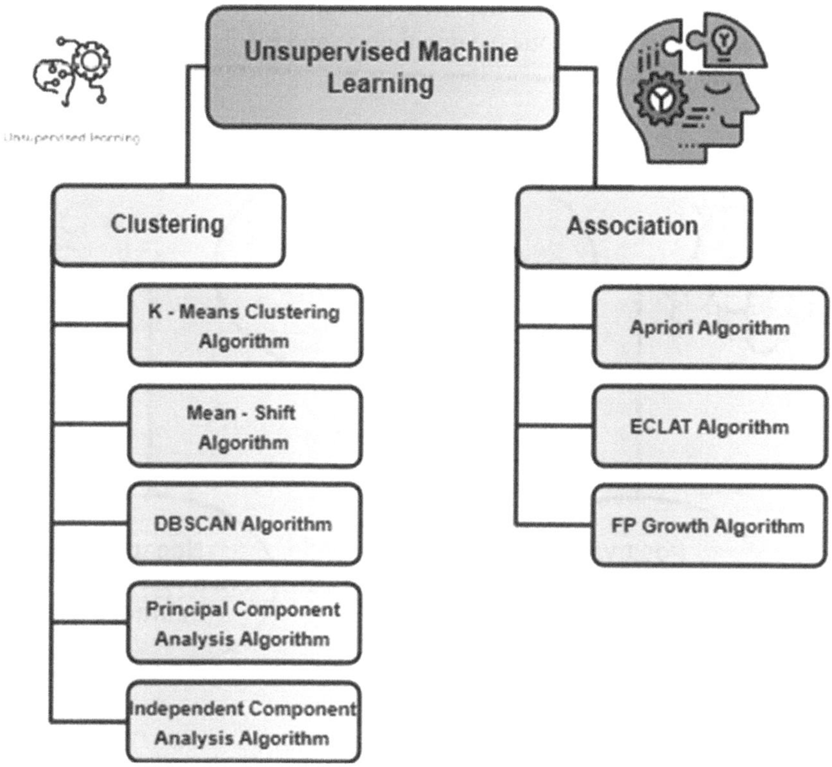

FIGURE 3.2 Types of unsupervised learning.

3.3.5 REINFORCEMENT LEARNING

A fundamental part of ML is RL, in which an agent uses information from its immediate environment to learn how to respond. The agent interacts with the environment through a series of actions, resulting in the acquisition of rewards or penalties based on the outcomes. The agent's primary objective is to devise a strategy that maximizes the total cumulative reward over its operational lifespan. In the field of RL, a popular paradigm is the Markov decision process (MDP). The structure of this learning framework consists of states, actions, and rewards, as well as a transition function that governs the impact of the agent's actions on the state of the environment. The agent's goal is to develop a policy that facilitates the transition between states and actions, allowing it to improve its overall utility over an extended period. RL, as depicted in Figure 3.3, has been applied in several areas such as robotics, game theory, and recommendation systems (RSs).

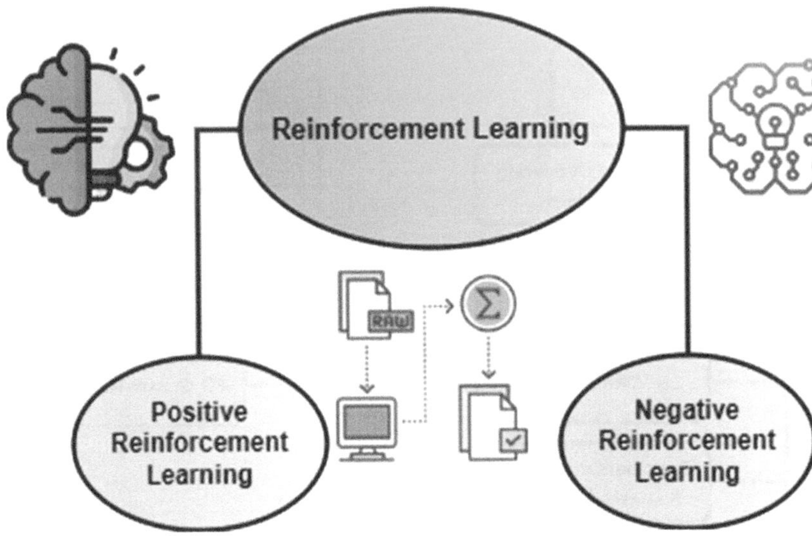

FIGURE 3.3 Types of reinforcement learning.

3.4 FUNDAMENTALS OF NATURAL LANGUAGE PROCESSING (NLP)

NLP is an interconnected discipline that includes software engineering, phonetics, and AI. The essential goal is to enable computers to understand, interpret, and communicate with human language in a way that corresponds to human perception.

3.4.1 INTRODUCTION TO NATURAL LANGUAGE PROCESSING (NLP)

Natural language processing, a truncation for NLP, is a multidisciplinary space that incorporates software engineering, etymology, and AI. The primary objective is to make it possible for computers to understand, translate, and interact with human language in a way that is comparable to human comprehension.

NLP plans to fill the gap between the intricacies of human language and the computational capacities of machines. This is achieved by coordinating computational approaches with measurable procedures. NLP is able to decipher the intricacies of natural language using these methods, obtaining meaning and structure from information that appears to lack organization [19].

3.4.2 KEY NATURAL LANGUAGE PROCESSING (NLP) TASKS

Tasks associated with the understanding and manipulation of human language are encompassed under the expansive domain of NLP. Numerous foundational NLP activities encompass:

Task	Description
Part-of-Speech Tagging	Assigning grammatical tags to each word in a sentence, such as noun, verb, adjective, or adverb.
Named Entity Recognition	Identifying and classifying named entities in a text, such as person names, organization names, locations, and dates.
Sentiment Analysis	Determining the sentiment or emotion expressed in a piece of text, such as positive, negative, or neutral.
Text Classification	Assigning predefined categories or labels to a piece of text based on its content or topic.
Machine Translation	Translating text from one language to another.
Text Summarization	Generating a concise summary of a longer piece of text, capturing the main points and key information.
Speech Recognition	Converting spoken language into written text.
Question Answering	Extracting specific information or facts from unstructured text and organizing it into structured formats.
Information Extraction	Extracting specific information or facts from unstructured text and organizing it into structured formats.
Dialogue Systems	Developing interactive systems that can engage in a conversation with users in natural language.

3.4.3 TOOLS AND LIBRARIES FOR NATURAL LANGUAGE PROCESSING (NLP)

NLP instances can be created utilizing a broad scope of devices and structures, ranging from basic text processing to robust AI models. NLP tools and libraries are outlined in the following list:

1. **NLTK:** It is a Python library that offers effective interfaces to more than 50 datasets and lexical resources. It also has text processing capabilities like tokenization, stemming, and parsing, making it useful for people who are just starting out with NLP.
2. **SpaCy:** The broadly utilized python NLP toolset, provides pre-trained models for various languages and many tasks like tokenization, part-of-speech (POS) tagging, NER, and dependency parsing. The reputation of this tool is established on its superior speed and effectiveness.
3. **Gensim:** It is an open-source topic modeling and similarity analysis package. It's usually used for word embedding (Word2Vec, FastText), topic modeling (LDA), and document similarity calculations.
4. **TextBlob:** It is a crucial Python library used for the processing of text-based information. POS tagging, noun phrase extraction, sentiment analysis, and text interpretation are just some of the common NLP tasks that can be carried out with the software's user-friendly and advanced interface [20].

3.5 THE INTERSECTION OF MACHINE LEARNING (ML) AND NATURAL LANGUAGE PROCESSING (NLP)

The fields of AI, ML, and NLP cooperate to examine the integral connection between information-driven calculations and etymological appreciation. The utilization of ML approaches empowers regular language handling models to learn examples, semantics, and logical data from huge datasets. The combination of these advances further develops etymological cycles, including feeling examination, named substance acknowledgment, and machine interpretation. The designs of neural networks, such as recurrent and transformer models, have revolutionized this intersection, as shown in Figure 3.4, making it possible to perform language processing that is both more complex and aware of its context. The combination of ML with NLP is essential to the progression of applications such as chatbots, language translation services, and the analysis of sentiment in social media [20].

3.5.1 HOW MACHINE LEARNING (ML) AND NATURAL LANGUAGE PROCESSING (NLP) COMPLEMENT EACH OTHER

Machine learning and NLP are two categories that are inextricably linked and complement each other in different ways. First and foremost, ML serves as the foundation for many NLP applications. Deep neural networks and other algorithms have revolutionized NLP by facilitating models such as transformer-based ones (e.g., BERT, GPT) to understand and generate human-like text. ML techniques are used in these models to process massive volumes of text input, learn patterns, and generate predictions.

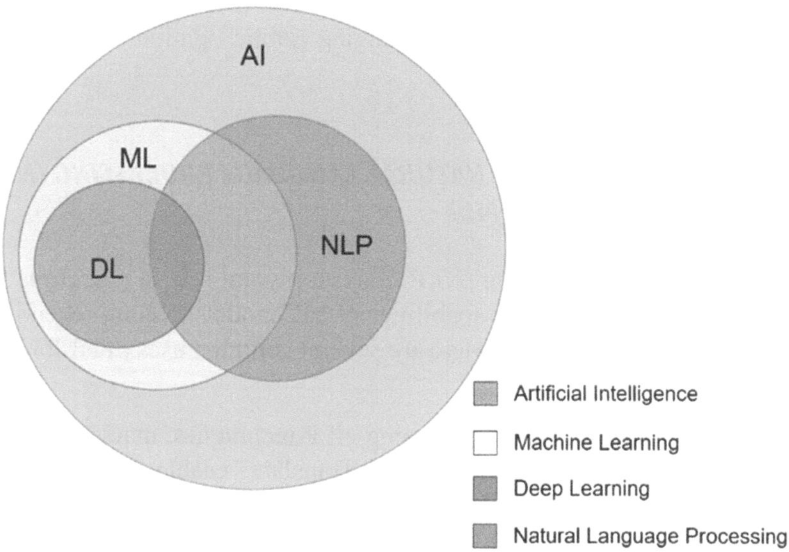

FIGURE 3.4 Relationship between artificial intelligence (AI), machine learning (ML), deep learning (DL), and natural language processing (NLP).

Natural language processing (NLP) on the other hand, improves ML by offering rich and organized data sources [21].

3.5.2 APPLICATIONS OF MACHINE LEARNING (ML) IN NATURAL LANGUAGE PROCESSING (NLP)

Machine learning (ML) plays a fundamental role in NLP by providing the tools and techniques to analyze and understand human language. Some key applications of ML in NLP include:

1. **Machine Translation:** ML models, specifically neural machine translation models, have greatly improved the accuracy and fluency of automatic translation systems like Google Translate.

2. **Speech Recognition:** Transforming spoken words into text is enabled by deep learning (DL), for example, by Siri and transcription services.
3. **Text Generation:** ML-based generative models such as GPT-3 produce coherent and contextually relevant text, which can be used in chatbots, content production, or creative writing.
4. **Sentiment Analysis:** It is crucial to use ML techniques to assess sentiment in text files for gauging public opinion and customer feedback.

3.5.3 APPLICATIONS OF NATURAL LANGUAGE PROCESSING (NLP) IN MACHINE LEARNING (ML)

Natural language processing (NLP) plays a pivotal role in numerous ML applications, enriching the capabilities of ML models to comprehend and work with human language. Here are several common uses of NLP within ML:

1. **Language Translation:** Using NLP techniques, machine translation models, such as Google Translate, enable cross-linguistic communication between languages.
2. **Speech-to-Text Conversion:** As a result of NLP, spoken language can be converted into written text, enabling voice assistants, transcription services, and voice-activated applications to operate.
3. **Text Summarization:** Automatic summarization models make use of NLP to produce concise summaries of lengthy texts, providing a mechanism for facilitating content curation and information retrieval (IR).
4. **Speech Synthesis:** Assistive technologies, voiceovers, and accessibility tools can benefit from NLP when it converts text into speech.
5. **Content Recommendations:** Personalized suggestions are produced through NLP-powered RSs. For example, when movie and product recommendations are analyzed based on user-generated content.

Machine learning (ML) applications benefit from NLP by effectively interacting with human language and finding applications across diverse fields [21].

3.6 CASE STUDIES

Certain case studies that demonstrate the convergence of ML and NLP in diverse applications are discussed as follows:

1. **Chatbots and Virtual Assistants:**
   i. **Case Study:** "Automated Customer Support."
   ii. **Description:** ML and NLP were applied synergistically to create an online chatbot that can assist clients independently. Consumers can navigate their purchase experience with this chatbot by answering common customer inquiries, resolving issues, and managing concerns. Consequently, human agents' workload was reduced, response times were accelerated, and customer satisfaction was enhanced.
2. **Language Translation:**
   i. **Case Study:** "Neural Machine Translation Advancements."
   ii. **Description:** ML and NLP techniques have been used by many internet-based translation services, including Google Translate. In this study, neural machine translation methods were used, which enhanced the quality of translation significantly by utilizing neural networks.
3. **Content Recommendation:**
   i. **Case Study:** "Personalized Content Recommendations on Streaming Platforms."
   ii. **Description:** Personalized recommendations are made based on user behavior, viewing histories, and text descriptions of material. As a result, higher user engagement and retention rates have been achieved through this methodology.
4. **Information Extraction:**
   i. **Case Study:** "News Aggregation and Summarization."
   ii. **Description:** News aggregation platforms use ML and NLP to extract key information from articles. A system like this identifies the most important entities within a news story, events that transpire, and sentiments that are portrayed.
5. **Healthcare and Medical Records:**
   i. **Case Study:** "Clinical Document Processing for Healthcare."
   ii. **Description:** The healthcare industry has used ML and NLP to analyze and extract crucial data from medical records,

physician notes, and research papers. Healthcare practitioners can utilize these technologies to obtain patient data quickly, streamline diagnoses, and enhance research effectiveness.

6. **Legal Industry:**
   i. **Case Study:** "Legal Document Analysis."
   ii. **Description:** A lot of legitimate records, agreements, and reports are analyzed by lawful associations and divisions utilizing AI (ML) and NLP. As well as advancing report search and revelation methods, NLP works with the extraction of pivotal sentences, the identification of significant legitimate points of reference, and the identification of critical lawful points of reference.
7. **Financial Services:**
   i. **Case Study:** "Fraud Detection in Financial Transactions."
   ii. **Description:** ML and NLP are essential for detecting and preventing fraud in the banking industry. Financial institutions must carefully examine transaction descriptions, interactions with customers, and consumer behavior in addition to the transactions themselves in order to identify and prevent fraud.

These case studies provide concrete examples of how ML and NLP converge to bring about advancements in various industries, from enhancing customer service to facilitating data-driven decision-making and automation [15, 22].

3.7 CONCLUSION

Besides, the convergence of AI, ML and NLP comprises a huge forward-moving step in our endeavors to limit the differentiation between human mental capacity and computational ability. It has become obvious from the examination of this connecting and helpful undertaking that it has enormous potential for change that will influence the global community in a significant manner. AI uses its capacity to extract valuable information from large datasets to provide us with predictive analysis and automation. Meanwhile, NLP enables machines to understand the complexity of human language while also producing it accurately and appropriately. A new era of exceptional creativity has emerged, reshaping industries and rethinking our relationship with technology [23–27].

KEYWORDS

- **artificial intelligence**
- **automation**
- **information extraction**
- **language translation**
- **machine learning**
- **natural language processing**
- **pre-trained models**
- **sentiment analysis**

REFERENCES

1. Harnal, S., Jain, A., Rathore, A. S., Baggan, V., Kaur, G., & Bala, R. (2023). Comparative approach for early diabetes detection with machine learning. In *2023 International Conference on Emerging Smart Computing and Informatics (ESCI)* (pp. 1–6). IEEE.
2. Kaur, G., Kaur, V., Sharma, Y., & Bansal, V. (2022). Analyzing various machine learning algorithms with SMOTE and ADASYN for image classification having imbalanced data. In *2022 IEEE International Conference on Current Development in Engineering and Technology (CCET)* (pp. 1–7). IEEE.
3. Ali, I., Mughal, N., Khand, Z. H., Ahmed, J., & Mujtaba, G. (2022). Resume classification system using natural language processing and machine learning techniques. *Mehran University Research Journal of Engineering & Technology, 41*(1), 65–79.
4. Pandiaraja, P., Boopesh, K. B., Deepthi, T., Laksmi Priya, M., & Noodhana, R. (2022, February). An analysis of document summarization for educational data classification using NLP with machine learning techniques. In *International Conference on Computing in Engineering & Technology* (pp. 127–143). Singapore: Springer Nature Singapore.
5. Lalitha, T. B., & Sreeja, P. S. (2023). Potential web content identification and classification system using NLP and machine learning techniques. *International Journal of Engineering Trends and Technology, 71*(4), 403–415.
6. Mahesh, B. (2020). Machine learning algorithms—A review. *International Journal of Science and Research (IJSR), 9*(1), 381–386.
7. Osisanwo, F. Y., Akinsola, J. E. T., Awodele, O., Hinmikaiye, J. O., Olakanmi, O., & Akinjobi, J. (2017). Supervised machine learning algorithms: Classification and comparison. *International Journal of Computer Trends and Technology (IJCTT), 48*(3), 128–138.
8. Khanum, M., Mahboob, T., Imtiaz, W., Ghafoor, H. A., & Sehar, R. (2015). A survey on unsupervised machine learning algorithms for automation, classification, and maintenance. *International Journal of Computer Applications, 119*(13).

9. Liddy, E. D. (2001). *Natural Language Processing.*
10. Nadkarni, P. M., Ohno-Machado, L., & Chapman, W. W. (2011). Natural language processing: An introduction. *Journal of the American Medical Informatics Association, 18*(5), 544–551.
11. Jones, K. S. (1994). Natural language processing: A historical review. *Current Issues in Computational Linguistics: In Honour of Don Walker*, 3–16.
12. Hristea, F., & Caragea, C. (2022). Preface to the special issue "natural language processing (NLP) and machine learning (ML)—Theory and applications." *Mathematics, 10*(14), 2481.
13. Le Glaz, A., Haralambous, Y., Kim-Dufor, D. H., Lenca, P., Billot, R., Ryan, T. C., & Lemey, C. (2021). Machine learning and natural language processing in mental health: Systematic review. *Journal of Medical Internet Research, 23*(5), e15708.
14. Tusar, M. T. H. K., & Islam, M. T. (2021). A comparative study of sentiment analysis using NLP and different machine learning techniques on US airline Twitter data. In *2021 International Conference on Electronics, Communications and Information Technology (ICECIT)* (pp. 1–4). IEEE.
15. Kulkarni, C. S., Bhavsar, A. U., Pingale, S. R., & Kumbhar, S. S. (2017). BANK CHAT BOT – An intelligent assistant system using NLP and machine learning. *International Research Journal of Engineering and Technology, 4*(5), 2374–2377.
16. Aggarwal, M., Khullar, V., Goyal, N., Singh, A., Tolba, A., Thompson, E. B., & Kumar, S. (2023). Pre-trained deep neural network-based features selection supported machine learning for rice leaf disease classification. *Agriculture, 13*(5), 936.
17. Harnal, S., Sharma, G., Malik, S., Kaur, G., Simaiya, S., Khurana, S., & Bagga, D. (2023). Current and future trends of cloud-based solutions for healthcare. In *Image Based Computing for Food and Health Analytics: Requirements, Challenges, Solutions and Practices (IBCFHA)* (pp. 115–136).
18. Kaur, V., Kumar, P., Kaur, G., & Kaur, A. (2023). Improved facial biometric authentication using MobileNetV2. In *2023 International Conference on Computational Intelligence and Sustainable Engineering Solutions (CISES)* (pp. 376–380). IEEE.
19. Kaur, G., Choudhary, P., Sahore, L., Gupta, S., & Kaur, V. (2023). Healthcare: In the era of blockchain. In *AI and Blockchain in Healthcare* (pp. 45–55). Singapore: Springer Nature Singapore.
20. Reddy, K., Reddy, M. A., Kaur, V., & Kaur, G. (2022). Career guidance system using ensemble learning. *Akhila and Kaur, Veerpal and Kaur, Gagandeep, Career Guidance System Using Ensemble Learning.*
21. Khullar, V., Tiwari, R. G., Agarwal, A. K., & Dutta, S. (2022). Physiological signals-based anxiety detection using ensemble machine learning. In *Cyber Intelligence and Information Retrieval: Proceedings of CIIR 2021* (pp. 597–608). Springer Singapore.
22. Arias-Barahona, M. X., Arteaga-Arteaga, H. B., Orozco-Arias, S., Flórez-Ruíz, J. C., Valencia-Díaz, M. A., & Tabares-Soto, R. (2023). Requests classification in the customer service area for software companies using machine learning and natural language processing. *PeerJ Computer Science, 9*, e1016.
23. Kaur, G., Goyal, S., & Kaur, H. (2021). Brief review of various machine learning algorithms. In *Proceedings of the International Conference on Innovative Computing & Communication (ICICC).*

24. Bindal, R., Sarangi, P. K., Kaur, G., & Dhiman, G. (2019). An approach for automatic recognition system for Indian vehicles numbers using k-nearest neighbors and decision tree classifier. *International Journal of Advanced Science and Technology, 28*(9), 477–492.
25. Kaur, V., Kaur, G., Dhiman, G., Bindal, R., & Mishra, M. K. (2020). Adaptability of machine learning in cryptography. *Solid State Technology, 63*(4), 2874–2880.
26. Sharifani, K., Amini, M., Akbari, Y., & Aghajanzadeh Godarzi, J. (2022). Operating machine learning across natural language processing techniques for improvement of fabricated news model. *International Journal of Science and Information System Research, 12*(9), 20–44.
27. Silvestri, S., Islam, S., Papastergiou, S., Tzagkarakis, C., & Ciampi, M. (2023). A machine learning approach for the NLP-based analysis of cyber threats and vulnerabilities of the healthcare ecosystem. *Sensors, 23*(2), 651.

CHAPTER 4

Navigating the AI Landscape: Architectures and Algorithms for Natural Language Processing

VIKAS SHARMA[1,*], SUNIL KUMAR[2], TARUN KUMAR VASHISHTH[3], MEGHNA[4], BHUPENDRA KUMAR[3], and KEWAL KRISHAN SHARMA[3]

[1]*Department of Computer Applications, SRM Institute of Science and Technology, Delhi NCR Campus, Ghaziabad, U.P., India*

[2]*School of Information Technology, Auro University, Gujrat, India*

[3]*School of Computer Science and Applications, IIMT University, Meerut, Uttar Pradesh, India*

[4]*School of Civil Engineering, IIMT University, Meerut, Uttar Pradesh, India*

**Corresponding author.*

ABSTRACT

In the ever-evolving landscape of artificial intelligence (AI), the domain of natural language processing (NLP) stands as a profound testament to the remarkable synergy between technology and human communication. This abstract introduces "Navigating the AI Landscape: Architectures and Algorithms for NLP," a comprehensive exploration of the intricacies, advancements, and challenges at the intersection of AI architectures and NLP algorithms. The quest for enabling machines to understand and generate human language has been a cornerstone of AI research for decades. This interdisciplinary endeavor has resulted in the development of various architectures and algorithms, each possessing distinct strengths and applications. *Navigating the AI Landscape* embarks on a journey through

Natural Language Processing and Artificial Intelligence: A Perspective Toward Current Trends, Challenges, and Applications. Vikas Khullar, Aryan Chaudhary, & Sunil Kumar (Eds.)

DOI: 10.1201/9781998511273-4

this intricate web of technologies, illuminating their inner workings and practical implications. While progress in NLP has been remarkable, it is not without challenges. Ethical considerations, bias mitigation, and interpretability remain at the forefront of discussions. This symposium provides a platform to examine these critical issues and to discuss emerging trends in AI fairness and transparency. Additionally, *Navigating the AI Landscape* recognizes the practical applications of NLP across diverse sectors, including healthcare, finance, customer service, and education. Through insightful case studies and practical demonstrations, participants will develop a comprehensive understanding of how AI architectures and algorithms are influencing various industries. In summary, this symposium aims to be a compass in the vast landscape of AI, guiding attendees through the architectures and algorithms that underpin the remarkable achievements in NLP. By facilitating discussions and sharing insights, it aspires to foster collaboration, ethical awareness, and innovation in the pursuit of harnessing AI for the betterment of society. Join us as we navigate the AI landscape and chart a course toward a more intelligent and language-aware future.

4.1 INTRODUCTION

4.1.1 THE EVOLUTION OF NATURAL LANGUAGE PROCESSING (NLP)

The evolution of NLP refers to the historical development and milestones in the field of NLP, which is a subfield of artificial intelligence (AI) that focuses on enabling computers to understand, interpret, and generate human language. NLP has a rich history dating back to the mid-20th century, and its evolution has been marked by significant advancements in technology, linguistic theory, and data-driven approaches. Let us explore the evolution of NLP in detail:

4.1.1.1 EARLY FOUNDATIONS (1950S–1960S)

- The origins of NLP can be traced back to the 1950s when computer scientists and linguists began to explore the idea of using computers to work with human language.
- One of the earliest and most influential developments during this period was the creation of machine translation systems, such as the Georgetown-IBM experiment, which aimed to automatically translate languages.

- Researchers attempted to develop rule-based systems to understand and process language, but they faced limitations due to the complexity of natural language.

4.1.1.2 SYMBOLIC ARTIFICIAL INTELLIGENCE (AI) AND RULE-BASED APPROACHES (1970S–1980S)

- During the 1970s and 1980s, NLP primarily relied on symbolic AI and rule-based approaches. Linguists and computer scientists worked on developing grammatical rules and formal language representations.
- Systems like SHRDLU, which could understand and respond to natural language commands in a limited domain, demonstrated early success in rule-based NLP.
- However, rule-based systems struggled with scalability and generalization to broader language understanding.

4.1.1.3 STATISTICAL METHODS (1990S–2000S)

- The late 1990s and early 2000s saw a shift towards statistical methods in NLP. Researchers started using large datasets and probabilistic models to address the shortcomings of rule-based systems.
- Hidden Markov Models (HMMs) and statistical machine translation (SMT) systems gained prominence during this era.
- The introduction of the Penn Treebank and the development of statistical parsers helped improve syntactic and grammatical analysis.

4.1.1.4 MACHINE LEARNING AND NEURAL NETWORKS (2010S–PRESENT)

- The 2010s marked a transformative period for NLP with the rise of machine learning (ML) and deep learning (DL) techniques.
- Word embeddings, such as Word2Vec and GloVe, revolutionized how computers understand the meanings of words and their relationships.
- The introduction of neural network architectures like recurrent neural networks (RNNs) and especially the transformer architecture

(as seen in models like BERT, GPT-3, and others) has dramatically improved language understanding and generation.

- Transfer learning (TL) became a prominent approach, allowing models pre-trained on large text corpora to be fine-tuned for specific NLP tasks, making NLP more accessible and effective.

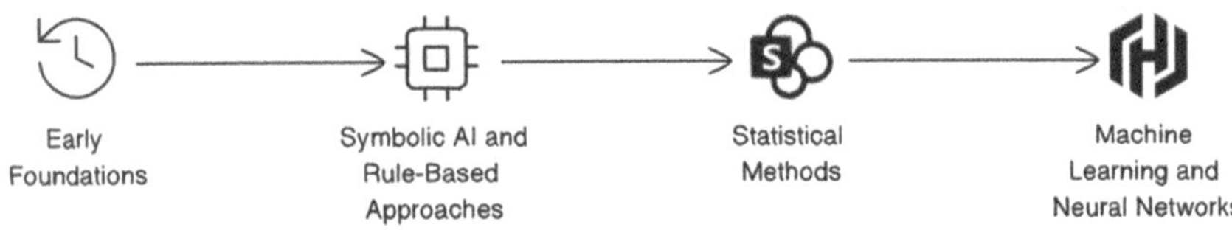

FIGURE 4.1 Evolution of natural language processing.

Hence, the evolution of NLP has transitioned from early rule-based approaches to statistical methods and, most significantly, to the era of DL and large language models (LLMs) (Figure 4.1). These advancements have made NLP an integral part of AI and have opened up numerous possibilities for applications across industries. NLP continues to evolve, with ongoing research and development driving its growth and impact.

4.1.2 SIGNIFICANCE OF NATURAL LANGUAGE PROCESSING (NLP) IN THE ARTIFICIAL INTELLIGENCE (AI) LANDSCAPE

The significance of NLP in the AI landscape is profound and multifaceted. NLP plays a critical role in advancing AI and has a wide range of applications that impact various industries and domains. Here's a detailed explanation of the significance of NLP in the AI landscape:

4.1.2.1 HUMAN-COMPUTER INTERACTION

- NLP enables more natural and intuitive interactions between humans and computers. Users can communicate with AI systems using spoken or written language, making technology more accessible.
- Applications include voice-activated assistants (e.g., Siri, Google Assistant), virtual keyboards, and speech-to-text interfaces.

4.1.2.2 INFORMATION RETRIEVAL AND SEARCH ENGINES

- NLP powers search engines to understand user queries better. Techniques like query understanding, document ranking, and semantic search deliver more accurate results.
- Google's BERT model, for instance, improved search results by understanding the context of words in queries.

4.1.2.3 CHATBOTS AND VIRTUAL ASSISTANTS

- NLP forms the core of chatbots and virtual assistants. These systems can understand user questions, provide answers, perform tasks, and engage in natural conversations.
- They are used in customer support, scheduling appointments, and automating routine tasks.

4.1.2.4 SENTIMENT ANALYSIS AND OPINION MINING

- NLP analyzes social media posts, reviews, and comments to determine public sentiment. This helps businesses understand customer feedback and make data-driven decisions.
- Companies use sentiment analysis to gauge brand perception and tailor marketing strategies.

4.1.2.5 LANGUAGE TRANSLATION

- Machine translation, powered by NLP, automatically translates text and speech between languages. This breaks down language barriers in communication and global business.
- Google Translate and other translation tools rely on NLP techniques.

4.1.2.6 FINANCE AND TRADING

- NLP analyzes financial news, reports, and social media sentiment to make investment decisions and predict market trends.

- Hedge funds and trading firms employ NLP for algorithmic trading strategies.

4.1.2.7 LEGAL AND COMPLIANCE

- NLP helps law firms and regulatory bodies review and extract insights from legal documents, contracts, and compliance records.
- Legal tech solutions use NLP for contract analysis and e-discovery.

4.1.2.8 CONTENT RECOMMENDATION

- Streaming platforms and e-commerce sites use NLP to understand user preferences and recommend content or products.
- Netflix's recommendation system (RS) relies on NLP for personalized content suggestions.

4.1.2.9 ACCESSIBILITY AND INCLUSION

- NLP technologies assist individuals with disabilities by providing speech-to-text, text-to-speech, and other accessibility features.
- Screen readers, speech recognition software, and communication aids use NLP.

4.1.2.10 EDUCATION

- NLP-powered tools automate grading, provide personalized learning experiences, and support language learning.
- Intelligent tutoring systems analyze student performance and adapt instructional content.

4.1.2.11 ETHICAL AND BIAS MITIGATION

- NLP is crucial for identifying and addressing biases in AI systems to ensure fair and ethical AI deployment.

- Bias detection models and guidelines promote fairness and transparency.

4.1.2.12 SECURITY AND FRAUD DETECTION

- NLP assists in identifying fraudulent activities by analyzing text data and detecting anomalies in communication patterns.
- Banks and cybersecurity firms use NLP for fraud detection and risk assessment.

4.1.2.13 GOVERNMENT AND PUBLIC SERVICES

- NLP is used for processing citizen inquiries, analyzing government documents, and improving public service interactions.
- Automated chatbots handle routine government queries and streamline services.

4.1.2.14 MULTILINGUAL COMMUNICATION

- NLP models that support multiple languages facilitate cross-cultural communication and information dissemination.
- Multilingual chatbots and translation services bridge language gaps.

4.1.2.15 CONTINUOUS LEARNING AND IMPROVEMENT

- NLP systems can learn and adapt over time through ML techniques, improving their performance as they encounter more data.
- Models like GPT-3 and BERT fine-tune themselves on massive datasets to become more proficient in understanding and generating human language.

These diverse applications demonstrate the pervasive impact of NLP across various industries and facets of daily life. As NLP technologies continue to advance, they will undoubtedly play an increasingly pivotal role in shaping the way we communicate, work, learn, and interact with the world around us.

4.2 FUNDAMENTALS OF NATURAL LANGUAGE PROCESSING (NLP)

The fundamentals of NLP are crucial for grasping the underlying concepts and principles of this interdisciplinary domain that connects computer science and linguistics. The subsections offer a thorough exploration of the foundational elements.

4.2.1 WHAT IS NATURAL LANGUAGE PROCESSING (NLP)?

4.2.1.1 DEFINING NATURAL LANGUAGE

- Natural language refers to the language that humans use for communication, such as English, Spanish, or Mandarin. It is characterized by its complexity, ambiguity, and the use of syntax, semantics, and pragmatics.
- NLP aims to enable computers to understand, interpret, and generate human language, allowing them to interact with humans in a way that is both meaningful and useful.

4.2.1.2 THE ROLE OF COMPUTERS IN PROCESSING LANGUAGE

- Computers, being deterministic machines, require structured data to operate efficiently. Natural language, being inherently unstructured, poses a challenge for computers to understand and process.
- NLP leverages computational techniques and algorithms to bridge the gap between human language and machine language. It involves tasks such as language understanding, language generation, and language translation (Figure 4.2).

4.2.2 KEY COMPONENTS OF NATURAL LANGUAGE PROCESSING (NLP)

The key components of NLP are given in Figure 4.3.

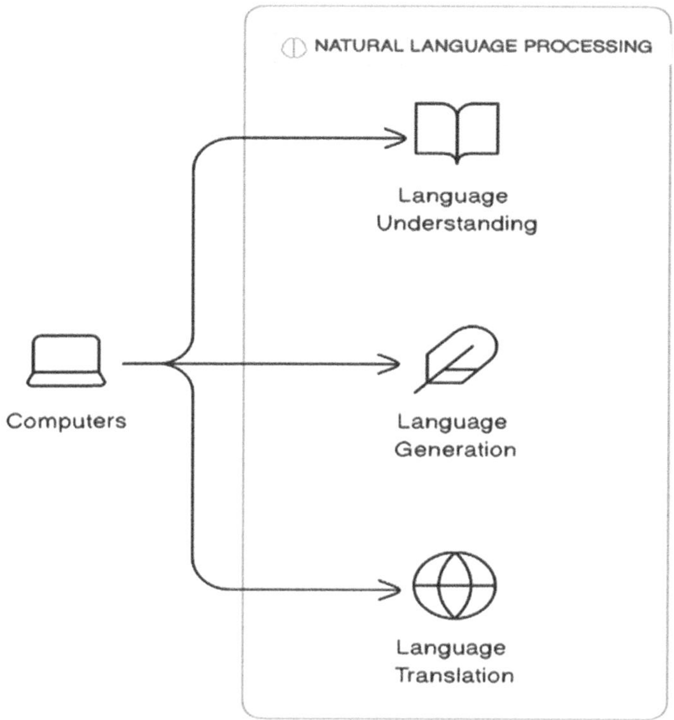

FIGURE 4.2 Role of computers in processing language.

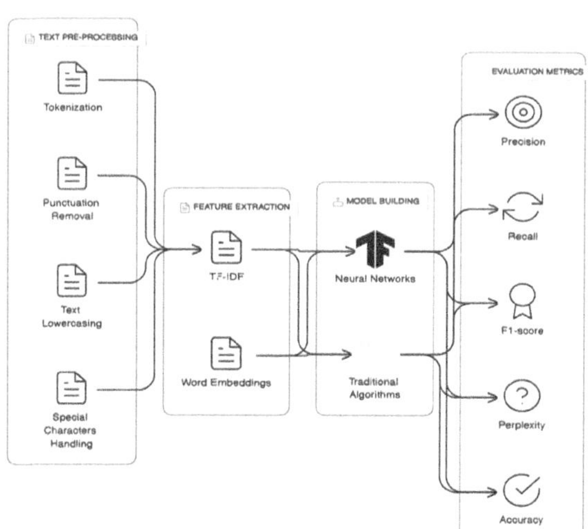

FIGURE 4.3 Components of natural language processing (NLP).

4.2.2.1 TEXT PRE-PROCESSING

- Text pre-processing involves cleaning and preparing raw text data for NLP tasks. It includes tasks like tokenization (breaking text into words or tokens), removing punctuation, converting text to lowercase, and handling special characters.
- Pre-processing is crucial for ensuring that the data is in a format that NLP algorithms can work with effectively.

4.2.2.2 FEATURE EXTRACTION

- Feature extraction involves converting text data into numerical or vector representations that ML models can use. Common techniques include TF-IDF and word embeddings (e.g., Word2Vec, GloVe).
- These features capture the semantics and relationships between words in a text, making it easier for algorithms to work with.

4.2.2.3 MODEL BUILDING

- Model building is the core of NLP, where ML models, such as neural networks and traditional algorithms, are trained to perform specific NLP tasks. Some common NLP tasks include sentiment analysis, named entity recognition (NER), and machine translation.
- Models are trained on annotated datasets and learn patterns and relationships within the data.

4.2.2.4 EVALUATION METRICS

- To assess the performance of NLP models, various evaluation metrics are used, depending on the specific task. Common metrics include accuracy, precision, recall, F1 score, and perplexity.
- Choosing the appropriate evaluation metric is essential to determine how well an NLP system is performing and whether it meets the intended objectives.

4.2.3 CHALLENGES IN NATURAL LANGUAGE PROCESSING (NLP)

4.2.3.1 AMBIGUITY

- Natural language is inherently ambiguous, with words and phrases often having multiple meanings depending on context. Resolving this ambiguity is a significant challenge in NLP.
- Techniques like contextual analysis and semantic disambiguation are used to address ambiguity in language understanding.

4.2.3.2 CONTEXTUAL UNDERSTANDING

- Understanding the context of a sentence or conversation is crucial for accurate NLP. Many NLP tasks, such as question answering and sentiment analysis, require a deep understanding of context.
- Recent advancements in NLP, including transformer-based models like BERT, have significantly improved contextual understanding.

4.2.3.3 MULTILINGUALISM

- NLP models need to work effectively in multiple languages to be globally useful. However, different languages have unique linguistic structures and challenges.
- Multilingual NLP models and cross-lingual techniques are developed to address multilingualism in NLP.

4.2.3.4 ETHICS AND BIAS

- NLP systems can inadvertently perpetuate biases present in training data, leading to unfair or biased outcomes. Addressing bias and ensuring ethical AI is a critical concern in NLP.
- Researchers and practitioners work on developing fairness-aware models and bias mitigation techniques to promote fairness and the ethical use of NLP systems.

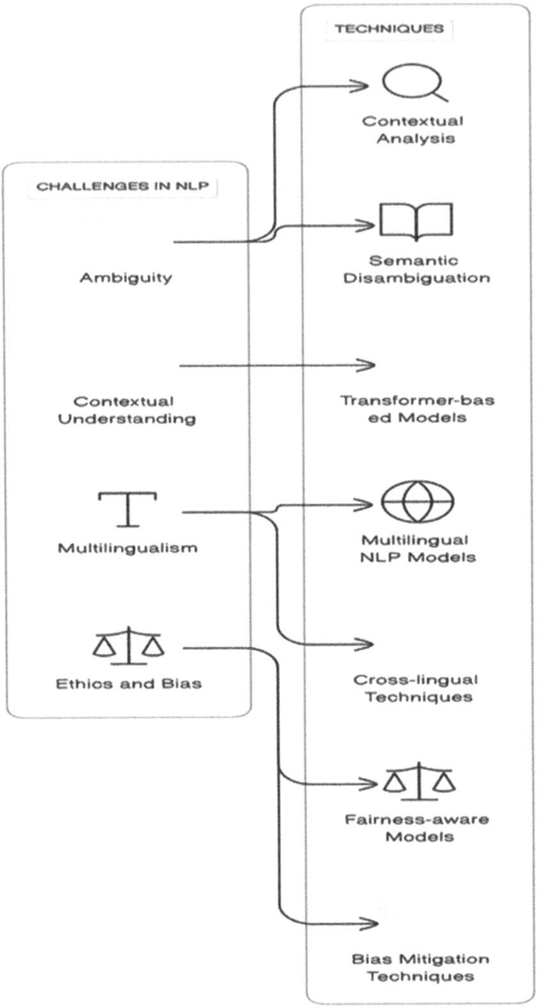

FIGURE 4.4 Challenges in natural language processing (NLP).

Hence, the fundamentals of NLP encompass understanding the nature of human language, the role of computers in processing it, key components like pre-processing, feature extraction, and model building, as well as addressing challenges such as ambiguity, contextual understanding, multilingualism, ethics, and bias (Figure 4.4). NLP continues to evolve, driven by ongoing research and the increasing demand for applications that leverage the power of language understanding and generation.

4.3 NATURAL LANGUAGE PROCESSING (NLP) ARCHITECTURES

NLP architectures refer to the different methodologies and techniques used in NLP to process, understand, and generate human language. These architectures have evolved over time, and each has its strengths and weaknesses. Here's a detailed explanation of various NLP architectures:

4.3.1 RULE-BASED APPROACHES

4.3.1.1 SYNTAX AND GRAMMAR RULES

- Rule-based NLP systems use predefined linguistic rules and grammatical structures to analyze and parse text. These rules define the relationships between words in a sentence.
- For example, a rule-based system might identify subject-verb-object (SVO) patterns in a sentence to extract meaning.

4.3.1.2 REGULAR EXPRESSIONS

- Regular expressions are patterns used to match and manipulate text. In NLP, they can be employed to find specific patterns, extract entities (e.g., email addresses, phone numbers), or perform text cleaning.
- Regular expressions are powerful but can be limited when dealing with complex linguistic nuances.

4.3.2 STATISTICAL AND MACHINE LEARNING APPROACHES

4.3.2.1 BAG OF WORDS (BOW)

- BoW represents text as a collection of words, ignoring the order and structure of the words in a document. It creates a vector where each dimension corresponds to a unique word in the corpus.
- BoW is often used for text classification tasks like spam detection, sentiment analysis, and document categorization.

4.3.2.2 WORD EMBEDDINGS (E.G., WORD2VEC, GLOVE)

- Word embeddings are dense vector representations of words in a continuous vector space. They capture semantic relationships between words.
- Word2Vec and GloVe are popular techniques that enable NLP models to understand word similarity, context, and analogies. They improve tasks like language understanding and machine translation.

4.3.2.3 RECURRENT NEURAL NETWORKS (RNNS)

- RNNs are a type of neural network architecture designed for sequence data, including text. They have feedback loops that allow them to maintain a hidden state to capture sequential dependencies.
- RNNs are used for tasks like text generation, speech recognition, and machine translation. However, they suffer from vanishing gradient problems for long sequences.

4.3.2.4 CONVOLUTIONAL NEURAL NETWORKS (CNNS)

- While commonly associated with computer vision, CNNs can also be applied to NLP tasks. In text processing, 1D convolutions are used to detect patterns and features within sequences of words.
- CNNs excel at tasks like text classification and sentiment analysis due to their ability to capture local patterns (Figure 4.5).

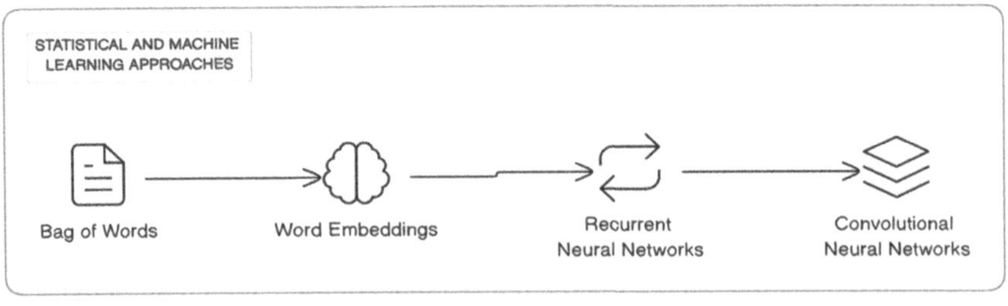

FIGURE 4.5 Statistical and machine learning approaches.

4.3.3 DEEP LEARNING ARCHITECTURES

4.3.3.1 LONG SHORT-TERM MEMORY (LSTM) NETWORKS

- LSTMs are a type of RNN designed to address the vanishing gradient problem. They can capture long-range dependencies in text data and are suitable for sequence-to-sequence tasks.
- LSTMs are commonly used in machine translation, speech recognition, and text summarization.

4.3.3.2 TRANSFORMER ARCHITECTURE

- The Transformer architecture, as introduced in the paper *Attention is All You Need*, has revolutionized NLP. This architecture relies on self-attention mechanisms to effectively process input data.
- Models like bidirectional encoder representations from transformers (BERT) and generative pre-trained transformer (GPT) have achieved state-of-the-art performance in various NLP tasks, including language understanding, generation, and translation (Figure 4.6).

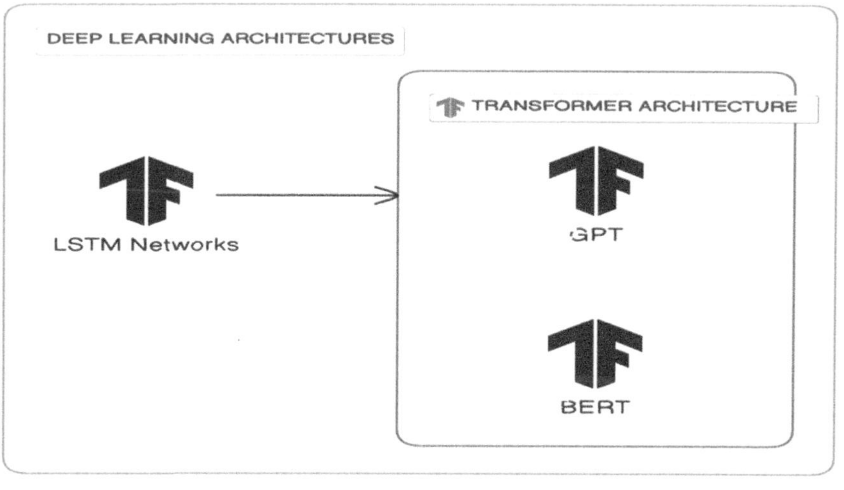

FIGURE 4.6 Deep learning architectures.

4.3.4 HYBRID APPROACHES

4.3.4.1 COMBINING RULE-BASED AND MACHINE LEARNING TECHNIQUES

- Hybrid approaches leverage the strengths of both rule-based and ML methods. They use linguistic rules and heuristics to preprocess text and guide the feature extraction or model-building process.
- For example, a rule-based system might identify entities like names and dates, which are then used as features in a ML model for information extraction.
- Hybrid approaches aim to overcome limitations in rule-based or statistical methods alone, offering more robust and accurate NLP solutions.

In summary, NLP architectures encompass a wide range of techniques and methodologies for processing and understanding natural language. While rule-based approaches rely on explicit linguistic rules, statistical and ML methods leverage data-driven approaches, and DL architectures like transformers have pushed the boundaries of NLP performance. Hybrid approaches combine the strengths of different techniques to address the complexities and challenges of NLP tasks. The choice of architecture often depends on the specific NLP task and the available data.

4.4 LITERATURE REVIEW

Hancock [1], in their article, aggregates prior research results demonstrating that shopper engagement technologies and data visualization tools facilitate immersive 3D experiences within user-generated digital virtual environments.

Junnu [2] offers a comprehensive survey of the pivotal role NLP plays in the healthcare landscape. The text outlines future trends, encompassing advancements in DL, multimodal data integration, explainable AI (XAI), and the potential for real-time NLP within clinical settings.

Chen et al. [3] provide a detailed account of research on four fundamental NLP tasks: information retrieval (IR), NER, literature-based discovery, and question answering. Additionally, they present work that directly addresses pandemic-related aspects through four more tasks: topic modeling, sentiment and emotion analysis, caseload forecasting, and misinformation detection.

Yang et al. [4] offer a comprehensive review encompassing AI-based NLP, its applications in healthcare and ophthalmology, future use cases, and potential deployment challenges.

Kang et al. [5] provide an extensive description of toolkits and procedural steps for using NLP as an analytical technique, along with a comprehensive examination of its advantages and disadvantages.

In their study, Álvarez-Carmona et al. [6] explored the significant advantages offered by AI, specifically NLP, in collecting and analyzing extensive user-generated content related to tourism services and products. They conducted a systematic review, providing valuable insights into the use of NLP in the tourism industry and research.

Wah et al. [7] presented the results of a workshop aimed at identifying applications, research challenges, and design considerations for high-performance computing and communications (HPCC) systems to support various applications.

Deng and Liu [8] commenced by offering an introductory overview of the fundamentals of NLP as an essential component of AI.

Wah et al. [7] introduced a novel approach for expediting neural architecture search in NLP through knowledge distillation, denoted as KD-NAS.

Danilevsky et al. [9] conducted a comprehensive survey on the contemporary status of XAI within the scope of NLP.

In this chapter by Chowdhary and Chowdhary [10], the challenges of NLP, advancements to date, NLP applications, components of NLP, and the grammatical requirements of the English language as needed by machines are presented.

Koroteev [11] provides an overview of the application of BERT, a widely used DL-based language model. The review explains BERT's workings, its primary applications in text analytics tasks, comparisons with similar models in various tasks, and outlines some proprietary models.

Galassi, Lippi, and Torroni [12] present a unified model for attention architectures in NLP, emphasizing their compatibility with vector representations of textual data.

4.5 NATURAL LANGUAGE PROCESSING (NLP) ALGORITHMS

NLP algorithms refer to the various computational techniques and methods used to process and extract meaningful information from natural language text. These algorithms are crucial for a wide range of NLP tasks. Let us delve into the details of these NLP algorithms.

4.5.1 TOKENIZATION AND TEXT PRE-PROCESSING

Text preprocessing techniques and their role in various NLP tasks are illustrated in Figure 4.7.

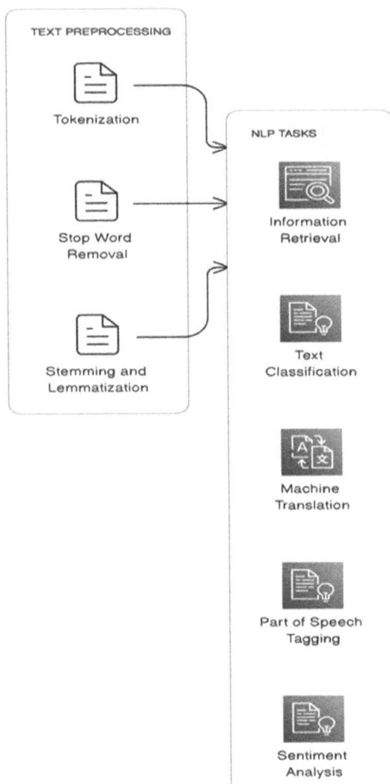

FIGURE 4.7 Tokenization and text pre-processing.

4.5.1.1 TOKENIZERS

- Tokenization is the process of breaking down a text into smaller units, called tokens. Tokens are typically words or subwords.

- Tokenizers split a sentence into individual words or subword units, making it easier for NLP models to understand and process text.
- Tization helps create structured input for tasks like sentiment analysis, part-of-speech (POS) tagging, and machine translation.

4.5.1.2 STOP WORD REMOVAL

- Stop words are common words like "and," "the," "in," and "is" that do not carry significant meaning on their own.
- Removing stop words from text can improve the efficiency of NLP algorithms and reduce the dimensionality of text data.
- This pre-processing step is essential for many tasks, including text classification and IR.

4.5.1.3 STEMMING AND LEMMATIZATION

- Stemming and lemmatization are techniques used to reduce words to their base or root form.
- Stemming involves chopping off prefixes or suffixes from words to obtain the root form (e.g., "running" becomes "run").
- Lemmatization, on the other hand, maps words to their dictionary or lemma form (e.g., "better" becomes "good").
- Stemming and lemmatization help in reducing the dimensionality of text data and improving the consistency of word representations.

4.5.2 NAMED ENTITY RECOGNITION (NER)

- NER is the process of identifying and classifying named entities in text, such as names of people, organizations, dates, locations, and more.
- NER algorithms use ML models, often trained on annotated data, to recognize and label these entities.
- NER is crucial for tasks like information extraction, event detection, and content recommendation.

4.5.3 SENTIMENT ANALYSIS

- Sentiment analysis, also known as opinion mining, determines the sentiment or emotional tone expressed in text, such as positive, negative, or neutral.
- It uses ML models to classify text as sentiment-positive, sentiment-negative, or sentiment-neutral.
- Applications include social media monitoring, customer feedback analysis, and brand sentiment tracking.

4.5.4 PART-OF-SPEECH TAGGING

- Part-of-speech tagging assigns grammatical tags (e.g., noun, verb, adjective) to each word in a sentence.
- It helps in understanding the grammatical structure of a sentence and disambiguating word meanings based on their context.
- Part-of-speech tagging is essential for syntactic parsing, machine translation, and grammar checking.

4.5.5 MACHINE TRANSLATION

- Machine translation algorithms translate text from one language to another automatically.
- Statistical methods, rule-based approaches, and neural machine translation models (like the transformer-based models) are used for this task.
- Machine translation has applications in global communication, content localization, and cross-language IR.

4.5.6 QUESTION ANSWERING SYSTEMS

- Question-answering systems are designed to provide concise and contextually relevant answers to user questions.
- These systems use various techniques, including IR, passage ranking, and NLU.
- Question-answering systems are used in chatbots, search engines, virtual assistants, and educational applications.

4.5.7 LANGUAGE GENERATION

- Language generation algorithms create human like text, often in response to a prompt or input.
- Techniques include rule-based templates, sequence-to-sequence models, and transformer-based models.
- Language generation is used in chatbots, content generation, and creative writing assistance.

4.5.8 COREFERENCE RESOLUTION

- Coreference resolution identifies when two or more words or phrases in a text refer to the same entity.
- It helps in understanding the relationships between entities in a text and ensures coherent discourse.
- Coreference resolution is important for tasks like text summarization and information extraction.

Natural language processing algorithms are foundational to the field of NLP, enabling computers to understand, extract information from, and generate human language text. These algorithms address a wide range of tasks, from basic text pre-processing to advanced tasks like sentiment analysis, machine translation, and coreference resolution. The choice of algorithm depends on the specific NLP task and the desired outcome.

4.6 ETHICAL CONSIDERATIONS IN NATURAL LANGUAGE PROCESSING (NLP)

Ethical considerations in NLP are critical, as NLP technologies have significant societal and ethical implications. This section delves into the ethical considerations in NLP in detail:

4.6.1 BIAS AND FAIRNESS

Sources of bias and strategies for mitigating bias in machine learning models are illustrated in Figure 4.8.

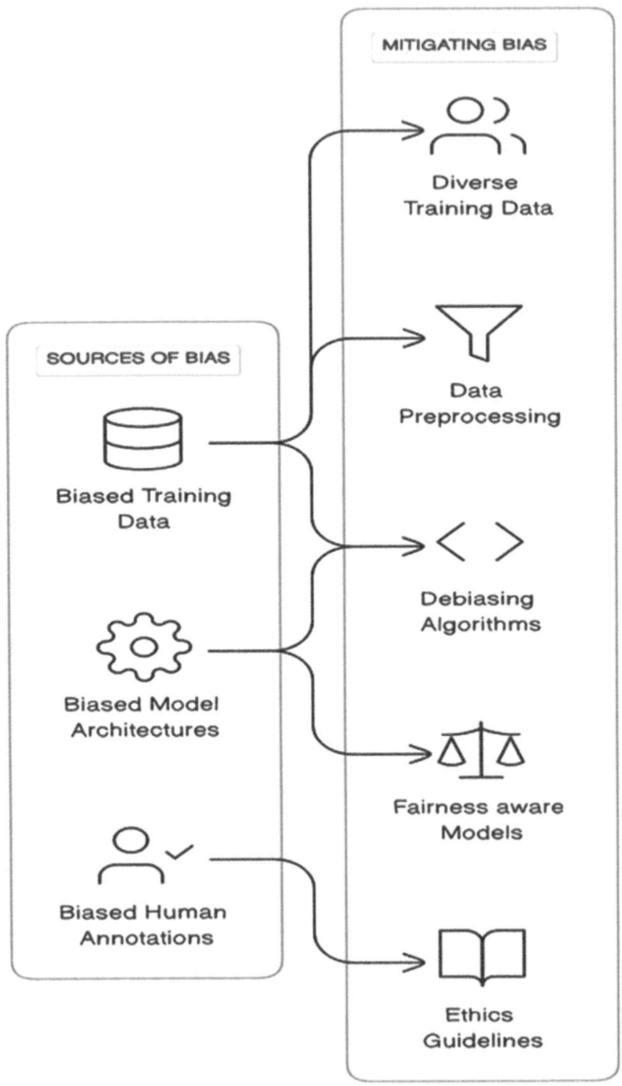

FIGURE 4.8 Bias and fairness in natural language processing (NLP).

4.6.1.1 SOURCES OF BIAS IN NATURAL LANGUAGE PROCESSING (NLP)

- Bias in NLP can arise from multiple sources, including biased training data, biased model architectures, and biased human annotations.

- Biased training data may reflect societal biases present in historical texts, user-generated content, or data collected from online sources.
- Model architectures that prioritize efficiency may inadvertently propagate biases present in the training data.
- Human annotators can introduce bias when labeling data, influenced by their own beliefs and perspectives.

4.6.1.2 MITIGATING BIAS

Mitigating bias in NLP is a complex challenge. Strategies include (Figure 4.9):

1. **Data Pre-Processing:** Identifying and removing biased or toxic content from training data.
2. **Debiasing Algorithms:** Developing algorithms that can reduce bias during training or post-processing.
3. **Diverse Training Data:** Ensuring training data represents a diverse range of perspectives and demographics.
4. **Fairness-Aware Models:** Designing models that explicitly incorporate fairness constraints.
5. **Ethics Guidelines:** Establishing clear guidelines and ethical standards for data collection, labeling, and model development.

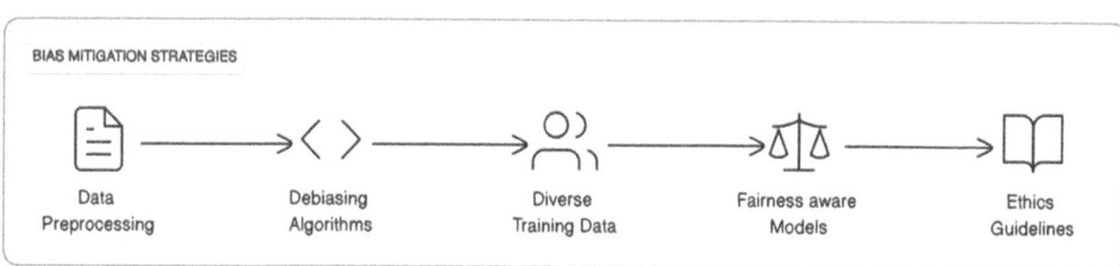

FIGURE 4.9 Bias mitigation strategies.

4.6.2 PRIVACY AND DATA HANDLING

4.6.2.1 DATA PRIVACY CONCERNS

- NLP often involves processing large volumes of personal and sensitive data, such as emails, messages, and medical records. Ensuring privacy is paramount.

- Privacy concerns include unauthorized access to personal information, data breaches, and the potential for misuse of user data.

4.6.2.2 GENERAL DATA PROTECTION REGULATION (GDPR) AND OTHER REGULATIONS

- The general data protection regulation (GDPR) in Europe and similar regulations globally mandate strict rules for handling personal data.
- NLP practitioners must comply with these regulations, which include obtaining informed consent for data usage, providing data transparency, and ensuring data security.
- NLP systems that process personal data must be designed to respect user privacy and comply with legal requirements.

4.6.3 TRANSPARENCY AND EXPLAINABILITY

4.6.3.1 MODEL INTERPRETABILITY

- NLP models can be highly complex, making it challenging to understand their decision-making processes. Model interpretability refers to the ability to explain why a model made a particular prediction or decision.
- Interpretability is important for ensuring that NLP systems are transparent and accountable, especially in critical applications like healthcare and legal domains.

4.6.3.2 EXPLAINABLE AI (XAI) IN NLP

- Explainable AI (XAI) is an area of research that focuses on developing techniques and methods to make AI models more interpretable.
- In NLP, XAI techniques may include generating human-readable explanations for model decisions, highlighting important features, and providing context for model outputs.
- XAI is vital for ensuring trust, understanding, and accountability in AI systems, particularly in scenarios where decisions impact individuals' lives.

Ethical considerations in NLP encompass addressing bias and fairness, ensuring data privacy and compliance with regulations, and enhancing transparency and explainability. These considerations are integral to responsible AI development and deployment, as they help mitigate harm, protect user rights, and build trust in NLP systems. NLP practitioners and researchers must actively work to address these ethical challenges to create technology that benefits society while minimizing risks and adverse impacts.

4.7 NAVIGATING THE NATURAL LANGUAGE PROCESSING (NLP) LANDSCAPE

Navigating the NLP landscape involves understanding how to make informed decisions when working with NLP technologies. It encompasses selecting the right NLP approach, building and training models, evaluating performance, and accessing resources for learning and development. A detailed explanation of each aspect is given in subsections.

4.7.1 CHOOSING THE RIGHT NATURAL LANGUAGE PROCESSING (NLP) APPROACH

Selecting the appropriate NLP approach is crucial, as different tasks and contexts may require different techniques. Consider the following factors:

1. **Task and Objective:** Start by defining the specific NLP task you want to address (e.g., sentiment analysis, machine translation, text summarization). Your choice of approach will depend on the task's nature and goals.
2. **Data Availability:** Assess the availability and quality of data. Some approaches may require large, labeled datasets, while others can work with limited data.
3. **Complexity:** Determine the complexity of the problem. Simple tasks may be handled with rule-based systems or basic ML models, while complex tasks may require DL approaches like transformers.
4. **Computational Resources:** Consider the hardware and infrastructure available for model training and deployment. DL models often require significant computational resources.

5. **Interpretability and Transparency:** If interpretability is a priority (e.g., in legal or healthcare applications), choose approaches that offer transparency, such as rule-based systems or models with XAI features.

4.7.2 BUILDING AND TRAINING NATURAL LANGUAGE PROCESSING (NLP) MODELS

Once you have chosen the right NLP approach, the next steps involve constructing and training your NLP models (Figure 4.10):

1. **Data Pre-Processing:** Prepare and clean your text data. This involves tasks like tokenization, stop word removal, and feature extraction.
2. **Model Selection:** Choose an appropriate NLP model architecture or algorithm based on your approach. Common choices include transformers, RNNs, and rule-based systems.
3. **Data Splitting:** Divide your dataset into training, validation, and test sets to evaluate model performance properly.
4. **Training and Fine-Tuning:** Train your model using the training data, adjusting hyperparameters and optimizing performance as needed. Fine-tuning pre-trained models is a common practice to leverage large-scale, pre-existing language models.
5. **Validation and Hyperparameter Tuning:** Use the validation set to fine-tune hyperparameters and prevent overfitting.
6. **Evaluation:** Evaluate your model's performance on the test set using appropriate metrics (e.g., accuracy, F1 score, BLEU score for machine translation).
7. **Regularization:** Apply regularization techniques like dropout or L1/L2 regularization to prevent model overfitting.

4.7.3 EVALUATING NATURAL LANGUAGE PROCESSING (NLP) PERFORMANCE

Evaluating NLP performance is critical to ensuring that your models are achieving their intended objectives. Consider the following evaluation practices:

1. **Task-Specific Metrics:** Use metrics relevant to the NLP task. For example, for text classification, accuracy or precision and recall are common metrics, while for machine translation, BLEU or ROUGE scores are used.
2. **Cross-Validation:** If data is limited, employ cross-validation techniques to obtain a more robust estimate of model performance.
3. **Bias and Fairness Assessment:** Assess your model for bias and fairness, particularly in applications involving sensitive data or decision-making.
4. **Human Evaluation:** In some cases, perform human evaluations to validate model outputs, especially for tasks that require nuanced understanding or subjective judgment.
5. **Continuous Monitoring:** Continuously monitor model performance and retrain models as necessary to adapt to evolving data and user needs.

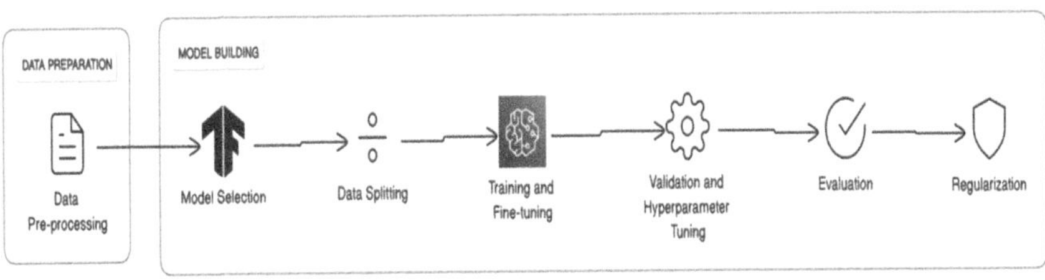

FIGURE 4.10 Building and training natural language processing (NLP).

4.7.4 RESOURCES FOR LEARNING AND DEVELOPMENT IN NATURAL LANGUAGE PROCESSING (NLP)

The field of NLP is constantly evolving, and staying updated is essential for success. Utilize various resources for learning and development:

1. **Online Courses and Tutorials:** Enroll in NLP courses and tutorials available on platforms like Coursera, edX, and YouTube.
2. **Research Papers:** Read NLP research papers published in top conferences like ACL, NAACL, and EMNLP to stay informed about the latest advancements.
3. **NLP Toolkits and Libraries:** Familiarize yourself with NLP toolkits like NLTK, spaCy, and libraries like TensorFlow and PyTorch.

4. **Community and Forums:** Join NLP communities on platforms like Reddit, Stack Overflow, and LinkedIn to ask questions and learn from peers.
5. **Open-Source Projects:** Contribute to open-source NLP projects and leverage pre-trained models available through projects like Hugging Face Transformers.
6. **Online Challenges:** Participate in NLP competitions and challenges, like Kaggle competitions, to gain practical experience.

Hence, navigating the NLP landscape involves a series of strategic decisions, from selecting the right NLP approach to building, training, and evaluating models. Staying informed about the latest developments in NLP through various resources is essential for continuous learning and development in this rapidly evolving field.

4.8 CASE STUDIES AND APPLICATIONS

Case studies and applications of NLP showcase the versatility and practicality of NLP in various domains. Let us explore these applications in detail in subsections.

4.8.1 NATURAL LANGUAGE PROCESSING (NLP) IN HEALTHCARE

1. **Clinical Text Analysis:** NLP is used to analyze clinical notes, medical records, and patient histories to extract valuable insights for diagnosis and treatment. It aids in identifying patterns, diseases, and potential risk factors.
2. **Medical Chatbots:** Healthcare chatbots powered by NLP assist patients in symptom checking, appointment scheduling, and medication reminders. They provide immediate access to medical information.
3. **Drug Discovery:** NLP helps in processing vast biomedical texts, research papers, and patents to identify potential drug candidates, drug interactions, and emerging treatments.
4. **Radiology Reports:** NLP can extract information from radiology reports, helping radiologists identify abnormalities and streamline reporting processes.

4.8.2 NATURAL LANGUAGE PROCESSING (NLP) IN CUSTOMER SUPPORT

1. **Chatbots and Virtual Assistants:** NLP-driven chatbots offer automated customer support, responding to inquiries, resolving issues, and guiding users through troubleshooting steps.
2. **Sentiment Analysis:** NLP is used to analyze customer feedback, reviews, and social media mentions to gauge customer sentiment and identify areas for improvement.
3. **Automated Email Responses:** NLP-powered email systems can understand and generate responses to customer emails, enhancing response times and efficiency.

4.8.3 NATURAL LANGUAGE PROCESSING (NLP) IN FINANCE

1. **Financial News Analysis:** NLP is employed to analyze news articles, press releases, and financial reports to assess their impact on stock prices and market trends.
2. **Risk Assessment:** NLP models help assess credit risk by analyzing applicant data, including bank statements and transaction history, for loan approvals.
3. **Fraud Detection:** NLP identifies unusual patterns and potentially fraudulent activities by analyzing transaction descriptions, customer communications, and other textual data.
4. **Algorithmic Trading:** NLP assists in parsing news and social media data for real-time market sentiment analysis, influencing trading decisions (Figure 4.11).

4.8.4 NATURAL LANGUAGE PROCESSING (NLP) IN SOCIAL MEDIA ANALYSIS

1. **Sentiment Analysis:** NLP is widely used to track and analyze sentiment in social media conversations, helping brands understand customer opinions and trends.
2. **Trend Detection:** NLP models identify emerging trends and hashtags on platforms like Twitter, enabling businesses to adapt marketing strategies.

3. **Social Listening:** Companies use NLP to monitor brand mentions, assess customer feedback, and manage online reputation.

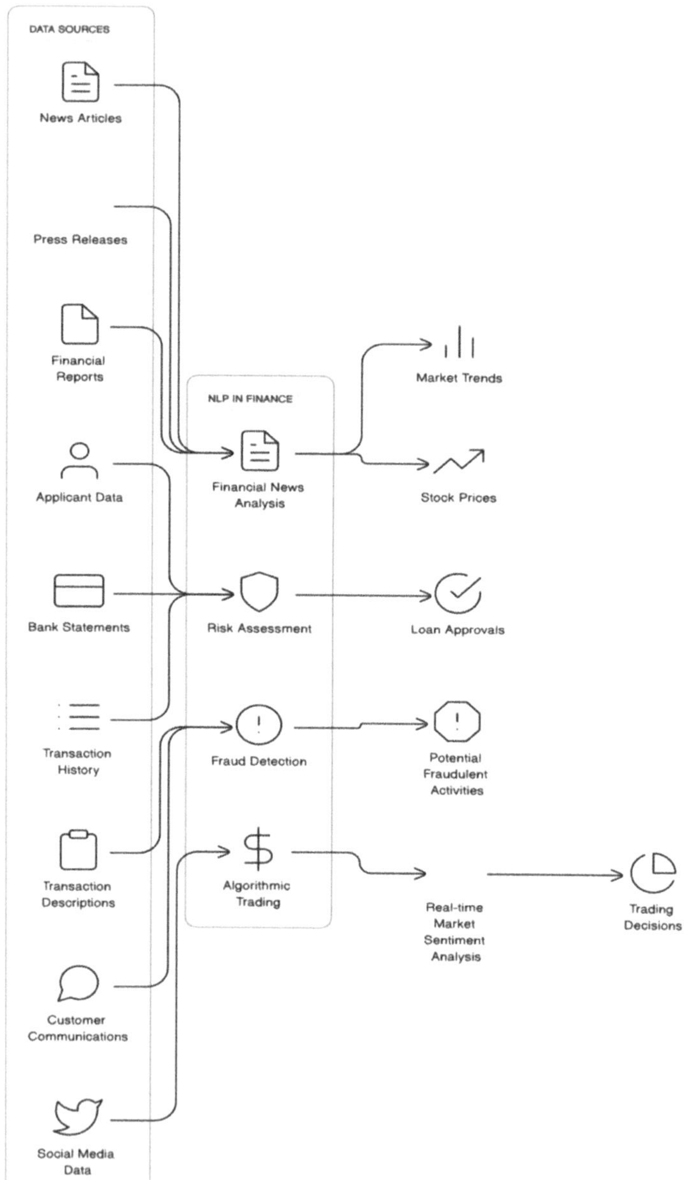

FIGURE 4.11 Natural language processing (NLP) in finance.

4.8.5 NATURAL LANGUAGE PROCESSING (NLP) IN LEGAL AND COMPLIANCE

1. **Legal Document Analysis:** NLP processes and extracts information from legal documents, contracts, and case law, facilitating legal research and compliance tasks.
2. **E-Discovery:** NLP assists in electronic discovery by quickly searching, analyzing, and categorizing large volumes of documents for legal proceedings.
3. **Regulatory Compliance:** NLP is used to ensure compliance with regulations by identifying risks, monitoring communications, and assessing compliance-related documents.

4.8.6 NATURAL LANGUAGE PROCESSING (NLP) IN E-COMMERCE

1. **Product Recommendations:** NLP analyzes user behavior, reviews, and preferences to generate personalized product recommendations, enhancing the shopping experience.
2. **Search Optimization:** NLP-driven search engines improve product search and query understanding, making it easier for customers to find what they need.
3. **Review Analysis:** NLP helps analyze customer reviews, extracting insights for product improvement and reputation management.

These case studies and applications illustrate how NLP technology is transforming industries by automating tasks, extracting insights from unstructured data, enhancing customer experiences, and aiding decision-making processes. NLP's ability to process and understand human language is invaluable in addressing diverse challenges across sectors.

4.9 FUTURE TRENDS IN NATURAL LANGUAGE PROCESSING (NLP)

The future of NLP holds exciting possibilities and advancements that will shape the way we interact with and benefit from language technologies. Here are detailed explanations of some key future trends in NLP:

4.9.1 ADVANCEMENTS IN MODEL ARCHITECTURES

NLP has seen rapid progress in model architectures over recent years, particularly with the emergence of transformer-based models like BERT, GPT-3, and T5.

1. **Continual Scaling:** We can expect the continual scaling of model sizes, leading to even larger and more powerful models. This will enable NLP systems to understand context better and generate more coherent and contextually relevant responses.
2. **Efficiency:** There will be a focus on developing more efficient and environmentally sustainable NLP models. Researchers will work on compressing large models while retaining their performance, making them accessible to a broader range of applications.
3. **Multilingual Models:** The development of multilingual models will continue, allowing NLP systems to handle multiple languages with greater proficiency. This will be especially important for global applications and cross-cultural communication.

4.9.2 MULTIMODAL NATURAL LANGUAGE PROCESSING (NLP)

Multimodal NLP involves the integration of text with other modalities such as images, videos, and audio for a richer understanding of content.

1. **Vision-Language Models:** Advances in models that understand both text and images will lead to enhanced capabilities in applications like image captioning, content moderation, and automatic generation of alt text for accessibility.
2. **Audio-Text Integration:** NLP models will become better at processing spoken language and audio data, enabling more accurate transcription services, voice assistants, and sentiment analysis from voice recordings.

4.9.3 NATURAL LANGUAGE PROCESSING (NLP) FOR LOW-RESOURCE LANGUAGES

Many languages have limited digital resources and NLP support. Future trends in NLP include addressing this linguistic imbalance.

1. **Zero-Shot and Few-Shot Learning:** NLP models will become more adept at zero-shot and few-shot learning, where they can perform tasks in languages with limited data by transferring knowledge from well-resourced languages.
2. **Cross-Lingual Transfer:** Techniques for transferring knowledge and models from high-resource to low-resource languages will become more sophisticated, fostering the development of NLP tools for marginalized languages.

4.9.4 NATURAL LANGUAGE PROCESSING (NLP) AND HUMAN-ARTIFICIAL INTELLIGENCE (AI) COLLABORATION

NLP systems will increasingly work alongside humans to enhance productivity, creativity, and problem-solving (Figure 4.12).

1. **Explainable AI (XAI):** Research will focus on making NLP models more interpretable and transparent. This will be crucial in domains where accountability and trust are essential, such as healthcare, law, and finance.
2. **Conversational AI:** NLP systems will become better at NLU and generation, enabling more fluid and human-like interactions in applications like customer service, virtual assistants, and education.
3. **AI-Augmented Creativity:** NLP models will aid in content creation, helping authors, marketers, and artists generate ideas, improve writing, and enhance creativity.

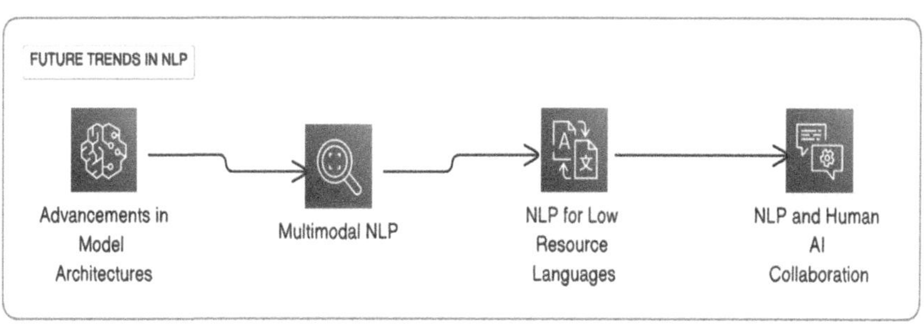

FIGURE 4.12 Future trends in natural language processing (NLP).

Hence, the future trends in NLP hold promise for more powerful and efficient model architectures, the integration of multiple modalities, improved support for low-resource languages, and enhanced collaboration between humans and AI. These advancements will shape the way NLP technologies are applied in various domains, making them more accessible, versatile, and beneficial to society.

4.10 CONCLUSION

In conclusion, the landscape of AI and its application to NLP has witnessed remarkable growth and transformation in recent years. This research paper has explored the intricate tapestry of NLP, delving into its fundamental concepts, architectural approaches, algorithmic strategies, ethical considerations, and real-world applications. The journey through this dynamic AI landscape has revealed that NLP is no longer confined to mere language understanding; it has evolved into a powerhouse of capabilities that can revolutionize industries and reshape human-computer interaction. From healthcare to customer support, finance to social media analysis, and beyond, NLP is finding applications in domains that touch every aspect of our lives. In navigating this landscape, we have emphasized the significance of ethical considerations. As NLP becomes increasingly integrated into society, addressing bias, ensuring privacy, and enhancing transparency and fairness are non-negotiable imperatives. The responsible development and deployment of NLP technologies are essential for achieving the full potential of AI while minimizing risks and upholding ethical standards. Furthermore, the future of NLP holds great promise, with advancements in model architectures, multimodal integration, support for low-resource languages, and a growing synergy between human intelligence and AI. These emerging trends are poised to unlock new horizons, making NLP more versatile, powerful, and beneficial to humanity [13–15].

KEYWORDS

- **artificial intelligence**
- **cross-lingual transfer**
- **explainable AI**

- **healthcare**
- **high-resource languages**
- **multimodal integration**
- **natural language processing**
- **zero-shot learning**

REFERENCES

1. Hancock, K. (2022). Cognitive artificial intelligence and predictive modeling algorithms, virtual navigation and geospatial mapping tools, and remote sensing data fusion techniques in the immersive metaverse environment. *Journal of Self-Governance and Management Economics, 10*(3), 40–55.
2. Junnu, M. (2023). Role of natural language processing in extracting insights from medical text data. *European Journal of Modern Medicine and Practice, 3*(9), 128–137.
3. Chen, Q., Leaman, R., Allot, A., Luo, L., Wei, C. H., Yan, S., & Lu, Z. (2021). Artificial intelligence in action: Addressing the COVID-19 pandemic with natural language processing. *Annual Review of Biomedical Data Science, 4*, 313–339.
4. Yang, L. W. Y., Ng, W. Y., Foo, L. L., Liu, Y., Yan, M., Lei, X., & Ting, D. S. W. (2021). Deep learning-based natural language processing in ophthalmology: Applications, challenges, and future directions. *Current Opinion in Ophthalmology, 32*(5), 397–405.
5. Kang, Y., Cai, Z., Tan, C. W., Huang, Q., & Liu, H. (2020). Natural language processing (NLP) in management research: A literature review. *Journal of Management Analytics, 7*(2), 139–172.
6. Álvarez-Carmona, M. Á., Aranda, R., Rodríguez-Gonzalez, A. Y., Fajardo-Delgado, D., Sánchez, M. G., Pérez-Espinosa, H., & Díaz-Pacheco, Á. (2022). Natural language processing applied to tourism research: A systematic review and future research directions. *Journal of King Saud University-Computer and Information Sciences, 34*(10), 10125–10144.
7. Wah, B. W., Huang, T. S., Joshi, A. K., Moldovan, D., Aloimonos, J., Bajcsy, R. K., & Weems, C. (1993). Report on workshop on high-performance computing and communications for grand challenge applications: Computer vision, speech and natural language processing, and artificial intelligence. *IEEE Transactions on Knowledge and Data Engineering, 5*(1), 138–154.
8. Deng, L., & Liu, Y. (2018). A joint introduction to natural language processing and to deep learning. In L. Deng & Y. Liu (Eds.), *Deep Learning in Natural Language Processing* (pp. 1–22). Springer.
9. Danilevsky, M., Qian, K., Aharonov, R., Katsis, Y., Kawas, B., & Sen, P. (2020). *A Survey of the State of Explainable AI for Natural Language Processing*. arXiv preprint arXiv:2010.00711.

10. Chowdhary, K., & Chowdhary, K. R. (2020). Natural language processing. *Fundamentals of Artificial Intelligence*, 603–649.
11. Koroteev, M. V. (2021). *BERT: A Review of Applications in Natural Language Processing and Understanding*. arXiv preprint arXiv:2103.11943.
12. Galassi, A., Lippi, M., & Torroni, P. (2020). Attention in natural language processing. *IEEE Transactions on Neural Networks and Learning Systems, 32*(10), 4291–4308.
13. Li, J., Liu, X., Zhang, S., Yang, M., Xu, R., & Qin, F. (2021). Accelerating neural architecture search for natural language processing with knowledge distillation and Earth Mover's Distance. In *Proceedings of the 44th International ACM SIGIR Conference on Research and Development in Information Retrieval* (pp. 2091–2095).
14. Vashishth, T. K., Sharma, V., Chaudhary, S., Panwar, R., Sharma, S., & Kumar, P. (2023). Advanced technologies and AI-enabled IoT applications in high-tech agriculture. In A. Khang (Ed.), *Handbook of Research on AI-Equipped IoT Applications in High-Tech Agriculture* (pp. 155–166). IGI Global.
15. Vashishth, T. K., Kumar, B., Sharma, V., Chaudhary, S., Kumar, S., & Sharma, K. K. (2023). The evolution of AI and its transformative effects on computing: A comparative analysis. In B. Mishra (Ed.), *Intelligent Engineering Applications and Applied Sciences for Sustainability* (pp. 425–442). IGI Global.

CHAPTER 5

Bridging Language Structure and Deep Learning Models: A Comprehensive Exploration of Natural Language Processing

ANURAG SHARMA*, DAVINDER SINGH, and NEHARIKA SHARMA

Department of Computer Science and Engineering, GNA University, Phagwara, Punjab, India

**Corresponding author.*

ABSTRACT

Natural language processing (NLP) is currently seen as a pillar of our digital age. Due to its expanding importance and widespread use, there is a need for thorough comprehension and ongoing research. This study delves further into NLP's complexities than it does on the surface. By exploring if common themes, like discussing the weather or analyzing peas, can provide a universal grammar rule that can produce all English sentences, we hope to grasp the basic structure of language. By doing this, we hope to ease the challenging process of teaching computers to comprehend and produce human language. This study addresses the dynamic world of deep learning (DL) models in NLP in addition to language structure. It provides a thorough evaluation of the numerous models and approaches used in various NLP applications. Understanding the advantages and disadvantages of various strategies is crucial for helping academics and practitioners select the models that are best suited for particular NLP tasks.

This work also acts as a knowledge bridge by bridging the gap between DL and NLP. It describes how DL approaches have transformed NLP,

Natural Language Processing and Artificial Intelligence: A Perspective Toward Current Trends, Challenges, and Applications. Vikas Khullar, Aryan Chaudhary, & Sunil Kumar (Eds.)

DOI: 10.1201/9781998511273-5

providing improved functionality for jobs like text analysis, translation, and chatbots. We clarify the profound influence of DL on the development of NLP, opening it up to a larger audience.

This work greatly contributes to the ongoing discussion on the development of language processing technology by offering insights into language structure and undertaking a comparative examination of DL models in NLP. For researchers, programmers, and companies wishing to use NLP and DL in their applications, it provides useful information.

5.1 INTRODUCTION

Natural language processing (NLP) has become a fundamental cornerstone of our ever-changing digital age, allowing robots to comprehend and interact with human language. The importance and pervasiveness of NLP have increased dramatically, requiring a thorough grasp and ongoing investigation of its complexities. In order to simplify the difficult effort of training computers to understand and produce human language, this research goes beyond the boundaries of NLP in an attempt to understand the underlying structure of language. The main goal of this research is to find a universal grammar rule that can produce any type of English sentence. The investigation takes its cues from seemingly uninteresting topics like weather discussion and pea analysis. The goal is to decode not just linguistic nuances but also the fundamentals of human communication. By doing this, we want to open up new avenues for NLP applications by facilitating more accurate and efficient machine translation. Moreover, this research is not a stand-alone project. It acknowledges its significant impact on the advancement of language processing technology and is closely entwined with the dynamic realm of deep learning (DL) models in NLP. These DL techniques have completely changed the field of NLP, improving its capacities for tasks like chatbot generation, translation, and text analysis. This paper bridges the gap between NLP and DL by explaining how these cutting-edge methods have increased NLP's usefulness and accessibility for a wider range of users and developers. In addition to its theoretical investigation, this study makes a useful contribution by providing an extensive assessment of the many models and approaches used in different NLP applications. Researchers and professionals can choose models that best fit their unique NLP tasks by being aware of the benefits and drawbacks of these various approaches. Thus, this work serves as an invaluable tool for researchers, developers, and companies looking to leverage DL and NLP in their applications.

We are ready to solve the mysteries of language structure and advance the rapidly developing area of NLP as we set out on this adventure through the complexity of NLP and DL.

5.2 BACKGROUND

NLP has come a long way since it was first developed in the 1950s [1], adjusting to new research paradigms and technological breakthroughs. This chapter offers a perceptive examination of the development of NLP, following its beginnings in syntax-driven processing to its current emphasis on semantics and context. Early on in the field of NLP research, syntax was given vital importance. It was widely accepted that understanding language's grammatical structure was essential to allowing computers to read and produce content that resembles that of humans. This syntactic-centric methodology was a product of the times in which it was used.

5.2.1 EXPANDING HORIZONS

An essential focus on syntax characterized the early years of NLP research. The general consensus was that understanding language's grammatical structure was essential to making computers able to understand and produce content that resembled that of humans. The syntactic-centric method was a product of its era as much as a need. During this period of expansion, NLP research experienced a paradigm shift, moving away from a narrow focus on grammatical structures towards a more holistic understanding of language. This evolution was driven, in part, by the realization that tasks such as question answering, information extraction, and topic modeling demanded a more nuanced approach. These tasks underscored the need to comprehend language beyond mere syntax, emphasizing the importance of capturing the richness of linguistic expression in its entirety.

5.2.2 FROM SYNTAX TO SEMANTICS

In recent years, the landscape of NLP research has undergone a transformative paradigm shift. The integration of DL techniques, empowered by access to expansive datasets, has enabled models to delve into the subtleties of language that transcend mere syntactic structures. This shift has broadened the focal point to encompass a spectrum of linguistic nuances, including sentiment analysis, opinion mining, and a profound exploration of semantics. This transformative phase heralds a new era in NLP, where the convergence of advanced technologies and expansive datasets has unlocked unprecedented capabilities for understanding and interpreting human language.

5.2.3 EARLY ATTEMPTS AT MACHINE TRANSLATION

The aspirational notion of developing a "translating machine" gained traction in the middle of the 1930s [2], as evidenced by the submission of two patents that sought to transform language translation. Georges Artsrouni was one of the first people to attempt this, and his method was to translate using a bilingual dictionary as the basis. He used a mechanism that involved using paper tape to map words directly from one language to another. Though inventive, Artsrouni's approach had a crucial flaw in that it was unable to account for the complexities of grammar. The foundation for a fundamental, if unfinished, method of language translation was established by Artsrouni's reliance on a bilingual dictionary for translation. At first, the idea of translating words between languages seemed promising as a step towards removing language barriers. But a major flaw in the equation turned out to be the absence of grammar. A key component of language structure is grammar, which consists of the rules and conventions that control how words are arranged to form coherent sentences. Artsrouni's method could not translate words coherently or accurately within their context if grammar was neglected during the translation process.

In addition to using a bilingual dictionary, Troyanskii went one step further and adopted an approach based on Esperanto, an internationally created language that is neutral and simple to learn. To tackle the intricacies of grammar in machine translation, Esperanto acted as a bridge. Troyanskii sought to develop a more sophisticated and precise translation system by utilizing an Esperanto-based methodology. Esperanto offered a framework that enabled a more methodical approach to handling grammatical nuances in translation because of its regular structure and simplified grammar. In order to achieve meaningful language translation, Troyanskii's method acknowledged the need to address grammar, which represented a conceptual advancement in the field. Both Artsrouni and Troyanskii faced the same challenge; in spite of these conceptual advances, their methods did not result in the creation of a useful, operational machine translation model. One significant barrier that eluded a simple, rule-based solution was the intricacy of language. The complexities of syntax, grammar, and context presented difficulties beyond the capabilities of the initial machine translation attempts. In hindsight, Artsrouni and Troyanskii's efforts set the groundwork for further research and advancement in the field of machine translation. Their conceptual breakthroughs emphasized the significance of taking grammar into account when translating languages, a notion that has been incorporated into contemporary NLP techniques. Even though their early efforts were unsuccessful, they provided insightful information that helped to advance the ongoing effort to overcome the difficulties involved in using computers to translate languages.

5.3 NATURAL LANGUAGE PROCESSING (NLP) CURVES

The advent of the Internet Age has brought about significant and swift changes to civilization, which we are currently witnessing more than before [1]. The passage implies that there is a persistent fear that even technologies that are actively expanding, adjusting, and introducing new developments may soon become outdated. The statement suggests that during the past 15 years, advances in NLP have not been as swift as those in certain other fields of technology. There is a growing awareness that NLP may not be able to keep up with the rate of technological advancement, and there is an underlying worry that despite continuous efforts to adapt and improve, it might be falling behind (Figure 5.1).

1. **Learning Curves:** In NLP, learning curves are frequently used to illustrate how a model performs over time or with progressively more training data. These curves usually depict the changes in a model's performance over training, such as accuracy or loss. If you have more than one model and they perform similarly on the task, their learning curves may overlap. Comparable learning rates or convergence behavior may be shown by overlapping learning curves.
2. **ROC Curves:** To assess how well categorization models are performing, receiver operating characteristic (ROC) curves are

frequently employed. For various classification thresholds, they show the trade-off between the true positive rate and false positive rate. The ROC curves of several models may overlap when used for a classification job, indicating comparable discriminatory power. A visual indicator that the models perform similarly across various decision criteria can be provided via overlapping ROC curves.

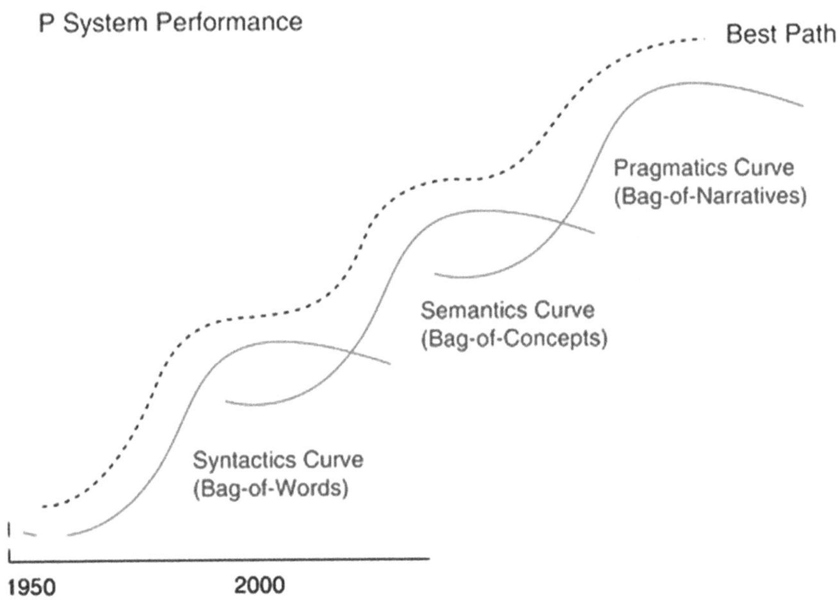

FIGURE 5.1 Evolution of system performance in natural language processing (NLP).

3. **Precision-Recall Curves:** In imbalanced datasets, precision-recall curves serve as an additional criterion of evaluation for classification models. The precision versus recall trade-off at various classification levels is depicted by these curves. The presence of overlapping accuracy-recall curves indicates that various models are attaining comparable recall and precision ratios, with no particular model outperforming the others in terms of performance.
4. **Word Embedding Spaces:** The trajectories of word vectors or similarity scores over time may be represented by curves in the context of word embeddings. Overlapping curves may suggest that particular words or concepts show comparable patterns in

their representations when you're displaying the evolution of word representations or relationships in an embedding space. This could be important to monitor changes in word context or semantic links over time.

5. **Activation Curves:** When examining neural network behavior in NLP, activation curves can be studied for individual neurons or layers. Overlapping activation curves may suggest that specific neurons or layers respond similarly across different inputs or tasks. This insight can provide valuable information about a network's capacity for generalization.
6. **Evaluation Metric Curves:** You may work with a variety of evaluation metrics, such as the F1 score for named entity recognition (NER) or the BLEU score for machine translation, depending on the particular NLP task at hand. Similar overall performance across multiple models may be indicated by overlapping curves for these metrics.

In the field of NLP, "overlapping NLP curves" can refer to different visual representations of model performance, learning behavior, or word representations. The particular interpretation you use will rely on the kind of curves you are using and the circumstances surrounding your NLP assignment. To make meaningful conclusions about the behavior or performance of the model, it is important to take into account the task at hand as well as the particular metrics or visualizations you are using.

5.4 NATURAL LANGUAGE PROCESSING (NLP) TOOLS

Tools for NLP are software elements made with the purpose of processing, interpreting, and analyzing human language. These tools use a variety of methods from artificial intelligence (AI), machine learning (ML), and computational linguistics to extract relevant information from speech or text [3]. Here are a few frequently utilized NLP tools:

5.4.1 THE NATURAL LANGUAGE TOOLKIT (NLTK)

The natural language toolkit, or NLTK, is a feature-rich Python programming language library for manipulating human language data. It offers a collection of resources and tools for work in computational linguistics and NLP. In both academia and business, NLTK is frequently utilized for a variety of language-related tasks.

5.4.1.1 KEY ATTRIBUTES OF NATURAL LANGUAGE TOOLKITS (NLTKS)

1. **Text Processing Utilities:** The NLTK contains modules for tasks like lemmatization, which reduces words to their dictionary form; stemming, which reduces text to words or sentences; and tokenization, which breaks text into words or sentences.
2. **Part-of-Speech Tagging:** To aid in comprehending the syntactic structure of text, NLTK is able to tag words in a sentence with their grammatical parts of speech.
3. **Parsing:** NLTK enables context-free grammar parsing, which allows users to examine the grammatical construction of sentences.
4. **Corpora and Resources:** NLTK is useful for linguistic research and language modeling because it offers access to a variety of language corpora and lexical resources.
5. **Machine Learning:** For tasks like NER, sentiment analysis, and text classification, NLTK integrates with ML algorithms.
6. **Community and Documentation:** NLTK is user-friendly for a wide range of NLP researchers, from novices to experts, thanks to its thriving community and comprehensive documentation.

5.4.2 SPACY

SpaCy is a powerful and effective NLP library made specifically for Python programmers. Here is a quick rundown of what SpaCy is:

5.4.2.1 GOAL AND LAYOUT

1. **Production Use:** SpaCy is ideal for real-world applications because it is specifically made to be used in production environments.
2. **Speed and Efficiency:** Due to its reputation for speed and efficiency, spaCy is a top option for jobs requiring rapid text processing in large quantities.

5.4.2.2 IMPORTANT CHARACTERISTICS

1. **Pre-Trained Models:** Pre-trained models for a number of languages are included with spaCy. These models can carry out tasks like dependency parsing, which determines the grammatical relationships between words in a sentence; part-of-speech (POS) tagging, which assigns grammatical categories to words; and NER, which recognizes entities like people, organizations, and locations. These models are trained on large annotated datasets.
2. **Tokenization:** SpaCy's tokenization is very effective and capable of handling complicated situations, such as punctuation and contractions.
3. **Recognition of Named Entities (NER):** When it comes to recognizing and categorizing named entities in text, spaCy shines. This entails identifying various entities, such as people, groups, places, times, and more.
4. **Part-of-Speech (POS) Tagging:** With spaCy, you can accurately tag parts of speech. Part of speech tagging is the process of assigning a grammatical category (such as a noun, verb, or adjective) to each word in a sentence.
5. **Reliance Parsing:** In spaCy, dependency parsing entails examining a sentence's grammatical structure by identifying the relationships between its words. This is essential for comprehending the syntactic connections found in a text.
6. **Language-Related Notes:** Text is annotated with linguistic data by spaCy, which facilitates the extraction of features and insights from the processed data.
7. **Adaptability:** In addition to having pre-trained models for a variety of NLP tasks, spaCy is meant to be extensible. If necessary, users can use domain-specific data to train their custom models.
8. **Easily Navigable API:** A user-friendly API offered by spaCy makes working with natural language data easier. Due to its thorough documentation, both novice and seasoned NLP practitioners can easily access it.

5.4.3 GENSIM

Python library called Gensim is used for document similarity analysis and topic modeling. Many different NLP tasks make extensive use of it, with the main objective being the processing and extraction of semantic information from large text corpora. The main functions and typical applications of Gensim are explained as follows:

5.4.3.1 VISIBLE MODELING

1. **Definition:** Finding abstract topics in a group of documents is possible with the help of a technique called topic modeling. Since each document is thought to contain a variety of topics, Gensim can recognize these topics by analyzing how words are used in different documents.
2. **Latent Dirichlet Allocation (LDA):** LDA is a well-liked probabilistic model for topic modeling that is implemented in Gensim. Within a given set of documents, LDA can identify topics and the word distributions that correspond to them.

5.4.3.2 ANALYSIS OF DOCUMENT SIMILARITY

1. **Vector Space Model:** Every document in Gensim is represented as a vector in a high-dimensional vector space. The similarity of vectors is then used to calculate document similarity.
2. **Cosine Similarity:** Gensim frequently gauges document similarity using cosine similarity. This metric, which ranges from –1 (totally dissimilar) to 1 (totally similar), computes the cosine of the angle formed by two vectors.

5.4.3.3 SUMMARY OF TEXT

1. **Extractive Summarization:** Important sentences or phrases are chosen from the source text to produce a succinct summary. Gensim can be used for this type of summarization.
2. **Sentence Similarity:** The most representative sentences for summarization can be found by measuring the similarity between sentences using Gensim's document similarity capabilities.
3. **Word Inserts:** Word2Vec Implementation: Word2Vec is a well-liked algorithm for learning word embeddings, and Gensim has

an implementation of it. Word embeddings capture the semantic relationships between words by representing words as continuous vector spaces.

4. **Semantic Similarity:** Based on the vector representations of words, Gensim's Word2Vec models can be used to calculate the semantic similarity between words.
5. **Efficiency and Scalability:** Gensim can handle huge text datasets and corpora because of its scalable and memory-efficient design.
6. **Easily Usable:** Gensim offers an intuitive API that makes training models and completing different NLP tasks easier. Because of its modular design, users can add more functionality and customize it.
7. **Combining with Different Libraries:** Gensim is a flexible option for a variety of applications because it is simple to integrate with other well-known NLP libraries and tools.

5.4.4 STANFORD NATURAL LANGUAGE PROCESSING (NLP) TOOLKIT

The Stanford NLP Group created the extensive set of tools and resources known as the Stanford NLP Toolkit, which is intended to make a variety of NLP tasks easier. The main attributes and features are explained as follows:

5.4.4.1 PART-OF-SPEECH (POS) TAGGING

1. **Definition:** POS tagging is the process of giving each word in a sentence a grammatical category (such as nouns, verbs, or adjectives).
2. **Apply Case:** Understanding a text's syntactic structure depends on POS tagging, which is also frequently the first step in a variety of other NLP tasks.
3. **Recognition of Named Entities (NER):** The identification and categorization of entities—people, groups, places, dates, and more—within a text is known as recognition.

4. **Use Case:** Knowledge graph construction, IR, and question-answering applications are made possible by the extraction of structured information from unstructured text using NER.

5.4.4.2 EMOTIONAL ANALYSIS

1. **Definition:** The goal of sentiment analysis, also known as opinion mining, is to ascertain the sentiment—whether positive, negative, or neutral—expressed in a text.
2. **Use Case:** Businesses and organizations can benefit from sentiment analysis's widespread application in understanding user opinions, customer feedback, and social media sentiment.

5.4.5 RELIANCE PARSING

1. **Definition:** Dependency parsing is the process of examining a sentence's grammatical structure by determining the syntactic connections between its words.
2. **Use Case:** Understanding the hierarchical structure of sentences is crucial for information extraction and machine translation, among other tasks. Dependency parsing helps with this.
3. **Settlement of Coreferences:** Coreference resolution is the process of figuring out which words or expressions in a text refer to the same thing.
4. **Use Case:** In order to comprehend the context, coreference resolution is necessary.
5. **Sentence Division and Tokenization:** Tokenization entails dividing a text into discrete words or tokens, whereas sentence splitting divides a text into sentences.
6. **Use Case:** These tasks serve as a foundation for other NLP tasks and are essential for additional text analysis and processing.
7. **Multilingual Assistance:** The Stanford NLP Toolkit is adaptable for managing text in diverse linguistic contexts because it supports a number of languages.
8. **Excellent Models:** Pre-trained models that have been developed with cutting-edge methods and are renowned for their accuracy and performance are included in the toolkit.

5.5 RECENT TRENDS IN DEEP LEARNING BASED NATURAL LANGUAGE PROCESSING (NLP)

The goal of natural language processing, or NLP, is to automatically analyze and represent human language using a collection of theoretically motivated computational techniques. NLP research has come a long way from the days of punch cards and batch processing, when it could take up to seven minutes to analyze a single sentence, to the modern day, typified by companies like Google. Millions of webpages can now be processed in a matter of milliseconds, demonstrating the remarkable advancement and effectiveness of NLP [1].

The importance of statistical NLP in handling challenging language-related tasks is discussed in this passage. The "curse of dimensionality" is acknowledged as a major barrier that statistical NLP had to overcome in its early stages of development when trying to learn joint probability functions for language models. Dealing with a large number of features or dimensions presents a challenge that makes computations complex and resource-intensive.

Researchers looked into a different strategy to get around this problem: teaching words to have distributed representations in a lower-dimensional space. Stated differently, the idea was to represent words in a lower-dimensional, more condensed space rather than a high-dimensional one, which could be computationally demanding. This alternate strategy, which was frequently carried out using methods like word embeddings, sought to Bengio et al. [4].

I can give you a summary of some of the most current developments in DL-based NLP as of January 2022 [5], when I last updated my knowledge. Please be aware, though, that the field is always changing and that further developments may occur after that date. The following trends extend until 2022:

1. **Domination of Transformer Architecture:** Transformer-based models have dominated, including generative pre-trained transformer (GPT) and bidirectional encoder representations from transformers (BERT). These models have produced state-of-the-art outcomes in machine translation, sentiment analysis, and language understanding, among other NLP tasks.
2. **Pre-conditioned Linguistic Models:** Pre-trained language models are now considered the norm. These models are optimized for

particular tasks after being trained on sizable corpora. By recognizing semantic relationships and contextual information within language, BERT, GPT, and their variations have raised the bar.

3. **Transferring Knowledge and Perfecting It:** TL is a growing field that involves adapting models that have been pre-trained on one task to another. This method is more efficient because it requires less extensive labeled data for each unique task.
4. **Differential NLP:** A growing area of focus is the integration of multimodal information, which combines textual and visual data. Interest in models that can comprehend and produce text, image, and occasionally audio content has grown.
5. **Model Efficiency Efforts:** Creating models that are more efficient is becoming popular. This includes developing faster and more compact models that can be deployed on edge devices, which is essential for IoT and mobile applications.
6. **Responsible and Ethical AI:** In NLP research, ethical considerations are becoming more and more important. Fair and impartial language model creation is receiving more attention as researchers and practitioners become more aware of the biases in training data and models.
7. **Learning on the Go and One-Time Acquisitions:** The goal of research has been to enable models to learn from fresh data continuously (continuous learning) and function well with few examples (few-shot learning). This holds significant importance in scenarios where models must adjust to dynamic environments and generate precise forecasts with a restricted amount of labeled data.
8. **Interpretability and Explain Ability:** Because DL models are inherently opaque, there is a growing need for techniques that reveal the reasoning behind the decisions made by the models. Research on explainable AI (XAI) is ongoing.

5.6 MODELING

Using natural language sentences as input, an NLP system can produce a type of class (for dividing tasks), a series of tags (for sequence-tagging tasks), or a different decision. In order to use neural NLP techniques, the following two major problems must be resolved:

1. Use a neural network to encode the natural language sentence, which is a string of words.
2. Produce a series of labels or an additional sentence in natural language.

Based on these two factors, a number of widely used neural NLP models will be introduced in this section, such as word root, sentence root, and order-to-order modeling. Terms in the input root are mapped into vectors within a continuous domain through word root. Complex connectivity, like RNNs, convolutional neural networks (CNNs), and self-attentive networks, can be used based on the word root for feature extraction, context-aware word root that takes into account the context information of the entire sentence, or sentence root that integrates all of the sentence's information. Sentence root can be used for sentence-level tasks like sentiment analysis and paraphrase classification, while context-aware word root can be used for sequential labeling tasks like named-entity recognition (NER) and POS tagging. In order to create a new sequence, sentence root can also be fed into a different RNN or self-attention network. This creates the encoder-decoder framework that underpins source-target sequence pair modeling.

5.6.1 WORD EMBEDDING AND SENTENCE EMBEDDING

In NLP, word and sentence embeddings are methods that let words and sentences be represented as vectors in a continuous vector space. Algorithms can comprehend and analyze natural language more effectively thanks to these representations, which capture semantic linkages and similarities.

1. **Word Embedding:** The definition of word embeddings is distributed representations of words in a continuous vector space, where words that have similar semantic properties are closer to one another.
   i. **Procedures:**
      a. **Word2Vec:** Created by Google, this program learns to predict words based on their context and represents words as dense vectors. Another well-liked technique that learns word representations by factorizing the word co-occurrence matrix is called global vectors for word representation (GloVe).
      b. **Quick Text:** FastText was introduced by Facebook and allows it to capture morphological information by extending word embeddings to the subword level.

c. **Apply Cases:** Many NLP activities, including machine translation, NER, sentiment analysis, and text classification, require word embeddings. They outperform conventional sparse representations by offering a dense representation of words that maintains semantic links.

2. **Sentence Embeddings:** Vector representations of complete sentences or phrases are known as sentence embeddings. They capture the meaning of a full string of words, as opposed to word embeddings.
   i. **Strategies:**
      a. **Averaging Word Embeddings:** To create a single vector that represents the complete sentence, just average the word embeddings of each word in the phrase.
      b. Doc2Vec, also known as Paragraph Vectors, is a Word2Vec extension that uses fixed-length vectors to represent whole texts or sentences.
      c. Using the hidden states as sentence embeddings, Infer-Sent trains a model to anticipate the next sentence in a given document.
      d. Google created the Universal Sentence Encoder, a model that encodes texts while preserving syntactic and semantic information.

5.6.1.1 CONTEXT-DRIVEN WORD ROOT

Term representations that are learned independently [7] of the particular content or context of the documents in which they occur are known as content-independent word embeddings. Conventional word embedding models, such as Word2Vec and GloVe, are content-independent in the sense that they train without taking into account the context or true meaning of words in a given document. Rather, they employ global patterns found in a sizable sample to represent word semantics.

5.6.1.2 RECURRENT NEURAL NETWORK DEPENDENT CONTEXT-AWARE WORD EMBEDDING

One kind of neural network architecture that works well for processing sequential data is called a RNN [8], which is helpful for capturing context-aware word embeddings. RNNs are able to simulate dependencies in sequential data, such as sentences, since they preserve hidden states that can record information from earlier time steps.

This is a condensed summary of the context-aware word embeddings that RNNs can do:

1. **Processing in Sequence:** The RNN processes each word in a sentence one at a time given the word order.
   i. **Hidden States:** The RNN keeps a hidden state at each time step that contains a summary of the data from the previous and current hidden states.
2. **Context Building:** The sequence's worth of information is stored in the hidden state, which functions as a kind of memory. This is especially helpful for capturing word dependencies and context.
3. **Output of Word Embedding:** A context-aware representation of the matching word in the sequence can be created using the RNN's output at each time step.

But conventional RNNs have drawbacks as well. For example, the vanishing gradient problem makes it challenging to capture long-term dependencies. To solve these problems, more sophisticated RNN variations were introduced.

5.6.1.3 WORD EMBEDDING WITH CONTEXT-AWARENESS USING SELF-ATTENTION

Word embedding with context-awareness using self-attention is a NLP technique that leverages self-attention mechanisms within Transformer architectures to capture rich contextual information for each word in a sequence. Unlike traditional approaches, self-attention allows words to dynamically weigh the importance of other words in the sequence, enabling the model to consider both local and global context. This bidirectional understanding improves the contextual representation of each word, making it aware of its surroundings in the sequence. Well-known models such as BERT use self-attention for pre-training on large corpora, which allows them to learn deeply contextualized embeddings. These embeddings are highly effective in a variety of NLP tasks, demonstrating their abilities.

5.6.1.4 CONVOLUTIONAL NEURAL NETWORK DEPENDENT ON CONTEXT-AWARE WORD EMBEDDING

Convolutional neural networks (CNNs) [9] are used in CNN-dependent context-aware word embedding to extract contextual information from word sequences. By convolving over input word embeddings, the CNN functions as a filter in this method, removing local patterns and features. The model can identify dependencies and nuances in the context by taking into account nearby words. CNNs perform well at capturing hierarchical features and are especially helpful for tasks like NER and sentiment analysis, where local context is important. CNNs provide a computationally efficient solution for context-aware word embeddings, making them appropriate for a variety of NLP tasks, even though they are not as naturally sequential as RNNs or self-attention-based models.

5.6.1.5 SENTENCE EMBEDDING

Sentence embedding is a NLP technique that uses continuous vector representations to represent entire sentences or phrases. Sentence embeddings attempt to capture the overall meaning and context of a complete sequence of words, in contrast to word embeddings, which concentrate on individual words. Sentence embeddings can be produced in a number of ways: by averaging word embeddings, by utilizing models like Doc2Vec, or by utilizing sophisticated methods like InferSent or the Universal Sentence Encoder. These embeddings are useful for tasks like document similarity, sentiment analysis, and clustering, where comprehending the general meaning of a sentence is crucial. They function as condensed representations that encode semantic information.

5.6.2 SOURCE-TARGET SEQUENCE PAIR MODELING

5.6.2.1 TASK OF SOURCE-TARGET SEQUENCE PAIR MODELING

One sequence after another modeling is a basic idea in machine translation and NLP since it entails converting an input sequence into an output sequence. An encoder processes the input sequence and extracts its contextual information, and a decoder creates the output sequence based on this context. This is the typical architecture of a model. This paradigm is adaptable and useful for a number of tasks, including text generation, summarization, and language translation. Sequence-to-sequence models have proven remarkably effective at capturing the natural structure of sequential data and have grown to be indispensable in enhancing language generation and understanding systems.

5.6.2.2 ENCODER AND DECODER FRAME WORK

A neural network architecture called the encoder-decoder framework is frequently used for a variety of sequence-to-sequence tasks, including text summarization and machine translation. An encoder in this framework transforms input sequences into a fixed-size context or representation. This context is then used by the decoder to produce an output sequence. This architecture enables the model to efficiently capture and translate data, making it particularly useful for tasks involving variable-length input and output sequences. By enabling the decoder to selectively focus on distinct segments of the input sequence during the generation of each element in the output, variations such as the attention mechanism further improve the performance of the framework. The encoder-decoder paradigm has shown itself to be adaptable, serving as the basis for models (Figure 5.2).

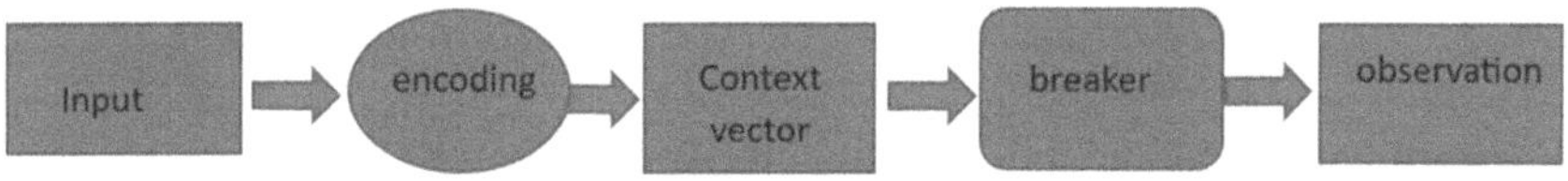

FIGURE 5.2 Encoder and decoder framework.

Here is the diagrammatic information for encoder and decoder framework.

An input sequence, like a sentence or an image, is fed into the encoder, which converts it into a context vector. A representation of the input sequence that preserves its structure and meaning is called a context vector. After that, the decoder creates an output sequence—like a translation or a caption—using the context vector.

Numerous neural network architectures, including CNNs and RNNs, can be used to implement the encoder and decoder. The particular architecture used will vary based on the task.

The encoder and decoder are described in further detail below:

1. **Encoding:** The input sequence must be processed by the encoder in order to create a context vector. An encoder is usually composed of a stack of layers in a RNN. One element at a time, each RNN layer processes the input sequence, updating its internal state in response to the current element and its previous state. The context vector is the top RNN layer's final state.
2. **Breaker:** One element at a time, the decoder is in charge of creating the output sequence. Usually, the decoder is just another stack of RNN layers. The context vector and the previous element of the output sequence are inputted into each RNN layer, which then outputs the subsequent element of the output sequence.
3. **Observation:** When creating the output sequence, the decoder can concentrate on various segments of the input sequence thanks to the attention mechanism. This can be useful for tasks like machine translation, where the decoder must focus on different parts of the source sentence to produce a target sentence that is both semantically and grammatically correct.

The encoder-decoder framework is a powerful and versatile architecture that can be used for a wide variety of sequence-to-sequence tasks. It has been shown to be effective for tasks such as machine translation, image captioning, and speech recognition.

5.6.2.3 ATTENTION-DRIVEN ENCODER AND DECODER SYSTEM

For applications like machine translation [10], the attention-driven encoder and decoder system is a potent NLP architecture. The input sequence is processed by an encoder in this framework, which results in a context vector that contains the important information. As the decoder creates the output, the attention mechanism enables it to selectively focus on various segments of the input sequence. The attention-based approach, in contrast to conventional sequence-to-sequence models, allows the model to give different elements in the input sequence different amounts of attention while it is decoding. The model performs better overall and can manage long-range dependencies better when tasks involve comprehending intricate relationships between input and output sequences, thanks to this dynamic allocation.

5.6.2.4 EVERY ENCODER AND DECODER FRAMEWORK IS ATTENTION BASED

A sophisticated architecture in NLP, the All-Attention-based Encoder-Decoder framework emphasizes the use of attention mechanisms for both encoding and decoding sequences. This framework enables the model to dynamically attend to all elements in the input sequence when generating each element in the output sequence, in contrast to traditional models that rely on fixed-size context representations. The model is better able to capture complex word relationships and dependencies in both the source and target languages, thanks to this bidirectional attention. This All-Attention paradigm is embodied by the Transformer model, which is represented by architectures such as BERT and GPT. It achieves impressive results in a range of language-related tasks, including summarization, machine translation, and language understanding. The framework is a mainstay of contemporary sequence-to-sequence modeling due to its adaptability and efficiency in managing long-range dependencies.

5.6.2.5 STATE OF THE ART MODELS IN NATURAL LANGUAGE PROCESSING (NLP)

NLP state-of-the-art models are the latest developments in the field, exhibiting remarkable performance on a variety of tasks. Models such as bidirectional encoder representations from transformers (BERT) and GPT are notable as of January 2022, when I last updated my knowledge. Google's BERT uses bidirectional attention mechanisms to extract contextual information, making it an excellent choice for tasks like sentiment analysis and question answering. Alternatively, GPT uses a generative approach, pre-training on a large corpus and demonstrating remarkable generation and understanding of language. In addition to pushing the envelope in terms of NLP performance, these models served as an inspiration for later architectures, which encouraged continued research and development in the search for ever more sophisticated language models.

5.6.2.5.1 Naïve bayes classifiers

An effective yet straightforward algorithm that is frequently used in NLP is the Naive Bayes classifier. It determines the likelihood of a specific category or label given the observed features based on Bayes' theorem. These characteristics frequently indicate the existence or frequency of words in a document in the context of NLP. Naive Bayes works well for text classification tasks like spam detection, sentiment analysis, and document categorization, despite its "naive" assumption of feature independence. When simplicity and real-time processing are essential, its computational efficiency and capacity to handle high-dimensional data make it a popular option.

5.6.2.5.2 Hidden Markov model (HMMs)

In NLP, HMMs are probabilistic models that are used to represent hidden states and sequential dependencies within a series of observations, like words in a sentence. An HMM, in the context of NLP, makes the assumption that there is a series of hidden states that are underlying, and each has an output that can be observed. The transitions between these hidden states, which stand in for latent linguistic structures, reveal the underlying syntactic or semantic relationships in the language. HMMs have been used for a variety of tasks where an understanding of language's sequential nature is essential, such as speech recognition and POS tagging. HMMs offer a fundamental framework for modeling sequential data in NLP, despite their simplicity.

5.6.2.5.3 Neural network

A neural network refers to a computational model that emulates the functionalities of biological neural systems within the human body. Networks perform in a similar manner and are composed of layers that consists of artificial neurons, often termed as nodes. These include an input layer that acknowledges data, one or multiple hidden layers that process information, and an output layer that makes forecasts or classifications. Connections between the neurons also known as synapse has a weight that changes during training to improve performance of the network model. The neurons perform activation functions upon their weighted inputs, thrusting diversity which is fundamental in trying to model sophisticated relations. After a through propagation, all the given inputs are processed, and the corresponding results or outputs are produced which upon comparison using a loss function are checked with the actual targets. Time and cost are deprecated within the single direction of answering the produced bases via back-propagation algorithms, ensuring that active synapses drives constant change through models such as gradient enlightenment. Fulfilling structures aids integrating an outlined view of an affiliate relating image recognizing, taught dialect handling, and texting the decision reasoning for making neural network an advanced technique.

5.6.2.5.4 BERT

NLP has greatly advanced thanks to the revolutionary BERT model. BERT is a Transformer-based architecture created by Google that takes into account both left and right context in a sequence to capture contextualized word representations. BERT can anticipate missing words in sentences thanks to its pre-training on large corpora with unsupervised learning objectives. This enables it to comprehend word relationships and meanings in both directions. With the help of this pre-training, BERT is able to produce context-aware embeddings, which have been shown to be very successful in a variety of NLP tasks, such as NER, text classification, and question answering. Because of BERT's success, many more transformer-based models have been created, and DL's use in natural language understanding (NLU) has gained traction.

5.7 EVOLUTION OF NATURAL LANGUAGE PROCESSING (NLP)

The development of NLP [11] is a story of technological progress over several decades. Starting in the 1950s and 1960s, NLP was primarily concerned with rule-based systems, which attempted to manually encode linguistic rules for language understanding. However, the field saw limited success due to computational limitations and the inherent complexities of natural language. Expert systems, which combined domain-specific knowledge with linguistic rules, were developed in the 1970s and 1980s, but they were only moderately successful. In the 1990s, NLP began to take on a more statistical approach, as probabilistic models and techniques like HMMs gained traction. This period laid the foundation for a more data-driven approach. With the development of ML techniques like SVMs and Maximum Entropy Models, the 2000s were a watershed decade for advancements in fields like machine translation and NER. But with the introduction of DL in the 2010s, NLP saw a true revolution. Complex linguistic patterns and context could be captured by neural network architectures, such as long short-term memory (LSTM) networks and RNNs. A breakthrough in the representation of words in continuous vector spaces and the capture of semantic relationships was made possible in 2013 with the release of Word2Vec and word embeddings. The true game-changer came along in 2018 when models based on transformers, such as BERT, achieved state-of-the-art performance in a variety of NLP tasks. The current state of NLP is defined by the prevalence of large pre-trained language models such as GPT-3, ethical issues related to AI, and an ongoing search for solutions to issues like bias and interpretability. NLP's history is a monument to unceasing innovation, with each new generation of researchers building on the achievements and shortcomings of the one before it to advance the field's understanding of human language to a more complex level.

5.8 INDUSTRIAL APPLICATIONS OF NATURAL LANGUAGE PROCESSING (NLP)

NLP strives to revolutionize human-to-machine interaction to the extent that conversing with a machine becomes as seamless as interacting with a human. The ongoing focus of NLP involves extracting meaning from unstructured data, facilitating its comprehension by machines. According to IDC's recent projection, the volume of data analyzed by cognitive systems is anticipated to surge 100-fold to reach 1.4 ZB by 2025, exerting a profound influence on numerous industries globally. Sectors such as robotics, healthcare, financial services, connected automobiles, and smart homes are poised to experience continued advancements through the integration of NLP. While NLP initially made its mark in the early 2000s with applications in machine translation for language conversion, it swiftly gained prominence in the customer service sector. One of the most prevalent applications of NLP in customer service is the implementation of "chatbots" or virtual assistants.

1. **Conversational System:** NLP conversational systems aim to create intelligent interfaces that enable smooth conversations between users and machines, marking a paradigm shift in human-computer

interaction. These systems interpret and react to natural language inputs by utilizing sophisticated algorithms, linguistic models, and ML techniques. Conversational AI has advanced remarkably over the years, from basic rule-based chatbots to complex systems with sentiment analysis, contextual understanding, and dynamic responses. RNNs and transformer models are two examples of neural network architectures that have made conversational agents more adept at understanding context and producing responses that seem human. Virtual assistants, social chat platforms, and customer service applications all make extensive use of these systems. By responding to user inputs, interpreting intent, and delivering pertinent and meaningful information, they seek to mimic the complexities of human conversation. The current state of NLP research in conversational systems is concentrated on improving user engagement and naturalness, as well as tackling issues such as managing ambiguity, context switching, and preserving cohesive conversations. Conversational systems have the potential to be extremely important in a number of fields, such as business, education, and healthcare, as technology develops further. They will also help to make human-machine interaction more natural and intuitive.

2. **Text Analytics:** One popular use of NLP is text analytics [12], which is the methodical analysis and derivation of significant insights from unstructured textual data. The importance of this field is growing across many industries as businesses realize how important it is to extract information from massive amounts of text. Textual content is processed, analyzed, and interpreted using NLP techniques, which makes tasks like entity recognition, sentiment analysis, and document classification possible. Text analytics, in its simplest form, uses NLP algorithms to convert unstructured text into structured data, enabling companies to extract useful information from sources like documents, social media, and customer reviews. Through the automated extraction of sentiments and patterns from text, businesses can improve customer experiences, make better decisions, and gain a competitive edge. The potential for further improving our capacity to extract meaningful insights from the ever-growing amounts of unstructured textual data in our digital world lies in the ongoing development of NLP and text analytics.

3. **Machine Translation:** Within the field of NLP, machine translation (MT) is a key application that is transforming cross-linguistic communication for both individuals and businesses. MT's main objective is to make it possible for text or spoken words to be automatically translated between languages, promoting intercultural communication and eradicating language barriers. MT has come a long way over the years, moving from rule-based systems to statistical methods and, more recently, gaining a great deal from the capabilities of neural network models, especially transformer architectures. Notable instances are DeepL and Google Translate, which use advanced algorithms to understand the subtleties of various languages and generate translations that are appropriate for the context. Despite ongoing difficulties with idiomatic expressions and cultural quirks, MT has emerged as a vital tool for cross-border cooperation, localizing content, and promoting international communication. Sophisticated NLP-driven machine translation systems will help us navigate a world of linguistic diversity with ease in the future as MT models are continuously improved and advanced technologies are integrated.
4. **Healthcare:** NLP has become a game-changer in the healthcare industry, enabling more accurate and efficient information extraction and analysis from clinical texts and patient records. With the use of NLP, healthcare professionals can extract insightful information from unstructured medical data, such as doctor notes, electronic health records (EHRs), and medical literature. This technology helps with clinical decision support by giving medical professionals quick access to relevant data so they can spot trends and make well-informed decisions. Improved patient care, illness diagnosis, and treatment planning are all made possible by the use of NLP in tasks like NER, sentiment analysis, and IR. Additionally, NLP applications can be used in medical coding, automating the process of obtaining billing codes from patient records and minimizing errors and administrative work. NLP's capacity to extract valuable information from enormous datasets holds enormous promise for improving patient outcomes, advancing medical research, and expediting healthcare procedures as the amount of data pertaining to healthcare grows.

5. **Automotive:** NLP is a game-changing technology with a wide range of applications in the automotive industry. The creation of voice-activated control systems for automobiles is one of the main uses. Natural language commands can be used by drivers to communicate with their cars through NLP, giving them hands-free control over features like climate control, music playback, and navigation. This leads to safer interactions and improves accessibility for drivers, which enhances the overall driving experience. Moreover, NLP is used in smart car assistants, which utilize conversational interfaces to provide drivers with real-time information and support. These assistants make driving easier and more intuitive by responding to voice commands, answering questions, and even carrying out simple tasks like sending messages or placing calls. NLP is anticipated to play a bigger role in the development of intelligent, conversational interfaces that promote smooth communication between drivers and their cars as the automotive industry moves closer to autonomous driving. This will ultimately improve both safety and convenience while driving.
6. **Finance:** Natural language processing, or NLP, has become a game-changer in the finance industry, transforming how banks examine and understand enormous volumes of textual data. Applications of NLP in finance are numerous and include everything from processing financial reports and customer interactions to sentiment analysis of news articles and social media. Financial analysts can evaluate risk factors, obtain important insights into market trends, and make well-informed investment decisions by utilizing NLP. In particular, sentiment analysis makes it possible to assess the sentiment of the market and the general public, giving traders and investors access to real-time data that helps them predict changes in the market. Furthermore, chatbots with NLP skills improve user experiences in customer interactions by comprehending and answering financial questions in a more conversational and natural way. NLP's capacity to sort through unstructured data and extract useful information is essential for ensuring compliance, optimizing financial processes, and ultimately fostering more informed and data-driven decision-making in the financial industry. With the development of NLP technologies, the field of finance stands to

benefit from their ability to solve intricate problems creatively and promote efficiency in financial operations.

7. **Manufacturing:** Two extremely promising application areas that will make use of NLP capabilities are robotics and process automation. Robots on factory floors can communicate with distant human operators via NLP and interpret commands for machine and product assembly and movement. In order to increase operational efficiency, industries have already started using NLP-based solutions to automate their process workflows.

KEYWORDS

- **chatbots and conversational AI**
- **comparative analysis of NLP models**
- **deep learning models**
- **language structure and grammar**
- **natural language processing**
- **text analysis and translation**

REFERENCES

1. Cambria, E., & White, B. (2014). Jumping NLP curves: A review of natural language processing research. *IEEE Computational Intelligence Magazine, 9*(2), 48–57. https://doi.org/10.1109/MCI.2014.2307227.
2. Johri, P., Khatri, S. K., Al-Taani, A. T., & Sabharwal, M. (2021). Natural language processing: History, evolution, application, and future work. In *Proceedings of the 3rd International Conference on Computing Informatics and Networks* (pp. 365–375).
3. Ginsberg, C. (2025). *8 Best Tools for Natural Language Processing in 2025*. Noble Desktop. Retrieved from: https://www.nobledesktop.com/classes-near-me/blog/best-natural-language-processing-tools (accessed on 13 March 2025).
4. Bengio, Y., Ducharme, R., Vincent, P., & Jauvin, C. (2003). A neural probabilistic language model. *Journal of Machine Learning Research, 3*, 1137–1155.
5. Weston, J., Bengio, S., & Usunier, N. (2011). Wsabie: Scaling up to large vocabulary image annotation. In *Proceedings of the International Joint Conference on Artificial Intelligence* (Vol. 11, pp. 2764–2770).

6. Deng, J., Dong, W., Socher, R., Li, L. J., Li, K., & Li, F. F. (2009). ImageNet: A large-scale hierarchical image database. In *Proceedings of the 2009 IEEE Conference on Computer Vision and Pattern Recognition* (pp. 248–255). IEEE.
7. Mikolov, T., Chen, K., Corrado, G., & Dean, J. (2013). *Efficient Estimation of Word Representations in Vector Space.* arXiv preprint arXiv:1301.3781.
8. Peters, M. E., Neumann, M., Iyyer, M., Gardner, M., Clark, C., Lee, K., & Zettlemoyer, L. (2018). Deep contextualized word representations. In *Proceedings of the 2018 Annual Conference of the North American Chapter of the Association for Computational Linguistics* (pp. 2227–2237).
9. Collobert, R., & Weston, J. (2008). A unified architecture for natural language processing: Deep neural networks with multitask learning. In *Proceedings of the 25th International Conference on Machine Learning* (pp. 160–167).
10. Bahdanau, D., Cho, K., & Bengio, Y. (2014). *Neural Machine Translation by Jointly Learning to Align and Translate*. arXiv preprint arXiv:1409.0473.
11. Turing, A. M. (1950). Computing machinery and intelligence. *Mind, 49*, 433–460.
12. Dale, R. (2017). The commercial NLP landscape in 2017. *Natural Language Engineering, 23*(4), 641–647.

CHAPTER 6

Artificial Intelligence Architectures and Algorithms in Natural Language Processing Ecosystems

SANGEETHA ANNAM* and VIKAS KHULLAR

Chitkara University Institute of Engineering and Technology, Chitkara University, Punjab, India

**Corresponding author.*

ABSTRACT

The goal of natural language processing (NLP), which is positioned midway between computational linguistics and computer science, is to transform spoken and written natural human languages into organized, mineable data. NLP can generate responses that resemble those of a human being or interpret textual data using statistical, linguistic, and artificial intelligence (AI) techniques. Within the AI field, NLP deals with creating and putting into practice algorithms and systems that can communicate with one another using natural language. Nowadays, the performance of NLP applications has increased to an unparalleled degree. This chapter will give you a general introduction and tutorial on AI architecture and its algorithms in NLP. Also, we outline how NLP has evolved historically, as well as a summary of some of the most notable aspects, such as architectures and algorithms of AI in NLP ecosystems. This chapter includes an overview of AI techniques applied to various NLP sub-problems. Then we go into the design of contemporary AI architectures and AI algorithms in NLP ecosystems, concluding with a summary.

Natural Language Processing and Artificial Intelligence: A Perspective Toward Current Trends, Challenges, and Applications. Vikas Khullar, Aryan Chaudhary, & Sunil Kumar (Eds.)

DOI: 10.1201/9781998511273-6

6.1 INTRODUCTION

Natural language involves both the application and learning of knowledge. People utilize language to exchange information, debate choices, and challenge views. Consequently, it is not unexpected that computational linguists began developing notions for representing knowledge as interactions between concepts in semantic networks as early as the 1950s and 1960s. Natural language processing (NLP) emerged as the nexus of linguistics and artificial intelligence (AI) in the 1950s. NLP was first separated from data information retrieval (IR), which rapidly indexes and searches vast amounts of text using highly scalable statistics-based approaches [1]. Three main issues need to be addressed to create computer programs that can comprehend natural language: the first is related to cognitive processes; the second is to the representation and interpretation of linguistic input; and the third is to general knowledge [2]. Due to its enormous quantity, lack of restrictions, and ambiguity, natural language posed two challenges to conventional parsing techniques that only relied on symbolic, human-crafted rules. Ultimately, NLP must derive meaning, or "semantics," from text. Formal grammars, which define the relationships between text units or speech elements like nouns, verbs, and adjectives, mostly deal with syntax. Grammars may be expanded to handle natural-language semantics by adding more rules and constraints and significantly increasing subcategorization.

A language is described as a system of rules or a collection of symbols that are integrated and used to transmit or broadcast information. NLP serves users who lack the time to acquire proficiency in new languages, as not all users may be well-versed in machine-specific language. NLP is a branch of AI and linguistics that focuses on teaching computers to comprehend words or sentences written in human languages. It can be divided into two categories: Natural language generation (NLG), which advances the task of understanding and producing text, and natural language understanding (NLU), or linguistics, which came into being to ease user workloads and fulfill the desire to communicate with computers in natural language [3]. The majority of NLP research in the literature thus far has been done by computer scientists; however, linguists, psychologists, philosophers, and other experts have also expressed interest. A fascinating feature of NLP is its contribution to the understanding of human language. The topic of NLP in computer communication is addressed by a variety of ideas and methods within the subject of NLP. The research-related tasks of NLP are: Automatic Summarization; Co-Reference Resolution; Discourse Analysis; Machine Translation; Morphological Segmentation; named entity recognition (NER); optical character recognition (OCR); part of speech tagging, etc.

6.1.1 DEVELOPMENT OF NATURAL LANGUAGE PROCESSING (NLP)

Natural language processing's primary goals include using a variety of tools, techniques, and algorithms to read, analyze, and manipulate data based on natural languages for the desired purpose. As a result, many academics have recently focused a great deal of emphasis on the creation of various tools and methodologies in the area of NLP and related fields of study. The stages of development are mentioned here: In the year 2001, the neural language model was developed; in 2008, multi-task learning came into existence; in 2013, word embedding and neural networks for NLP were developed; from 2014 to 2015, the sequence-to-sequence model and attention mechanism models were developed; and finally, in the year 2018, the pretrained language model was developed in NLP [4].

6.1.2 NATURAL LANGUAGE PROCESSING (NLP) COMPONENTS

NLP may be divided into two categories: NLG and NLU. NLP advances the process of text generation and comprehension. The major components of NLP are: (i) Generator and Speaker; (ii) Component and level of representation; and (iii) Application. The broader classification of NLP is depicted in Figure 6.1.

NLU makes it possible for computers to comprehend natural language and analyze it by removing concepts, entities, emotions, keywords, and other elements. In customer service applications, it's utilized to comprehend issues that consumers report, either orally or in writing. The branch of linguistics known as phonology deals with the orderly arrangement of sound. The word's constituent components stand for the smallest meaning units, or morphemes. Morphemes are the starting point for morphology, which is the nature of words. Both humans and NLP algorithms decipher word meanings in lexical systems.

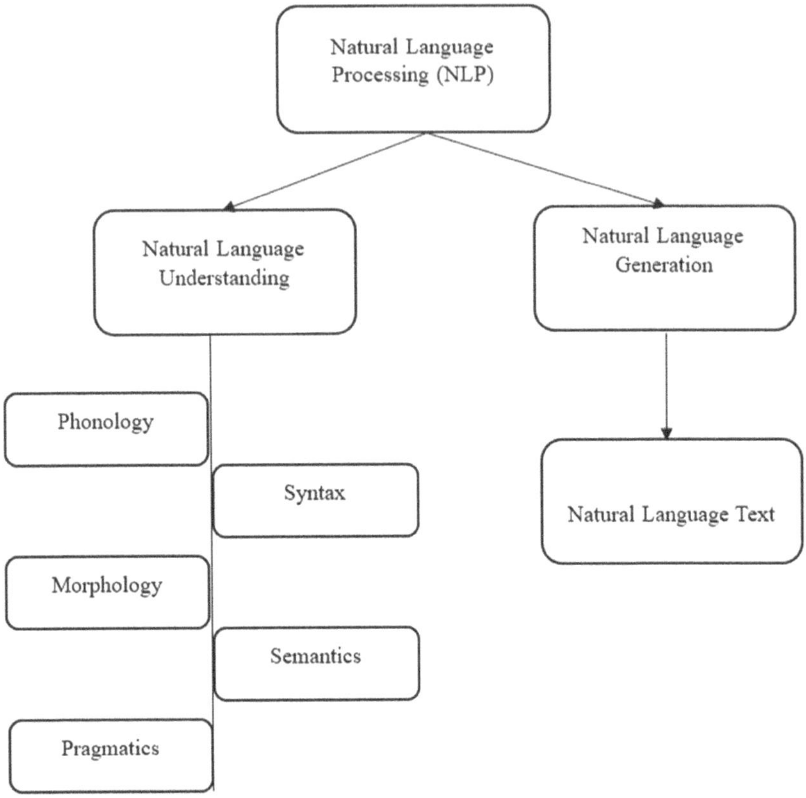

FIGURE 6.1 Natural language processing (NLP) broader classification.

Following lexical-level PoS tagging, words are categorized into phrases, which are subsequently grouped into clauses, and finally, at the syntactic level, phrases are concatenated to generate sentences. By dissecting the sentence's grammatical structure, it highlights the proper construction of a sentence. A phrase that displays the structural dependencies between words is the result of this level. By analyzing the logical structure of a phrase and selecting the most pertinent terms to comprehend how various concepts or words interact with one another, semantic processing ascertains the potential meanings of a given sentence. Discourse analyzes logical structure by establishing links between words and sentences to guarantee coherence [5]. The pragmatic level concentrates on information or content that originates from sources other than the document's substance. It addresses the inferences made by the speaker and the listener [6].

The act of creating meaningful words, sentences, and paragraphs from an internal model is known as NLG, or NLG. It is a component of the processing of natural languages and consists of four stages: goal identification, goal planning (based on scenario analysis and accessible communication sources), and goal realization as text. It is the antithesis of comprehension [7]. Applications for NLP include machine translation, question answering, information extraction, email spam detection, summarization, and more. After that, we go over a few topics and the pertinent research that has been done in those areas.

1. **Machine Translation:** The number of people using the Internet who do not speak English has skyrocketed over the past several decades worldwide, meaning that a significant portion of the data being fed into the Internet is not in English. Businesses in global marketplaces, particularly those that are new and expanding, have leaped to translate their materials into a number of languages in order to remain competitive. To automate translations, businesses have been releasing new apps and enhancing machine learning (ML) NLP models.
2. **Text Categorization:** Large volumes of data, such as official papers, military fatality reports, market data, newswires, and so forth, are fed into categorization systems, which then allocate the data to specified categories or indexes.
3. **Automatic Summarization:** When we must retrieve a particular, crucial piece of knowledge from a vast body of knowledge, information overload becomes a serious issue. When used to provide a summary of a news article or a collection of blog entries, automatic summarization is particularly pertinent as it minimizes repetition from many sources and maximizes the variety of material gathered.
4. **Sentiment Analysis:** It is a very helpful tool for social media monitoring since it gives us a general idea of how the public feels about certain issues. Retailers utilize NLP tools, including sentiment analysis, to extract thoughts and views from online content. They then incorporate these insights into the design and presentation of their merchandise to attract and retain consumers. The domains of customer service and market research benefit from this approach.
5. **Spam Filtering:** It functions through text classification, and in more recent times, anti-spam filtering and text categorization have benefited from the use of several ML approaches.

6. **Information Extraction:** Finding interesting terms in textual material is the focus of information extraction. In several applications, extracting entities—like names, locations, events, dates, times, and prices—is an effective method of condensing the data pertinent to the user's requirements.
7. **Summarization:** In the current digital era, information overload is a genuine problem, and our ability to access and acquire knowledge already outpaces our comprehension of it.

6.1.3 NATURAL LANGUAGE PROCESSING (NLP) PROCESS

The processes in the NLP pipeline are designed to help read and comprehend human language. The first step involves sentence segmentation, followed by word tokenization, stemming, lemmatization, stop word analysis, dependency parsing, and part-of-speech (POS) tagging.

1. **Tokenization:** The act of dividing a text paragraph into more manageable sections, such as words or phrases, is known as tokenization. One unit used as the foundation of a phrase or paragraph is called a token. This gives the computer a more manageable way to handle words.
2. **Stop Word Removal:** Eliminating superfluous words that do not contribute to the content of the text (word clutter) to leave just the words that convey the most information.
3. **Lemmatization/Stemming:** When a term is reduced to its most basic form, its meaning is more easily understood.
4. **Part-of-Speech Tagging:** This involves identifying a word's type and grouping it into nouns, pronouns, prepositions, verbs, etc.

6.2 ROLE OF NATURAL LANGUAGE PROCESSING (NLP) AND ARTIFICIAL INTELLIGENCE (AI) IN NATURAL LANGUAGE PROCESSING (NLP) ECOSYSTEMS

To accomplish clinical and research goals, an ideal ecosystem would allow clinical NLP researchers and developers to quickly test alternative components, assess and compare performance on benchmark datasets, and adapt NLP techniques to user applications [8]. The healthcare system has benefited greatly from digitization, which has improved patient safety, care quality, cost savings, and efficiency in the delivery of healthcare services [9]. The major classifications of AI in NLP consist of: Language Understanding; Question and Answering; Arithmetic; Common Sense Reasoning; Translation; Code Completion; General Knowledge; Reading Comprehension; Summarization; Local Inference Chains; Dialogue; and Language Understanding. A broad definition of AI is the study of artificial computer systems that can accomplish difficult tasks at a human level or with human-like skills. Although the area includes human encoding of expert knowledge and a range of symbol manipulation systems, the bulk of methodologies and techniques used by the research discussed here are related to ML. Generally speaking, machine translation involves using a statistical engine like Google Translate to translate sentences between languages. The difficulty with machine translation technology is not so much in translating words as it is in maintaining sentence meaning, syntax, and tenses.

Artificial intelligence enables computers to mimic human behaviors [10]. ML techniques are used to realize AI. Deep learning (DL) is a subject of AI that uses algorithms inspired by the function of the human brain. NLP's main goal is to bring computational linguistics and AI together so that people and machines can communicate naturally. A program has to understand syntax (grammar), semantics (phrase meaning), morphology (tense), and pragmatics (conversation) in order to interact with people. By applying AI and ML to deduce meaning from language and provide context, NLP enables computer programs to understand unstructured information in a manner similar to that of human brains.

Artificial intelligence [11] is broadly classified into major categories: (i) ML; (ii) NLP, further classified into Speech and Text Processing; (iii) Computer Vision; and (iv) Robotics and Manufacturing Optimization, etc. NLP is capable of handling enormous amounts of unstructured digital information, including comments on social media, emails, reports, newspaper articles, and forum postings, among many more types. This facilitates the process of deriving conclusions from data and making wise choices more quickly. Human language is broken down into discrete parts and analyzed as distinct concepts or thoughts one at a time in NLP. After that, it makes connections between them and searches for context, which enables it to decipher the meaning and tone of the input. The two steps of NLP are algorithm creation and data preprocessing.

6.2.1 ADVANTAGES AND LIMITATIONS OF NATURAL LANGUAGE PROCESSING (NLP) IN ARTIFICIAL INTELLIGENCE (AI)

Prior to the creation of NLP technology, codes or computer languages were used by humans to interact with computers. The procedure of communicating is easier now. NLP made it possible for computers to comprehend spoken and written human language, which improved communication. NLP's benefits [12] include: (i) writing summaries of lengthy documentation; (ii) correcting data entry errors; (iii) enabling voice assistants, like Alexa or Siri, to understand voice commands; (iv) identifying customer sentiment in comments and messages to improve user experience; (v) making use of employee labor by taking over some tasks; chatbots can replace customer service; and (vi) executing analytics that were formerly unattainable due to data volume.

Even with the rapid advancement of technology, NLP has drawbacks and restrictions. The primary difficulties with NLP are:

1. **Lack of Context:** Human communication is rarely precise and ordered, unlike the orders we give computers. It is frequently ambiguous and full of terms that, in the absence of context, a computer cannot comprehend. Natural speech is a challenge for NLP algorithms since it contains context, slang, and different accents.
2. **Recognizing Voice Tone:** People utilize rhetorical devices including sarcasm, irony, and exaggeration. Although speech recognition microphones are getting better, they are still unable to distinguish between different voice tones. The emphasis placed on a word in a phrase can alter its meaning, and a word itself might have more than one meaning. All of these are minute variations that NLP systems may overlook.
3. **Errors in Writing or Speaking:** Text analysis issues might arise from mispronounced, misspelled, or improperly used words. Although a writer can mitigate this issue by employing proofreading tools to eliminate particular problems, those systems are unable to comprehend the goal of being entirely error-free.
4. **Language Evolution and Changes:** Language is similar to a living thing. It has no set rules and is always expanding and changing. Since computers "like" to obey instructions, NLP algorithms may become outdated very quickly due to the unpredictable nature of language changes.

6.2.2 ARTIFICIAL INTELLIGENCE (AI) ARCHITECTURE IN NATURAL LANGUAGE PROCESSING (NLP)

The architectural flow of NLP, represented in Figure 6.2, shows the following stages: Data is collected from the user interface, and words are segmented automatically. After segmentation, the keywords are collected and extracted. The query handling and its responses are managed accordingly.

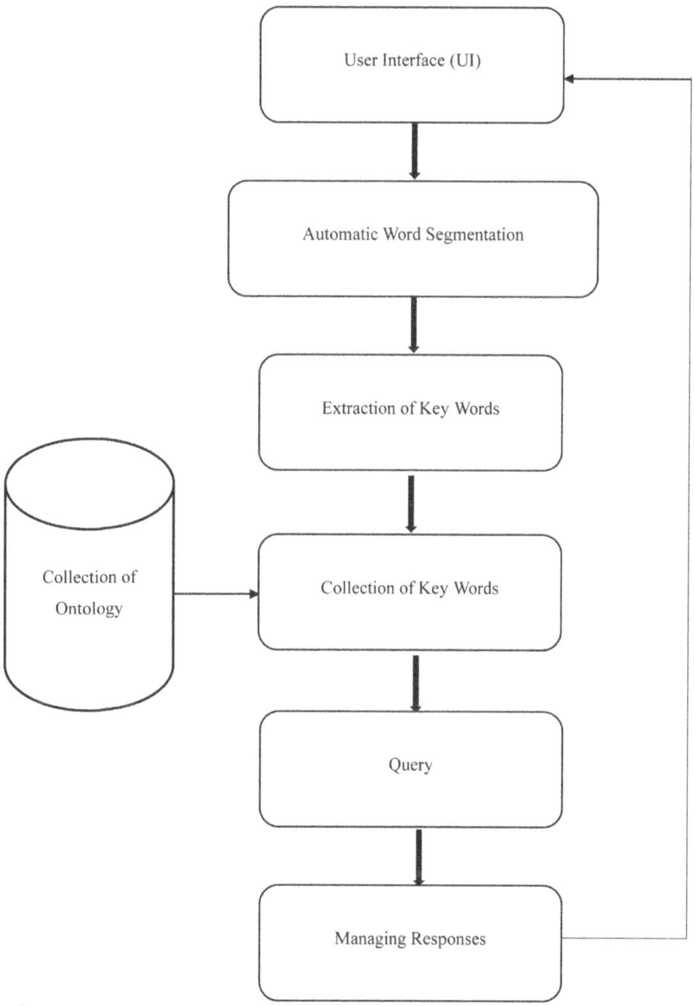

FIGURE 6.2 Natural language processing (NLP) for artificial intelligence (AI) architecture.

6.2.3 ARTIFICIAL INTELLIGENCE (AI) ALGORITHMS IN NATURAL LANGUAGE PROCESSING (NLP)

1. **Large-Scale Data:** Big data is defined as anything that can be measured, recorded, and classified. Currently, over 80% of the data is accessible in its raw format. Big data is derived from data that is kept in both corporations and large organizations. Examples include personnel data, business acquisition and sale documents, corporate transactions, prior organizational records, social media, etc. Even though the language used by humans is ambiguous and unstructured, making it difficult for computers to understand, NLP may be used to harness the enormous amount of unstructured data and look for growing patterns to better understand the information contained in the data. With the use of big data, it may resolve important business-related issues, regardless of the industry in which they operate—retail, healthcare, business, or finance.
2. **Log Analysis and Log Mining:** Tokenization, stemming, lemmatization, parsing, and other methods are among those used to transform log messages into a structured format. After logs are accessible in a well-documented format, they are subjected to log analysis and mining in order to extract valuable information and knowledge. An example of this would be a server failure-related error log.
3. **Pattern Recognition:** This is one method that filters out messages by matching messages from logs with messages kept in a pattern book.
4. **Deep Learning Models:** Words are represented as "one-hot" encoded vectors using a rule-based methodology. Semantic representation is secondary to syntactic representation in the conventional approach. The BoW model can't tell certain situations apart. The Figure 6.3 shows the usage of DL models [10] in AI.

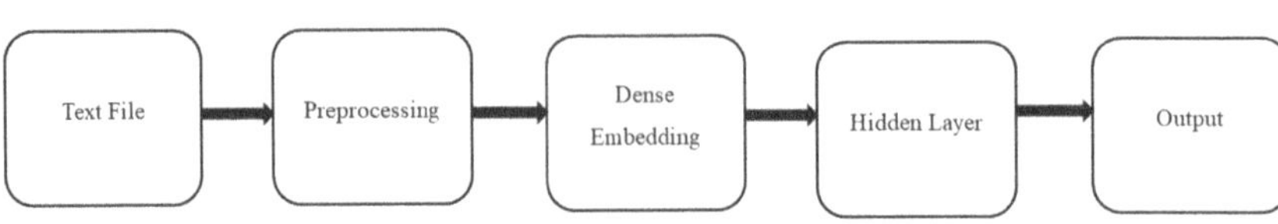

FIGURE 6.3 Deep learning process.

Table 6.1 shows the different models/applications of DL applied to the NLP process.

TABLE 6.1 Applications of NLP

Applications of NLP	Machine Learning / Deep Learning Algorithms
Tokenization	Neural network
Entity Recognition	
Intent Extraction	
Part of Speech tagging	
Image Captioning	Recurrent neural networks
Machine Translation	
Question Answering System	
Sentiment Analysis	Recursive neural networks
Paraphrase and Object Detection	
Parsing Sentences	
Sentence Classification	Convolutional neural network
Spam Detection	
Semantic relation extraction	
Terms; Topics; Sentiment; and Semantic analysis	Support vector machine, K-nearest neighbor, Naïve bayes, Boosting algorithm, Random Forest, Adaboost, Log R, XGBoost, Bagging, Stochastic gradient descent, and ensemble methods.

5. **Text Normalization:** To standardize the format of disparate messages, log messages are normalized. This is carried out in situations when distinct log messages use different wording, but they all have the same understanding and originate from different operating systems or applications.
6. **Automated Text Classification and Tagging:** Sorting and labeling distinct log messages entails classifying the messages according to different keywords so that they may be analyzed later.
7. **Artificial Ignorance:** It's a method of removing boring log messages using ML algorithms. It may also be used to identify anomalies in how systems typically operate.

6.2.4 BENEFITS OF ARTIFICIAL INTELLIGENCE (AI) ALGORITHMS IN NATURAL LANGUAGE PROCESSING (NLP)

Because there are so many subtle differences in language, the computer science discipline of NLP faces many difficult challenges. NLP entails breaking down language into semantic components, such as sentences and elements of speech. Using techniques focused more on linguistic-conceptual linkages than textual analysis, NLP, a branch of AI, organizes and extracts meaningful information from unstructured text. The most important uses for NLP are as follows:

1. **In Health Care Systems [9]:**
   i. **Electronic Health Record (EHR):** Medical practice was revolutionized by EHR systems, which allowed for speedier treatment starts, error removal, and efficient diagnosis. Patient care was enhanced by NLP's capacity to analyze raw information from medical records more quickly and efficiently. Researchers may find it simpler to obtain and compile clinical data with the use of EHR. Predicting illness development, complications, death, early diagnosis, disease prevention, and other outcomes can all be done using this.
   ii. **Clinical Decision Support Systems (CDSS):** CDSS, a medical selection assistance system designed to enhance healthcare delivery by improving clinical decisions with centered medical knowledge, affected the individual's information and other fitness information. It is one of the applications that NLP also offers. A conventional CDSS is software that is intended for simple clinical decision-making. It works by first picking a patient's suggested symptoms, which are compared to an electronic clinical knowledge base.
   iii. **Dragon Medical One:** A voice recognition program called Dragon Nuance is essential to the electronic medical record-keeping process since it allows medical personnel to dictate directly into the device. Every report has 99% accuracy in word recognition and may be dictated. Using natural language commands, format, edit, and browse notes quickly and effectively to maximize processes. Any device may be used as a microphone to access this program.

iv. **Clinical Trial Matching:** The method of matching clinical trials is based on predetermined parameters, including age, gender, medical history, and the stage of a certain ailment that will be examined. It is managed by researchers and medical investigators, and it is also known as observational or interventional studies. NLP may be utilized in clinical trials to improve patient outcomes, support doctors' and researchers' decision-making, reduce costs, and increase overall productivity. NLP may be utilized in clinical trials to improve patient outcomes, support doctors' and researchers' decision-making, reduce costs, and increase overall efficiency.
v. **Internet of Medical Things:** With unique IDs, the internet of things (IoT) [13] manages a wide range of networked computing devices, machines, objects, people, and animals. It can send data over the network without the need for human interaction. It includes the monitoring and control systems—such as thermostats and IoT-enabled HVAC and thermostatic equipment—that enable homes to be smart.
vi. **Vision Therapy Software:** This is the first website for visual treatment powered by AI worldwide. CureSee is a software vision treatment program that uses AI to address conditions like convergence insufficiency, amblyopia, and squint eye. It is a surgical in nature procedure that strives to help people with these issues because lens implants, glasses, or surgery are not effective treatments for these conditions.
vii. **Detection of Alzheimer's Disease:** A person with Alzheimer's disease loses their ability to think clearly and remember things accurately. The development of Alzheimer's disease is attributed to alterations in the brain that may ultimately lead to the death of brain cells. The changes include the buildup of abnormal protein clusters that impede cell-to-cell communication and amyloid plaque buildup.

Artificial intelligence may help with Alzheimer's disease since it is able to recognize it. Early-stage Alzheimer's disease may even be beneficial to the condition. ML, a subset of AI, may be able to recognize various diseases and mitigate their consequences through the use of AI. MRI equipment based on DL compares images of Alzheimer's disease with other disorders. DL refers to an AI function that simulates the way the human brain gathers information and creates patterns for use in making decisions. In order to proceed, the MRI image of a sick person is rotated in DL, which ultimately leads to patient analysis. Alzheimer's disease can be totally cured, but AI can lessen its severe consequences.

2. **In Legal Artificial Intelligence [14]:** Legal artificial intelligence (LegalAI) is the use of AI technologies, particularly NLP, to address legal domain problems. LegalAI has gained a lot of interest recently among both AI researchers and attorneys since it helps the legal system by relieving legal professionals of a burdensome paperwork backlog. While NLP scholars focus primarily on data-driven and embedding approaches, legal practitioners frequently consider how to solve tasks using rule-based and symbol-based methods.
   - i. **Legal Judgment Prediction (LJP):** One of Legal AI's most important jobs is LJP, particularly for the Civil Law system. The facts and the legislative articles are used to determine the judgment outcomes in the civil law system. Legal penalties won't be imposed until after a person has broken the law's list of forbidden behaviors. The primary focus of the LJP duty is predicting the outcome of a judgment based on the facts described in a case as well as the information included in the statutory articles of the Civil Law system.
   - ii. **Similar Case Matching:** Similar case matching (SCM) has emerged as a crucial area of study for Legal AI in order to forecast the outcomes of judgments more accurately in the Common Law system. SCM focuses on identifying pairs of related situations, while there are other ways to define similarity. SCM necessitates modeling the relationships between instances using data at various granularities, such as the element, event, and fact levels.
   - iii. **Legal Question-Answering:** Legal question answering (LQA) is a common use of Legal AI that seeks to provide answers to inquiries in the legal field. Offering trustworthy and excellent legal consultation services to non-experts is one of the most significant aspects of the work of legal professionals. The format of questions in LQA varies; some focus on explaining legal ideas, while others may analyze particular situations. Additionally, there are significant differences in the way that

professionals and non-professionals phrase queries, particularly when elucidating concepts unique to a certain subject.

3. **In Management Research:** The capacity of NLP to automatically interpret and analyze human discourse is causing it to gain popularity in management research. NLP research in several management disciplines focuses on a range of study subjects and analytical techniques since textual data are created with company activities on a ubiquitous basis. In particular, NLP is frequently used in information systems and marketing research to record consumer attitudes and actions. Other fields, including operations management, finance, and accounting, usually concentrate on obtaining detailed information from company filings, including annual reports and patent data. Scholars in the fields of information systems, marketing, and operations management have embraced DL algorithms for analytical purposes. These algorithms consider the context of words, allowing computers to express words based on their semantic worth and therefore "understand" material more effectively.
4. **In Accounting and Finance:** Scholars of accounting primarily study various aspects of corporate performance, such as firm worth. As the Internet and social media continue to grow quickly, consumers are using these online channels to voice their thoughts about businesses. The data and the feelings that go along with it may have an impact on the purchasing decisions of other clients and the investment decisions of investors, which may affect a company's worth. Differentiated influencing techniques are presented by tweeters or commentators and investment types during this process.

   The predictive influence of news and announcement linguistic qualities, such as tone, mood, frequency, and atypical word flow, on consumer response and company sales, or reverse causality, has traditionally been the subject of discussion among finance researchers. They have also looked at how this predictive impact differs across developed and emerging markets on a national level.
5. **In Information Systems:** Organization research and consumer behavior are the most studied areas in the field of information systems with relation to NLP application. Research on organizations has included topics such as innovation, R&D investment, and more. It has been demonstrated that new entry threats in the rapidly evolving information technology sector significantly influence the

decisions made by firms, particularly when it comes to R&D expenditure and the features of the industry that set the mechanism apart.

6. **In Marketing, Management, and Strategy:** Relevant NLP marketing research seeks to enhance sales performance, preserve brand reputation, and improve customer experience. Businesses pay close attention to how customers behave when it comes to social media opinion-sharing, online searching, and communication with other users. Compared to PCs, the introduction of smartphones allows consumers to instantly write brief text messages and convey emotion at any time [15]. Studies on strategic management have mostly concentrated on new and established businesses. It is crucial for newcomers to plan how to best position themselves in product marketplaces. Numerous NLP techniques have been used by operations management researchers on a variety of subjects, including inventory management, sales forecasting, product management, and operational efficiency [16–18].

6.3 CONCLUSION

NLP is a significant area of AI. It uses DL, special NLP techniques, and traditional ML algorithms to create models based on a range of study fields in order to comprehend and extract information from texts. The emergence of big data, cloud computing, and DL has led to the widespread application of NLP in a number of fields, including management research. NLP makes people's lives easier and significantly increases productivity and accuracy at work.

KEYWORDS

- **AI algorithms in NLP**
- **artificial intelligence**
- **big data and cloud computing**
- **deep learning**
- **machine learning**
- **natural language processing**

REFERENCES

1. Nadkarni, P. M., Ohno-Machado, L., & Chapman, W. W. (2011). Natural language processing: An introduction. *Journal of the American Medical Informatics Association, 18*(5), 544–551.
2. Chowdhary, K., & Chowdhary, K. R. (2020). Natural language processing. In *Fundamentals of Artificial Intelligence* (pp. 603–649).
3. Khurana, D., Koli, A., Khatter, K., & Singh, S. (2023). Natural language processing: State of the art, current trends and challenges. *Multimedia Tools and Applications, 82*(3), 3713–3744.
4. Olivetti, E. A., et al. (2020). Data-driven materials research enabled by natural language processing and information extraction. *Applied Physics Reviews, 7*(4).
5. Eisenstein, J. (2019). *Introduction to Natural Language Processing.* MIT Press.
6. Otter, D. W., Medina, J. R., & Kalita, J. K. (2020). A survey of the usages of deep learning for natural language processing. *IEEE Transactions on Neural Networks and Learning Systems, 32*(2), 604–624.
7. Gatt, A., & Krahmer, E. (2018). Survey of the state of the art in natural language generation: Core tasks, applications and evaluation. *Journal of Artificial Intelligence Research, 61*, 65–170.
8. Garvin, J. H., et al. (2017). An evolving ecosystem for natural language processing in the Department of Veterans Affairs. *Journal of Medical Systems, 41*, 1–9.
9. Silvestri, S., Islam, S., Papastergiou, S., Tzagkarakis, C., & Ciampi, M. (2023). A machine learning approach for the NLP-based analysis of cyber threats and vulnerabilities of the healthcare ecosystem. *Sensors, 23*(2), 651.
10. Torfi, A., Shirvani, R. A., Keneshloo, Y., Tavaf, N., & Fox, E. A. (2020). *Natural Language Processing Advancements by Deep Learning: A Survey.* arXiv preprint arXiv:2003.01200.
11. Chowdhary, K. R. (2020). *Fundamentals of Artificial Intelligence.* Springer.
12. Vajjala, S., Majumder, B., Gupta, A., & Surana, H. (2020). *Practical Natural Language Processing: A Comprehensive Guide to Building Real-World NLP Systems.* O'Reilly Media.
13. González García, C., Núñez Valdéz, E. R., García Díaz, V., Pelayo García-Bustelo, B. C., & Cueva Lovelle, J. M. (2019). A review of artificial intelligence in the Internet of Things. *International Journal of Interactive Multimedia and Artificial Intelligence, 5.*
14. Zhong, H., Xiao, C., Tu, C., Zhang, T., Liu, Z., & Sun, M. (2020). *How Does NLP Benefit the Legal System: A Summary of Legal Artificial Intelligence.* arXiv preprint arXiv:2004.12158.
15. Wolf, T., et al. (2020). Transformers: State-of-the-art natural language processing. In *Proceedings of the 2020 Conference on Empirical Methods in Natural Language Processing: System Demonstrations* (pp. 38–45).
16. Kang, Y., Cai, Z., Tan, C. W., Huang, Q., & Liu, H. (2020). Natural language processing (NLP) in management research: A literature review. *Journal of Management Analytics, 7*(2), 139–172.

17. Verma, K., et al. (2019). Latest tools for data mining and machine learning. *International Journal of Innovative Technology and Exploring Engineering, 8*(9), 18–23.
18. Lilhore, U. K., et al. (2022). Hybrid model for detection of cervical cancer using causal analysis and machine learning techniques. *Computational and Mathematical Methods in Medicine, 2022.*

CHAPTER 7

Classification of Real and Fake News Using Machine Learning and Deep Learning Techniques

NOMAN ISLAM[1,*], GHAZALA SHAFI SHEIKH[1], MAFAZA MOHI[2], LAIBA NADEEM[2], and RUQQAIYA AZIZ[2]

[1]*Iqra University, Karachi, Pakistan*

[2]*NED University, Karachi, Pakistan*

**Corresponding author.*

ABSTRACT

This paper employs various machine and deep learning (DL) models for fake news detection. The ever-evolving news industry, coupled with online social media platforms, has presented a challenge to consumers in identifying whether they are consuming authentic or false information. Fake news can be specifically designed to manipulate the mindset of users for their advantage. To address this challenge, this paper has attempted to simulate various machine and DL models to empower today's readers and equip them with a platform that examines the truthfulness of news and the authenticity of such news publication outlets. The machine learning (ML) models that have been tested in this research are logistic regression, support vector machine (SVM), decision tree, random forest, KNN, Naïve Bayes, and passive-aggressive classifiers. For DL models, long short-term memory (LSTM), Bi-LSTM, gated recurrent unit (GRU), convolutional neural network (CNN), and bi-directional encoded representation of transformers (BERT) have been utilized. To our knowledge, this is the most

Natural Language Processing and Artificial Intelligence: A Perspective Toward Current Trends, Challenges, and Applications. Vikas Khullar, Aryan Chaudhary, & Sunil Kumar (Eds.)

DOI: 10.1201/9781998511273-7

comprehensive comparison involving seven ML and six DL classifiers. The models have been tested based on accuracy, precision, recall, and F1 scores. BERT has been found to provide the maximum accuracy of around 97% on test data. Other DL models have also provided accuracy of around 90%.

7.1 INTRODUCTION

Fake news is a representation of information that deliberately puts forward misinformation or non-credible pieces of information, spread through both conventional printing media and recent online news platforms. The rising increase in such news for the defamation of certain political leaders and political plans has left the general public in confusion. These malicious entities have now gained a bulk of the readers' attention and have become successful in molding their perceptions to an extreme extent.

Social media, without even the concern of spreading hoax news, acts as a prime entity in the dissemination of information, or more specifically, fake news, and has made such news reach millions and thousands of users. These pieces of information, with the help of vast social networks, travel to even the most remote areas of the globe. Sometimes, over time and across places, the information loses its authenticity.

The notion of fake news dates back to the beginning of the 19th century, used by journalists to falsify the claims of contemporary newspapers and magazines. However, the term gained popularity with the tweets and speeches of the former American president. As mentioned above, fake news has existed for a long time; however, its aftermaths are now more visible and menacing. With deceptive words and eye-catching terminology used, online users can get trapped in a vicious set of beliefs. This set of beliefs can be used to fulfill the malicious objectives of certain political propaganda.

Social media is considered the least trusted source for news globally, as a result of the 2016 presidential elections. A survey in the spring of 2020 found that almost 60% of citizens aged 16–24 in the United Kingdom relied on social media to consume information relevant to the Coronavirus, of which 59% had encountered fake news on the matter. Around 52% of Americans claim to encounter fake news online regularly [1].

Meanwhile, around 25–30% of French citizens, mostly in the age bracket of 15–18 years, have used social media as their primary medium for information regarding COVID-19, making these young consumers more at risk of such fallacious pieces of information. At such a tender age, these consumers are not mature enough and usually pass it on without any verification, putting even the shrewdest consumers at risk. Unfortunately, there is no quick fix to eradicate the false news appraisal or even to raise awareness for educating online users to identify the news before circulating it to others.

Governments all around the globe, including Pakistan, are working hard to capture these perpetrators or at least thwart the proliferation of such misinformation through social media platforms. Yet, the problem is getting worse with every passing day. Not only are governments involved, but social media giants like Facebook removed around 7 million posts purported to be fake during the second quarter of 2020, primarily related to the Coronavirus pandemic. The removal of such misinformation was an indirect way to save thousands of lives. Meanwhile, the giant also removed around 22.5 million posts containing hate speech and around 7.7 million posts that promoted extremist organizations. These statistics are evidence of the risks that a piece of fake news can pose, showing that governments and social media platforms around the world need to revise their agendas and tools to combat such fake appraisals [2].

The main goal of this paper is to identify fake news using contemporary approaches. We examine various popular approaches for fake news detection. This includes machine learning (ML) and DL classifiers. Among popular ML classification techniques, the paper tested logistic regression, SVMs, decision trees, random forests, k-nearest neighbors, Naïve Bayes, and passive-aggressive classifiers. The DL models such as long short-term memory (LSTM), GRU, bi-directional LSTM, convolutional neural networks (CNN), and bi-directional encoded representations of transformers (BERT) were tested. The rest of the sections of the paper present the proposed approach. It starts with a brief overview of relevant literature. The methodology is presented afterward. This is followed by results and conclusions.

7.2 LITERATURE REVIEW

The topic of fake news detection has gained a lot of attention in recent years [3]. A comprehensive survey on the topic can be seen in Shu et al. [4], Zhou et al. [5]. An overview of ML approaches for fake news detection has been provided in Manzoor and Singla [6], Sirikulviriya and

Sinthupinyo [7], Kevric et al. [8]. According to Afroz et al. [9], linguistic features change when someone writes fake news. A discussion on fake news in the context of the 2016 US election has been presented in Allcott and Gentzkow [10]. An interesting discussion on what fake news is in the context of levels of facticity and deception has been presented in Tandoc et al. [11].

A number of approaches have been tested for fake news detection in the literature. Pérez-Rosas et al. [1] has identified linguistic differences between fake and legitimate news. Khanam et al. [2] has applied various ML classifiers such as random forest, SVM, linear regression, and k-nearest neighbor to classify fake news. Various preprocessing steps have been performed using the NLTK library, and for classification, scikit-learn was used. In Ahmad et al. [12], the authors proposed a linguistic inquiry and word count feature set along with an ensemble of classification techniques for fake news identification. DL has also been used in various studies for fake news detection. It is a branch of ML based on artificial neural networks with many hidden layers that can autonomously learn the feature representation from data. In this direction, Abedalla et al. [13] presented various approaches based on LSTM, CNNs, and bi-directional LSTM. An unsupervised approach based on a Bayesian network has been presented in Yang et al. [14].

Fake news detection in other languages has also gained interest. In Meesad [45], fake news detection in the Thai language has been presented based on several well-known ML algorithms, and LSTM has been found to provide the best accuracy.

Several studies have investigated the effect of culture on the spread of fake news [16]. It has been found that age, and not gender or education, has an impact on the spread of fake news. In Islam et al. [17], Ternion, a three-pronged solution to fake news detection, has been presented. The author's credibility is considered, as well as the stance of the news; that is, cosine similarity between the title and content of the news is also considered for identifying fake news. Finally, ML algorithms were applied for the detection of fake news. A comparison of various ML and DL approaches for fake news detection has been done in Mishra et al. [18].

From the literature, it is found that ML and DL are the dominant approaches for fake news detection. However, the current literature has not considered recent natural language approaches such as BERT for fake news detection. Hence, this paper presents a comparison of recent DL approaches.

7.3 METHODOLOGY

The following section depicts the ML pipeline for fake and real news. Using the models, a tool is deployed for detecting fake news. The pipeline starts with the acquisition of data from various resources, that is, the dataset collection phase. That dataset was extracted from Gist GitHub Repository [19]. The dataset is then preprocessed (cleaning, imputation, standardization/normalization, etc.). After the data is cleaned, it is processed to select features. The dataset with defined features is then split into respective training and testing samples. Finally, the classifier is trained on the dataset. For ML models, the term frequency-inverse document frequency (TfIDF) vectorizer is used. For DL models such as LSTM, CNN, and GRU, an embedding layer is used to get the vectorized representation of input text. For BERT, the default word piece tokenizer is used. Figure 7.1 illustrates the detailed workflow of the proposed system methodology. The strategy is based on conducting various experiments on the dataset using various ML and DL algorithms described in the previous section of the paper. Finally, the best-performing model is selected, and details of it are documented.

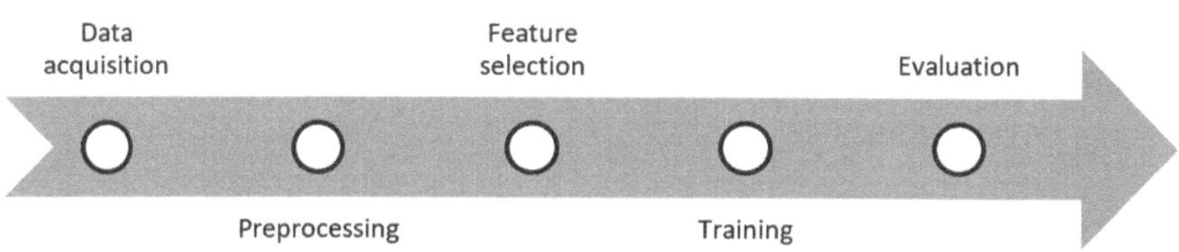

FIGURE 7.1 Proposed machine learning pipeline.

7.3.1 DATASET DETAILS

Figure 7.2 shows the dataset. The dataset contains details such as title (news headline), text (description/news), subject (news, politics, left-news, US news, government, and Middle East), and date (publishing date, on-air date).

The dataset is first pre-processed using noise removal and imputation techniques. After the pre-processing phase, exploratory data analysis and data visualization are performed. Python libraries such as Seaborn and Pyplot from Matplotlib were used.

	title	text	subject	date
0	Donald Trump Sends Out Embarrassing New Year'...	Donald Trump just couldn t wish all Americans ...	News	December 31, 2017
1	Drunk Bragging Trump Staffer Started Russian ...	House Intelligence Committee Chairman Devin Nu...	News	December 31, 2017
2	Sheriff David Clarke Becomes An Internet Joke...	On Friday, it was revealed that former Milwauk...	News	December 30, 2017
3	Trump Is So Obsessed He Even Has Obama's Name...	On Christmas day, Donald Trump announced that ...	News	December 29, 2017
4	Pope Francis Just Called Out Donald Trump Dur...	Pope Francis used his annual Christmas Day mes...	News	December 25, 2017

FIGURE 7.2 Snapshot of the dataset.

Figures 7.3 and 7.4 show the graphical and tabular representation of all the obtained insights regarding the subject column.

News	**9050**
politics	**6841**
left-news	**4459**
Government News	**1570**
US_News	**783**
Middle-east	**778**
Name: subject, dtype: int64	

FIGURE 7.3 Count of different subject-based news in the dataset.

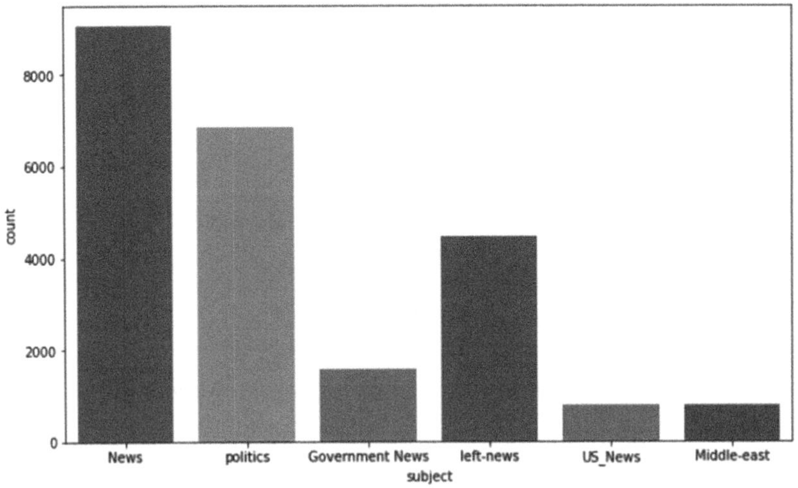

FIGURE 7.4 Depiction of count of different news based on their subject.

After obtaining an accurate count of subjects based on the news, a particular text visualization library named Word Cloud is deployed to get the most frequently appearing list of words. Before feeding the data to print the word cloud, the text data is first joined with its publisher details. The data is then fed to visualize the word cloud. Figure 7.5 shows the word cloud:

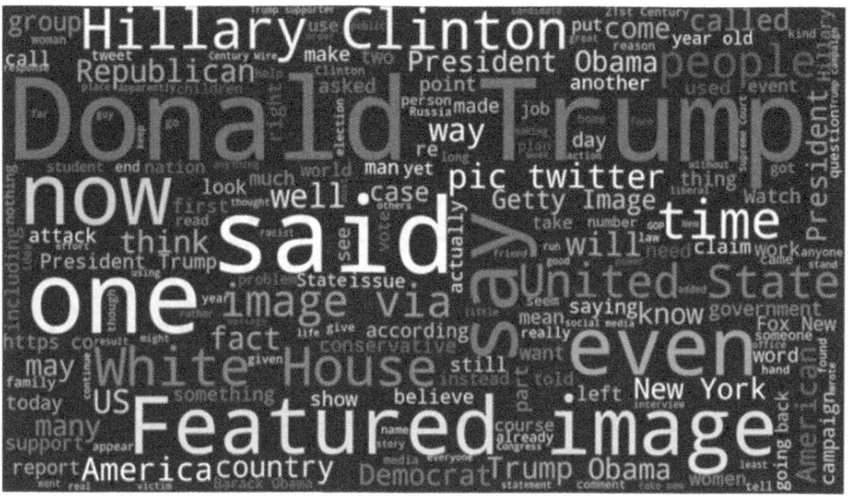

FIGURE 7.5 Word cloud illustrating the most frequent words in fake news.

The next step is to examine the dataset containing true news using a word cloud. The words occurring most frequently can be depicted in Figure 7.6.

7.3.2 PREPROCESSING

At first, the number of unknown publishers is counted from the dataset. The records containing unknown publishers are then dropped to get the cleaned data, that is, text with its publisher details. The next step is to use the NLTK library to perform POS tagging on the textual data for feature selection. The data is further cleaned by removing extra spaces from the text using the kgptalkie library from the preprocessor package of Python. Then, to split each word and form the list of words, Gensim, a library from Python, is deployed. This forms a vector of words containing all words enlisted in the array of words. Next, the technique of tokenization is applied to the word vector. The tokenizer assigns each word a unique index of its own. Afterward, the length of each row is counted (the number of words in each row). For a better understanding, the count of words in each row is depicted using a histogram in Figure 7.7.

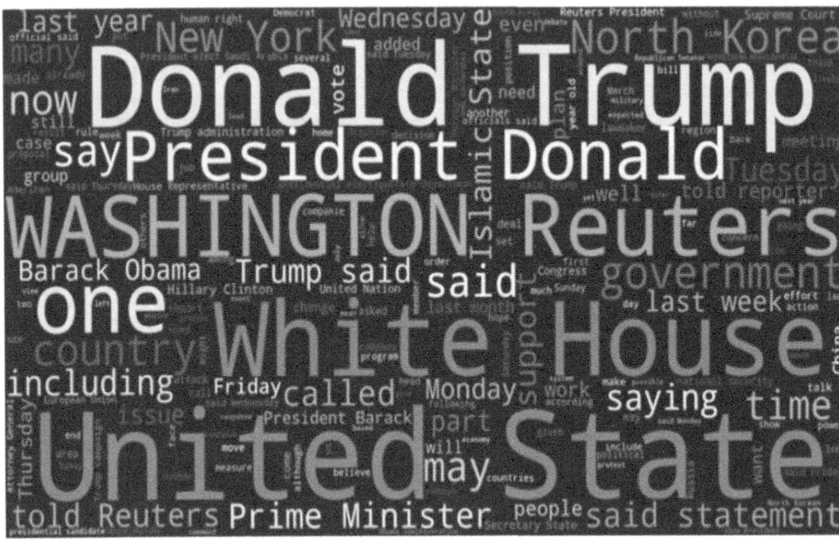

FIGURE 7.6 Frequency of each word in real news.

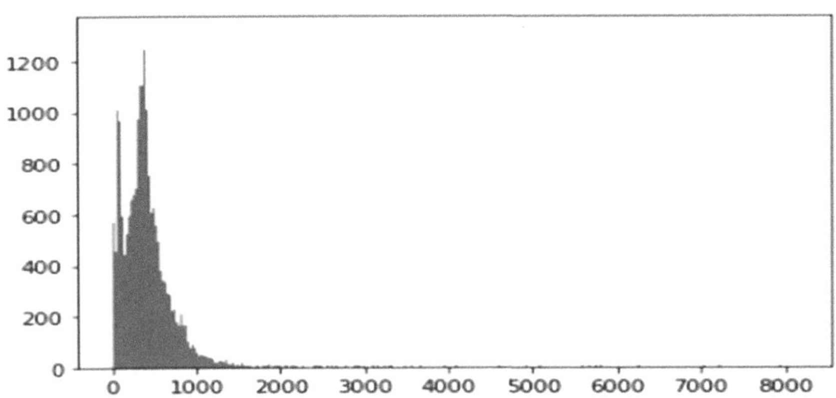

FIGURE 7.7 Histogram of count of words in each row.

After this, the padded sequence and weight matrix are utilized to prepare the data for word embedding. Word embedding is the most commonly used visual representation of document vocabulary. Through this, we have captured the context of a word in a document, semantic and syntactic similarity, and the relation of each word with others.

After all this, the dataset is now prepared to be fitted into various models. For that, the data is then split into a training and test set by a 70:30 ratio (i.e., 70% of the data is reserved for the training examples and the remaining 30% for the testing examples), and the model is then set for prediction.

7.3.3 CLASSIFICATION

Various ML models have been used for classification. This section presents an overview of the classifiers used along with their implementation details.

7.3.3.1 MACHINE LEARNING CLASSIFIERS

Various ML classifiers were trained. For this purpose, the text is first vectorized using the TF-IDF vectorizer in sklearn. The TF-IDF considers the frequency of the term in documents in proportion to the presence of the same term in other documents. If t is the term, d is the current document, and D is the collection of documents, then:

$$\text{TF-IDF}(t, d, D) = \text{tf}(t,d) * \text{idf}(t,D) \tag{7.1}$$

This paper considers the following classifiers. Unless stated, the parameters were kept to default value:

1. **Logistic Regression:** It is used for binary classification and employs the SoftMax function to squeeze the output between 0 and 1.
2. **Support Vector Machine:** It uses the kernel technique to find a high dimensional decision boundary while being in lower dimension.
3. **Decision Trees:** It tests features at various nodes of the tree to reach to final class of the data.
4. **Random Forest:** It uses a collection of trees ensemble for classification.
5. **Naïve Bayes:** It is based on Bayesian theorem to find the posterior probability based on prior and conditional probabilities.

6. **K-NN:** For an item to be classified, it chooses the labels of a set of neighbors and based on them, classifies a data item.
7. **Passive Aggressive Classifier:** It is based on remaining passive for correct classification while aggressive to misclassification.

7.3.3.2 DEEP LEARNING CLASSIFIERS

Deep learning is an extension of ML that is based on a deep network of neuron layers that learns increasingly complex representations of the data. The following classifiers were used. Figure 7.8 shows the implementation details of these classifiers.

1. **Multilayer Perceptron:** It is based on multiple layers of neurons. The neurons of one layer are connected to all the neurons of the next layer through weighted connections, which are then processed further with the help of an activation function. The first layer is the input layer, the last layer is the output layer, and in between, there are multiple hidden layers.
2. **Convolutional Neural Network:** It is based on convolution and max pooling operations. The convolution is based on a set of filters/windows that slide over the data points, and a dot product of the items overlapping with the window and the corresponding window coefficients is taken. Max pooling is based on dividing the data into small groups, and the maximum of each group is taken. The objective of both convolution and max pooling is to find the local features from the data, which are then classified using a fully connected layer.
3. **Long Short Term Memory:** It is based on a recurrent model that generates its output based not only on the current input but also on the previous output of the network. To avoid the vanishing gradient problem, the network also maintains a cell state.
4. **Gated Recurrent Unit:** It is similar to LSTM, but with a simplified model as it doesn't have an output gate.
5. **Bidirectional LSTM:** It is based on scanning the sequence from both left to right and right to left. It is expected to provide better accuracy because of its bi-directional nature.

   For each of the above models, the input is first converted into sequences and padded to the maximum size using TensorFlow's text vectorizer. They are then passed through an embedding layer to vectorize them into integer values.

6. **Bidirectional Encoded Representation of Transformers:** It is based on the transformer model that comprises multi-head attention blocks. The novelty of the transformer is that it can process input in parallel and also attend to various parts of the text. BERT adds to the novelty further by using a word piece tokenizer that can reduce the vocabulary.

Layer (type)	Output Shape	Param #
text_vectorization (TextVectorization)	(None, 100)	0
embedding_2 (Embedding)	(None, 100, 8)	20000
flatten (Flatten)	(None, 800)	0
dense_3 (Dense)	(None, 64)	51264
dense_4 (Dense)	(None, 32)	2080
dense_5 (Dense)	(None, 1)	33

Total params: 73,377
Trainable params: 73,377
Non-trainable params: 0

a. Multilayer perceptron neural network

Model: "sequential"

Layer (type)	Output Shape	Param #
text_vectorization (TextVectorization)	(None, None)	0
embedding (Embedding)	(None, None, 8)	20000
conv1d (Conv1D)	(None, None, 16)	400
global_max_pooling1d (GlobalMaxPooling1D)	(None, 16)	0
dense (Dense)	(None, 1)	17

Total params: 20,417
Trainable params: 20,417
Non-trainable params: 0

b. Convolutional neural network (CNN)

Model: "sequential"

Layer (type)	Output Shape	Param #
text_vectorization (TextVectorization)	(None, 100)	0
embedding (Embedding)	(None, 100, 8)	20000
lstm (LSTM)	(None, 16)	1600
dense (Dense)	(None, 1)	17

Total params: 21,617
Trainable params: 21,617
Non-trainable params: 0

c. Long short term memory (LSTM)

Model: "sequential"

Layer (type)	Output Shape	Param #
text_vectorization (TextVectorization)	(None, 100)	0
embedding (Embedding)	(None, 100, 8)	20000
gru (GRU)	(None, 16)	1248
dense (Dense)	(None, 1)	17

Total params: 21,265
Trainable params: 21,265
Non-trainable params: 0

d. Gated recurrent unit (GRU)

Model: "sequential_1"

Layer (type)	Output Shape	Param #
text_vectorization (TextVectorization)	(None, 100)	0
embedding_1 (Embedding)	(None, 100, 8)	20000
bidirectional (Bidirectional)	(None, 32)	3200
dense_1 (Dense)	(None, 1)	33

e. Bidirectional LSTM

input_ids (InputLayer)	[(None, 150)]	0	[]
token_type_ids (InputLayer)	[(None, 150)]	0	[]
tf_bert_model (TFBertModel)	TFBaseModelOutputWithPoolingAndCrossAttentions(last_hidden_state=(None, 150, 768), pooler_output=(None, 768), past_key_values=None, hidden_states=None, attentions=None, cross_attentions=None)	108310272	['attention_mask[0][0]', 'input_ids[0][0]', 'token_type_ids[0][0]']
dropout_37 (Dropout)	(None, 768)	0	['tf_bert_model[0][1]']
dense (Dense)	(None, 200)	153800	['dropout_37[0][0]']
dropout_38 (Dropout)	(None, 200)	0	['dense[0][0]']
dense_1 (Dense)	(None, 1)	201	['dropout_38[0][0]']

Total params: 108,464,273
Trainable params: 108,464,273
Non-trainable params: 0

f. Bidirectional encoded representation of transformer (BERT) model

FIGURE 7.8 Details of the deep learning models used.

7.4 RESULTS

This paper presents a comparison of various models for fake news identification. The data has been gathered from a well-known website that is already labeled as fake or trustworthy news. The data is split into training and testing datasets in the ratio of 70–30. For convenience, fake news is labeled as 1 and genuine news as 0. The models were tested on the scikit-learn and TensorFlow frameworks. The next few paragraphs present the results based on performance measures like accuracy, precision, recall, and F1 score.

7.4.1 MACHINE LEARNING CLASSIFIERS

Figure 7.9 shows the performance of various ML classifiers. An accuracy of around 93% is achieved by the decision tree classifier. Additionally, the other parameters are reasonably good.

	precision	recall	f1-score	support
FAKE	0.89	0.95	0.92	935
REAL	0.95	0.89	0.92	966
accuracy			0.92	1901
macro avg	0.92	0.92	0.92	1901
weighted avg	0.92	0.92	0.92	1901

(a) Logistic regression

	precision	recall	f1-score	support
FAKE	0.89	0.95	0.92	935
REAL	0.95	0.89	0.92	966
accuracy			0.92	1901
macro avg	0.92	0.92	0.92	1901
weighted avg	0.92	0.92	0.92	1901

(b) Support vector machine

	precision	recall	f1-score	support
FAKE	0.90	0.96	0.93	935
REAL	0.96	0.90	0.93	966
accuracy			0.93	1901
macro avg	0.93	0.93	0.93	1901
weighted avg	0.93	0.93	0.93	1901

(c) Decision tree classifier

	precision	recall	f1-score	support
FAKE	0.90	0.90	0.90	935
REAL	0.90	0.90	0.90	966
accuracy			0.90	1901
macro avg	0.90	0.90	0.90	1901
weighted avg	0.90	0.90	0.90	1901

(d) Random forest classifier

	precision	recall	f1-score	support
FAKE	0.80	0.76	0.78	935
REAL	0.78	0.81	0.79	966
accuracy			0.79	1901
macro avg	0.79	0.79	0.79	1901
weighted avg	0.79	0.79	0.79	1901

(e) Gaussian Naïve Bayes classifier

	precision	recall	f1-score	support
FAKE	0.52	1.00	0.69	935
REAL	1.00	0.12	0.22	966
accuracy			0.55	1901
macro avg	0.76	0.56	0.45	1901
weighted avg	0.77	0.55	0.45	1901

(f) k-neighbor classifier

FIGURE 7.9 Performance of various machine learning classifiers.

We investigated the passive-aggressive classifier in detail, as it has recently gained a lot of attention. Figure 7.10 shows the performance of the classifier. The ROC curve is also plotted. An accuracy of 94% is achieved, with a similar precision and recall score.

	precision	recall	f1-score	support
FAKE	0.93	0.94	0.94	935
REAL	0.94	0.93	0.94	966
accuracy			0.94	1901
macro avg	0.94	0.94	0.94	1901
weighted avg	0.94	0.94	0.94	1901

(a) Classification report

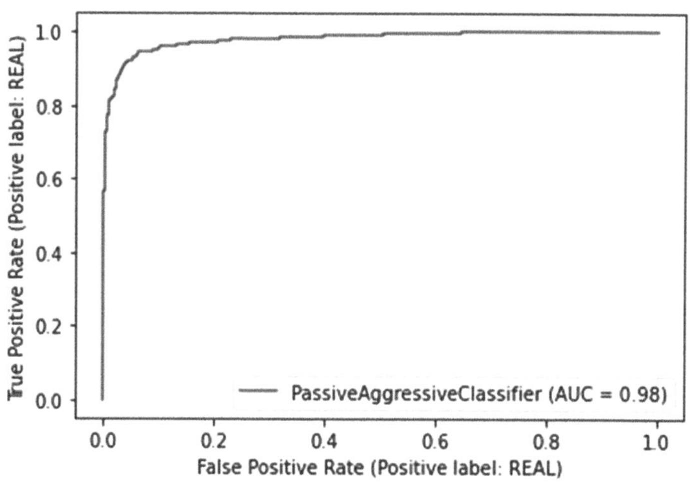

(b) ROC curve

FIGURE 7.10 Results of passive aggressive classifier.

7.4.2 DEEP LEARNING CLASSIFIERS

Figure 7.11 shows the performance of the multilayer perceptron. From the accuracy report in Figure 11(a), around 90% accuracy, precision, recall, and F1 score are achieved. The training and validation accuracy and loss graphs are shown in Figures 7.11(b) and 7.11(c).

```
              precision    recall  f1-score   support

           0       0.90      0.90      0.90       935
           1       0.90      0.90      0.90       966

    accuracy                           0.90      1901
   macro avg       0.90      0.90      0.90      1901
weighted avg       0.90      0.90      0.90      1901
```

(a) Accuracy report

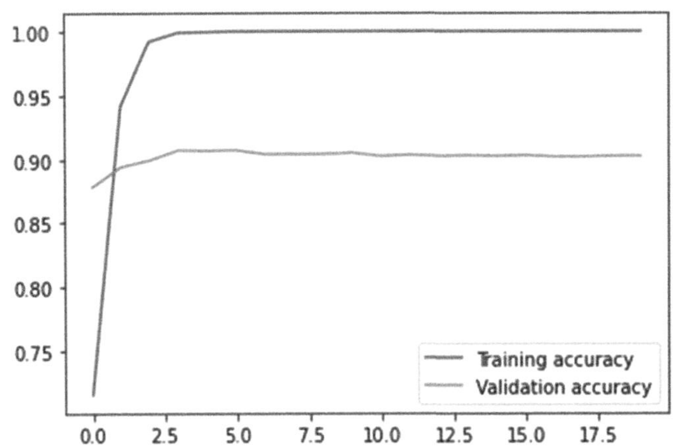

(b) Training vs. validation accuracy

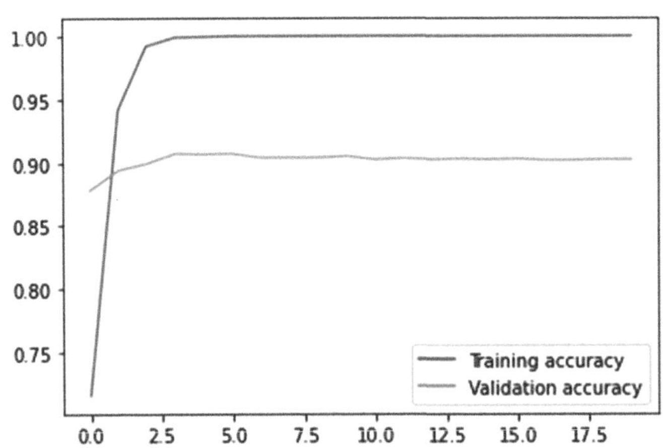

(c) Training vs. validation loss

FIGURE 7.11 Performance of multilayer perceptron.

Figures 7.12–7.16 show the performance of various DL classifiers trained for 20 epochs. As can be seen, BERT provides the best accuracy among these classifiers.

	precision	recall	f1-score	support
0	0.90	0.92	0.91	966
1	0.92	0.89	0.90	935
accuracy			0.91	1901
macro avg	0.91	0.91	0.91	1901
weighted avg	0.91	0.91	0.91	1901

(a) Accuracy report

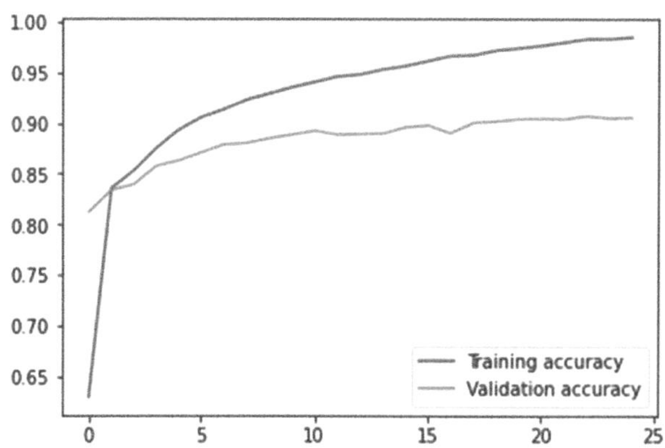

(b) Training Vs validation accuracy

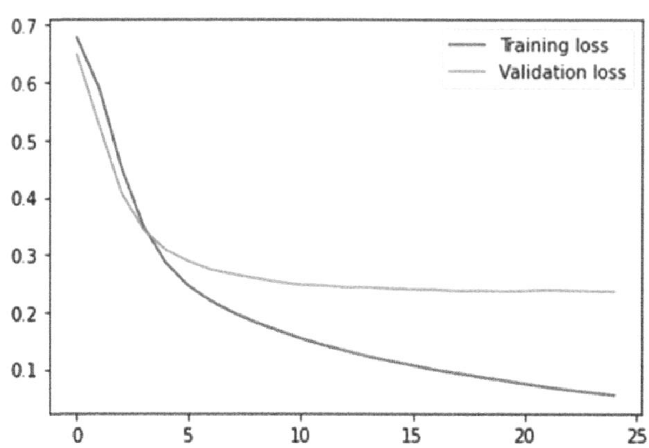

(c) Training Vs validation loss

FIGURE 7.12 Performance of convolutional neural network.

	precision	recall	f1-score	support
0	0.88	0.90	0.89	935
1	0.90	0.89	0.89	966
accuracy			0.89	1901
macro avg	0.89	0.89	0.89	1901
weighted avg	0.89	0.89	0.89	1901

(a) Classification report

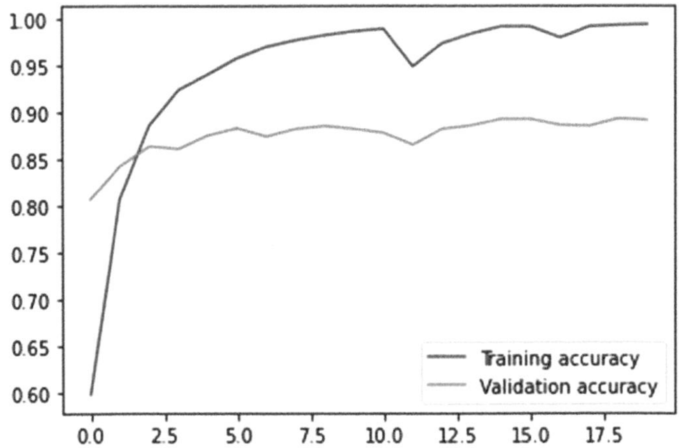

(b) Training Vs validation accuracy

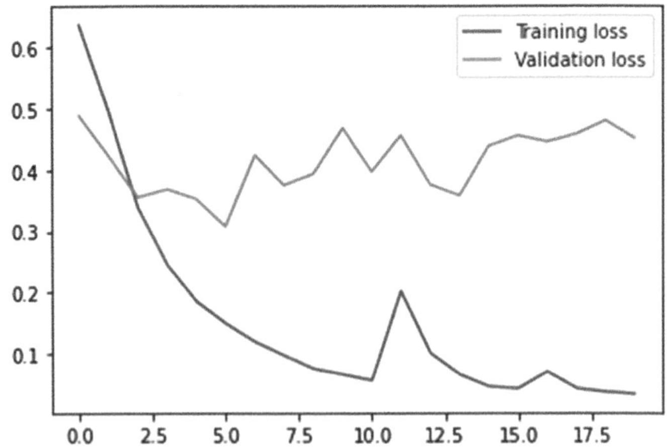

(c) Training Vs validation loss

FIGURE 7.13 Performance of LSTM.

	precision	recall	f1-score	support
0	0.89	0.88	0.88	935
1	0.88	0.89	0.89	966
accuracy			0.89	1901
macro avg	0.89	0.89	0.89	1901
weighted avg	0.89	0.89	0.89	1901

(a) Accuracy report

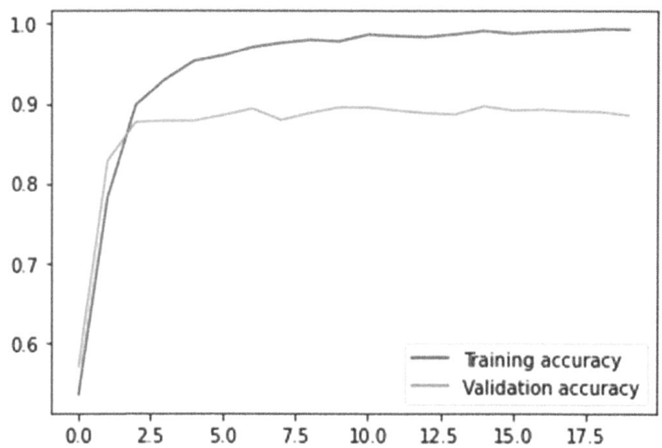

(b) Training Vs validation accuracy

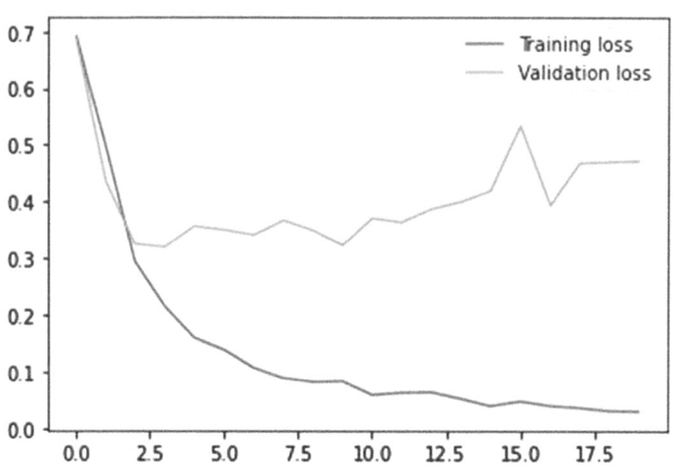

(c) Training Vs validation loss

FIGURE 7.14 Performance of gated recurrent unit (GRU).

	precision	recall	f1-score	support
0	0.87	0.90	0.89	935
1	0.90	0.87	0.89	966
accuracy			0.89	1901
macro avg	0.89	0.89	0.89	1901
weighted avg	0.89	0.89	0.89	1901

(a) Classification report

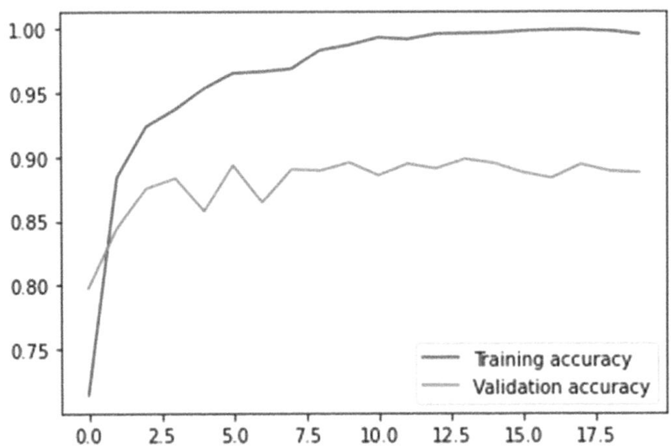

(b) Training Vs validation accuracy

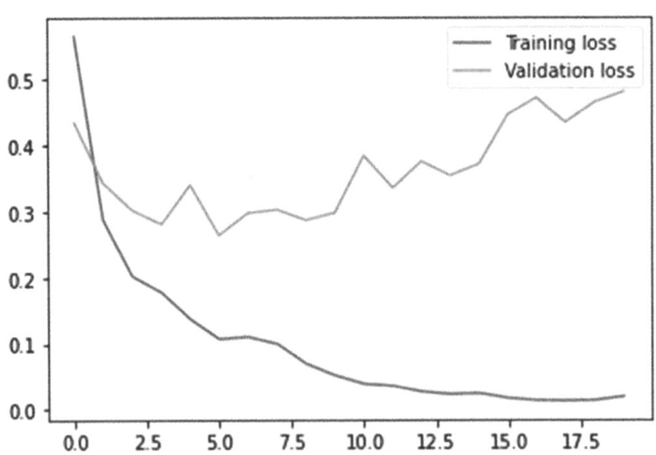

(c) Training Vs validation loss

FIGURE 7.15 Performance of bidirectional long short-term memory (LSTM).

	precision	recall	f1-score	support
0	0.97	0.97	0.97	639
1	0.97	0.97	0.97	628
accuracy			0.97	1267
macro avg	0.97	0.97	0.97	1267
weighted avg	0.97	0.97	0.97	1267

(a) Classification report

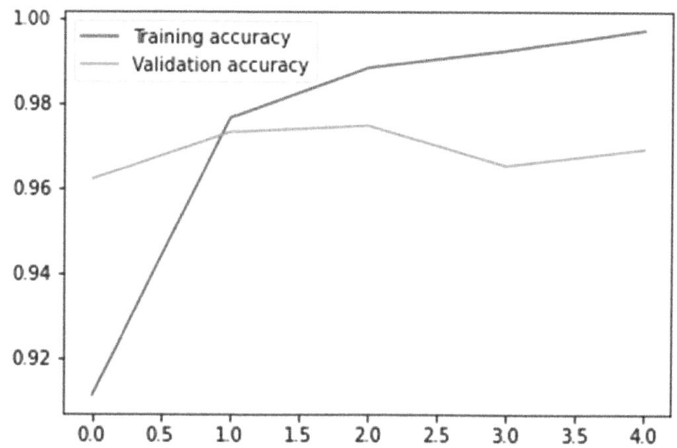

(b) Training Vs validation accuracy

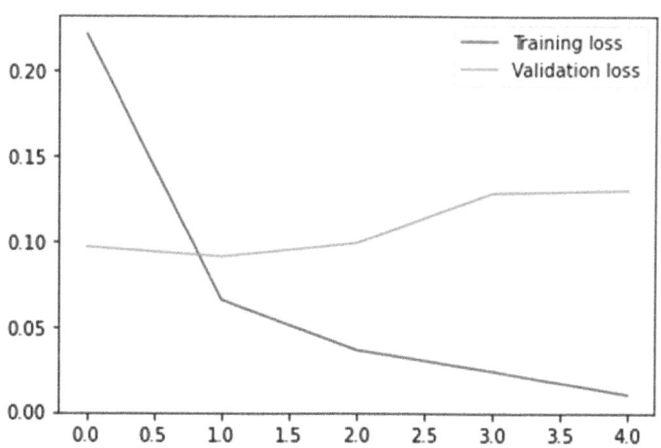

(c) Training Vs validation loss

FIGURE 7.16 Performance of bi-directional encoded representation of transformers (BERT) classifier.

Finally, Table 7.1 compares the performance of the various DL classifiers for the maximum validation accuracy, precision, recall, and F1 score of these classifiers. From the table, it can be seen that BERT provides the highest accuracy on the data with 97.47%. The second in the list is the passive-aggressive classifier with an accuracy of 94% (Figure 7.10).

TABLE 7.1 Comparison of Deep Learning Classifiers for Fake News Detection

Model	Accuracy % (Best)	Precision %	Recall %	F1 %
Multilayer perceptron	90.64	90	90	90
CNN	90.69	91	91	91
LSTM	89.11	89	89	89
GRU	89.64	89	89	89
Bi-directional LSTM	89.79	89	89	89
BERT	97.47	97	97	97

7.5 CONCLUSION

The majority of tasks in today's world are performed online. Newspapers, which were once read in the form of hard copies, are gradually being shifted to online news sites. In addition, there is a growing trend of social media platforms such as Facebook and Twitter being used for the consumption of news. Although many ML algorithms have been used for the identification of fake and genuine news, the discrimination of false news from genuine news is becoming more cumbersome due to the dynamic and variable nature of features present in fake news on social media networks. In this paper, we have used different ML and DL classifiers for fake news detection. To our knowledge, this is the most comprehensive comparison involving seven ML and six DL classifiers. From the comparative study, it has been found that BERT provides the most accuracy, precision, and recall. After that, the passive-aggressive classifier provides an accuracy of around 94%. Other DL models such as CNN, LSTM, and GRU also provide accuracy around 90%. The proposed work can be extended in various directions. The paper has not tested pretrained word embeddings such as Word2Vec or GloVe embedding. In addition, the BERT model can be tested after making a few modifications. Innovations such as residual connections and batch layer normalizations can also be tried for gaining further improvements. Finally, the proposed models should be tested on other datasets [20, 21].

KEYWORDS

- **deep learning**
- **fake news detection**
- **machine learning**
- **natural language processing**
- **recurrent neural network**
- **social media and news authenticity**

REFERENCES

1. Pérez-Rosas, V., Kleinberg, B., Lefevre, A., & Mihalcea, R. (2017). *Automatic Detection of Fake News*. arXiv preprint arXiv:1708.07104.
2. Khanam, Z., Alwasel, B. N., Sirafi, H., & Rashid, M. (2021). Fake news detection using machine learning approaches. In *IOP Conference Series: Materials Science and Engineering* (Vol. 1099, No. 1, p. 012040). IOP Publishing.
3. Wang, W. Y. (2017). *"Liar, Liar Pants on Fire": A New Benchmark Dataset for Fake News Detection.* arXiv preprint arXiv:1705.00648.
4. Shu, K., Sliva, A., Wang, S., Tang, J., & Liu, H. (2017). Fake news detection on social media: A data mining perspective. *ACM SIGKDD Explorations Newsletter, 19*(1), 22–36.
5. Zhou, X., Zafarani, R., Shu, K., & Liu, H. (2019). Fake news: Fundamental theories, detection strategies, and challenges. In *Proceedings of the Twelfth ACM International Conference on Web Search and Data Mining* (pp. 836, 837).
6. Manzoor, S. I., & Singla, J. (2019). Fake news detection using machine learning approaches: A systematic review. In *2019 3rd International Conference on Trends in Electronics and Informatics (ICOEI)* (pp. 230–234). IEEE.
7. Sirikulviriya, N., & Sinthupinyo, S. (2011). Integration of rules from a random forest. In *International Conference on Information and Electronics Engineering* (Vol. 6, pp. 194–198).
8. Kevric, J., Jukic, S., & Subasi, A. (2017). An effective combining classifier approach using tree algorithms for network intrusion detection. *Neural Computing and Applications, 28*(1), 1051–1058.
9. Afroz, S., Brennan, M., & Greenstadt, R. (2012). Detecting hoaxes, frauds, and deception in writing style online. In *2012 IEEE Symposium on Security and Privacy* (pp. 461–475). IEEE.
10. Allcott, H., & Gentzkow, M. (2017). Social media and fake news in the 2016 election. *Journal of Economic Perspectives, 31*(2), 211–236.
11. Tandoc, E. C., Jr., Lim, Z. W., & Ling, R. (2018). Defining "fake news": A typology of scholarly definitions. *Digital Journalism, 6*(2), 137–153.

12. Ahmad, I., Yousaf, M., Yousaf, S., & Ahmad, M. O. (2020). Fake news detection using machine learning ensemble methods. *Complexity, 2020.*
13. Abedalla, A., Al-Sadi, A., & Abdullah, M. (2019). A closer look at fake news detection: A deep learning perspective. In *Proceedings of the 2019 3rd International Conference on Advances in Artificial Intelligence* (pp. 24–28).
14. Yang, S., Shu, K., Wang, S., Gu, R., Wu, F., & Liu, H. (2019). Unsupervised fake news detection on social media: A generative approach. In *Proceedings of the AAAI Conference on Artificial Intelligence* (Vol. 33, No. 1, pp. 5644–5651).
15. Meesad, P. (2021). Thai fake news detection based on information retrieval, natural language processing, and machine learning. *SN Computer Science, 2*(6), 1–17.
16. Rampersad, G., & Althiyabi, T. (2020). Fake news: Acceptance by demographics and culture on social media. *Journal of Information Technology & Politics, 17*(1), 1–11.
17. Islam, N., Shaikh, A., Qaiser, A., Asiri, Y., Almakdi, S., Sulaiman, A., Moazzam, V., & Babar, S. A. (2021). Ternion: An autonomous model for fake news detection. *Applied Sciences, 11*(19), 9292.
18. Mishra, S., Shukla, P., & Agarwal, R. (2022). Analyzing machine learning-enabled fake news detection techniques for diversified datasets. *Wireless Communications and Mobile Computing, 2022.*
19. Gist GitHub repository. (2022). Retrieved from: https://gist.github.com/catyselman/ (accessed on 13 March 2025).
20. Jain, A., & Kasbe, A. (2018). Fake news detection. In *2018 IEEE International Students' Conference on Electrical, Electronics and Computer Science (SCEECS)* (pp. 1–5). IEEE. https://doi.org/10.1109/SCEECS.2018.8546944.
21. Tapaswi, N., & Jain, S. (2012). Treebank-based deep grammar acquisition and part-of-speech tagging for Sanskrit sentences. In *2012 CSI Sixth International Conference on Software Engineering (CONSEG)* (pp. 1–4). IEEE.

CHAPTER 8

Unleashing the Power of Big Data: Information Retrieval and Text Mining Strategies

VIKAS SHARMA[1,*], SUNIL KUMAR[2], TARUN KUMAR VASHISHTH[3], KEWAL KRISHAN SHARMA[3], BHUPENDRA KUMAR[3], and MEGHNA[4]

[1]*Department of Computer Applications, SRM Institute of Science and Technology, Delhi NCR Campus, Ghaziabad, U.P., India*

[2]*School of Information Technology, Auro University, Gujrat, India*

[3]*School of Computer Science and Applications, IIMT University, Meerut, Uttar Pradesh, India*

[3]*School of Civil Engineering, IIMT University, Meerut, Uttar Pradesh, India*

**Corresponding author.*

ABSTRACT

The majority of tasks in today's world are performed online. Newspapers, which were once read in the form of hard copies, are gradually being shifted to online news sites. In addition, there is a growing trend of social media platforms such as Facebook and Twitter being used for the consumption of news. Although many machine learning (ML) algorithms have been used for the identification of fake and genuine news, the discrimination of false news from genuine news is becoming more cumbersome due to the dynamic and variable nature of features present in fake news on social media networks. In this paper, we have used different ML and deep learning (DL) classifiers for fake news detection. To our knowledge, this is the most comprehensive comparison involving seven ML and

Natural Language Processing and Artificial Intelligence: A Perspective Toward Current Trends, Challenges, and Applications. Vikas Khullar, Aryan Chaudhary, & Sunil Kumar (Eds.)

DOI: 10.1201/9781998511273-8

six DL classifiers. From the comparative study, it has been found that BERT provides the most accuracy, precision, and recall. After that, the passive-aggressive classifier provides an accuracy of around 94%. Other DL models such as CNN, LSTM, and GRU also provide accuracy around 90%. The proposed work can be extended in various directions. The paper has not tested pretrained word embeddings such as Word2Vec or GloVe embedding. In addition, the BERT model can be tested after making a few modifications. Innovations such as residual connections and batch layer normalizations can also be tried for gaining further improvements. Finally, the proposed models should be tested on other datasets.

8.1 INTRODUCTION

In the digital age, the exponential growth of data has ushered in an era often characterized as "Big Data." This proliferation of data encompasses a myriad of sources, including textual information, which permeates documents, social media, emails, and web content. As organizations grapple with the sheer complexity and volume of this textual data, the need for effective strategies to harness its potential has never been more pressing. This research paper explores the pivotal role of information retrieval (IR) and text mining strategies in unleashing big data's power.

The advent of big data has ushered in a paradigm shift, changing data into a strategic asset for businesses, governments, and researchers alike. It has fundamentally altered the way organizations operate, offering unprecedented opportunities for informed decision-making, personalized user experiences, and innovative research endeavors. However, the sheer volume and unstructured nature of textual data pose significant challenges. Without the means to efficiently extract, process, and interpret this textual information, organizations risk drowning in an ocean of unutilized potential. IR and text mining emerge as the linchpins in overcoming these challenges. These multidisciplinary fields draw upon computer science, linguistics, data analytics, and information science to equip organizations with the tools and techniques needed to navigate the vast landscape of textual data effectively. IR facilitates the rapid and precise retrieval of relevant information from massive text collections, ensuring that organizations can locate the proverbial "needle in the haystack" when searching for critical insights. Text mining, on the other hand, empowers organizations to transform unstructured text into structured data, thereby unlocking valuable patterns, sentiments, and knowledge hidden within (Figure 8.1).

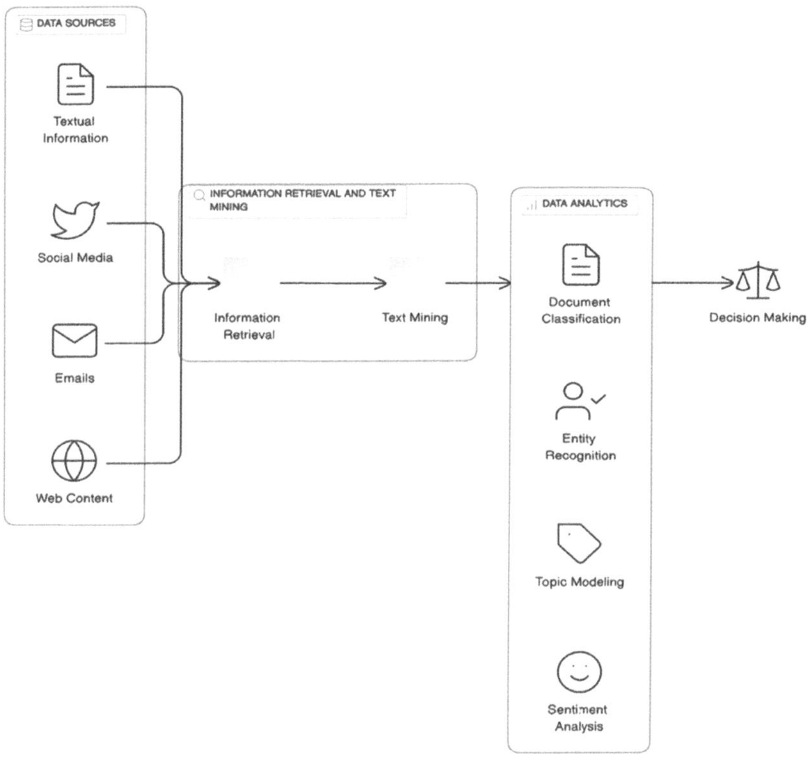

FIGURE 8.1 Big data: data analysis in digital age.

The research paper will delve into the fundamental concepts that underpin IR and text mining strategies, elucidating the intricacies of these techniques. It will explore the evolving landscape of Big Data and the unique challenges posed by textual data within this context. Furthermore, the paper will provide an in-depth examination of the tools, technologies, and methodologies available to organizations seeking to harness Big Data's power through IR and text mining. Beyond the technical aspects, this research paper will also delve into the strategies organizations can employ to extract actionable insights from the textual data deluge. From sentiment analysis to topic modeling, and from entity recognition to document classification, a plethora of techniques will be dissected and discussed within the broader context of real-world applications.

Ultimately, this chapter seeks to provide a comprehensive understanding of how IR and text mining strategies are not merely tools but rather the keys that unlock the transformative potential of Big Data. By shedding light on these essential techniques, we aim to equip organizations, researchers, and practitioners with the knowledge and insights necessary to navigate the ever-expanding landscape of textual data and harness its power for informed decision-making, innovation, and societal advancement.

8.2 LITERATURE REVIEW

Tien [1] introduces the concept of Big Data, which refers to datasets too huge for traditional tools to acquire, access, analyze, or apply efficiently within a reasonable timeframe.

Bizer et al. [2] examine the opportunities and challenges presented by Big Data in the context of the Semantic Web, Semantic Technologies, and database communities.

Chen, Chiang, and Storey [3] conduct an analysis of current research in business intelligence and analytics (BI&A). Their work delves into identifying the challenges and opportunities in BI&A research and education.

In their comprehensive paper, Bilal et al. [4] bridge interdisciplinary gaps by conducting an extensive review of literature in statistics, data mining, warehousing, machine learning (ML), and analytics of big data. The paper assesses the recent adoption of big data in the construction industry while exploring its upcoming strengths in various domain-specific sub-areas within the field.

In their paper, Ibekwe-Sanjuan and Geoffrey [5] provide a high-level analysis of the impacts of big data on knowledge organization (KO) and knowledge organization systems (KOSs). They engage with ongoing debates in the KO community regarding universal bibliographic classifications and thesauri in the context of the web and explore the epistemological and methodological considerations in data-driven inquiry.

In their paper, Wu et al. [6] introduce a HACE theorem that defines the key aspects of the big data revolution. They also put forward a big data processing model, considering the perspective of data mining.

In their paper, Wang et al. [7] provide an extensive overview of Big Data, covering topics such as its concepts, characteristics, processing paradigms, and decision-making techniques. They also delve into the practical use of Big Data in the realm of social science and discuss the existing issues and potential future directions in this field.

In their work, Pramanik et al. [8] provide a comprehensive examination of the technical aspects of Big Data classification, including the fundamentals like the necessity, types, patterns, phases, and approaches.

In their paper, Li et al. [9] start by providing a concise introduction to the concept, characteristics, and applications of "Big Data." They then proceed to summarize and analyze the various types of data relevant to the three primary phases of product lifecycle management (PLM): the initial phase, center phase, and final phase.

In their research, Bharathi [10] presents a comprehensive big data information security risk spectrum, which includes 25 precisely defined risk factors organized into seven constructs, offering a prioritized and ranked assessment.

Skënduli, Biba, and Ceci [11] conduct an extensive literature review of distributed parallel data-intensive ML algorithms applied to huge data. The reviewed algorithms span various ML categories, encompassing unsupervised learning, supervised learning, semi-supervised learning, and DL.

In their paper, Chen and Zhang [12] provide an insightful exploration of Big Data, delving into its applications, opportunities, and challenges. They also offer an overview of techniques and technologies employed to address Big Data-related issues.

8.3 THE LANDSCAPE OF BIG DATA

The landscape of massive data is a critical section in the research paper that sets the stage for understanding the role of IR and text mining strategies. In this section, you will define and contextualize big data, discuss the sources of textual data, and emphasize the importance of unstructured textual data in the realm of analytics of big data. Let's delve into each aspect in detail:

8.3.1 DEFINING BIG DATA WITH THE 4VS

The 4Vs—Volume, Variety, Velocity, and Veracity—serve as a framework for characterizing the essential characteristics of Big Data. These dimensions highlight the immense scale, diversity, speed, and data quality challenges associated with Big Data, setting the stage for understanding the complexities and opportunities that arise when dealing with vast and dynamic datasets in today's data-driven world.

8.3.1.1 VOLUME

1. **Definition:** Volume refers to the sheer amount of data that constitutes Big Data. It is characterized by sets of data that are too extensive to be managed, processed, or analyzed using traditional data management procedures.
2. **Scale:** Big Data can range from terabytes to petabytes, exabytes, and even yottabytes. Massive scale is beyond the capacity of typical databases and storage systems.
3. **Examples:** Massive datasets include those generated by social media platforms (millions of tweets, posts, and images daily), scientific research (large-scale simulations and experiments), and IoT (e.g., sensor data from smart cities) (Figure 8.2).

8.3.1.2 VARIETY

1. **Definition:** Variety represents the diversity of data formats and types within huge Big Data. It encompasses structured, semi-structured, and unstructured data, each with distinct characteristics.
2. **Structured Data:** This includes data that fits neatly into traditional relational databases, such as tables with rows and columns. Examples are customer databases, transaction logs, and financial records.
3. **Semi-Structured Data:** Semi-structured data does not conform to robust database structures but has some extent of organization. Common formats include XML and JSON, often used in web data and document storage.
4. **Unstructured Data:** Such information lacks a predefined structure and includes text, images, audio, video, and more. It is often in natural language and is challenging to analyze without appropriate tools.

8.3.1.3 VELOCITY

1. **Definition:** Velocity refers to the speed at which data is originated, collected, and must be processed in Big Data scenarios. It's about the real-time or near real-time nature of information.

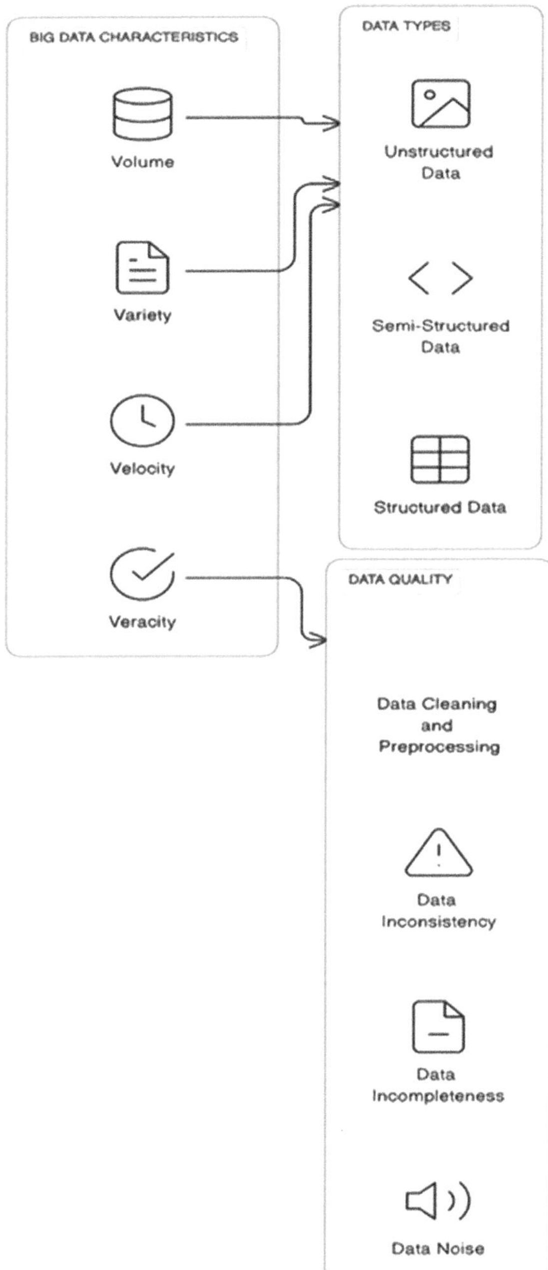

FIGURE 8.2 The four Vs—volume, variety, velocity, and veracity—in massive data.

2. **Real-Time Data:** Some data sources, such as social media, IoT devices, and financial markets, produce data at high speeds, sometimes in milliseconds. Organizations need to capture and analyze this data quickly to make informed decisions.
3. **Example:** In online retail, user activity data, including clicks, views, and purchases, is generated in real time and used to personalize product recommendations instantly.

8.3.1.4 VERACITY

1. **Definition:** Veracity deals with the consistency and trustworthiness of data in Big Data environments. It acknowledges that not all data is clean, accurate, and dependable.
2. **Data Quality Issues:** Veracity includes challenges like data noise (irrelevant or erroneous data points), data incompleteness (missing values), and data inconsistency (conflicting information).
3. **Data Cleaning and Preprocessing:** To derive meaningful insights from Big Data, organizations often need to invest significant effort in data cleaning and preprocessing to enhance data quality.
4. **Importance:** Poor data quality can lead to incorrect analysis and decision-making, emphasizing the importance of addressing veracity issues in Big Data projects.

8.3.2 SOURCES OF TEXTUAL DATA

Emphasizing the ubiquity of textual data across these diverse sources underscores its role as a valuable and pervasive source of information and insights. In today's digital landscape, textual data is instrumental in shaping decision-making, understanding user behavior, tracking trends, and gaining competitive advantages across various industries and domains. This data's availability and accessibility have made it a cornerstone of data-driven strategies and analytics.

8.3.2.1 SOCIAL MEDIA PLATFORMS

1. **Examples:** Social media platforms like Twitter, Facebook, Instagram, LinkedIn, and more provide a massive volume of textual

data. Users generate text through posts, comments, messages, and interactions.

2. **Data Types:** Social media data can include text, images, videos, hashtags, and mentions. Textual data, in the form of user-generated content, is particularly valuable for sentiment analysis, trend identification, and understanding user behavior.
3. **Importance:** Social media data reflects public sentiment, opinions, trends, and emerging issues, making it invaluable for market research, brand monitoring, and real-time analysis.

8.3.2.2 BLOGS AND ONLINE FORUMS

1. **Examples:** Blogs, discussion forums (e.g., Reddit, Stack Overflow), and Q&A websites (e.g., Quora) are rich sources of textual data. Users engage in discussions, ask questions, and provide detailed responses.
2. **Data Types:** Textual data on these platforms consists of user-written articles, forum threads, comments, and answers to questions. It often includes valuable expertise and domain-specific knowledge.
3. **Importance:** Blogs and forums offer insights into niche topics, technical troubleshooting, and community-driven knowledge sharing. Analyzing this data can help identify emerging trends and address user concerns.

8.3.2.3 NEWS ARTICLES AND RESEARCH PAPERS

1. **Examples:** News websites, scientific journals, and digital libraries host vast repositories of textual data in the form of articles, reports, and research papers.
2. **Data Types:** Textual data includes news articles, academic papers, reports, and reviews. This content covers a wide range of topics and domains.
3. **Importance:** News and research data are critical for staying informed, conducting academic research, and tracking developments in various fields. Text mining can help extract insights, trends, and sentiments from this data.

8.3.2.4 EMAILS AND ORGANIZATIONAL COMMUNICATION

1. **Examples:** Within organizations, emails, chat messages, and other communication channels generate significant textual data. This data pertains to internal discussions, project updates, and external correspondence.
2. **Data Types:** Textual data in emails and messages includes conversations, attachments, meeting notes, and more. It often contains crucial business information.
3. **Importance:** Analyzing internal communication can improve team collaboration, identify communication patterns, and support decision-making. It can also aid in compliance and legal investigations.

8.3.2.5 CUSTOMER REVIEWS AND FEEDBACK

1. **Examples:** E-commerce platforms, review websites (e.g., Yelp, TripAdvisor), and mobile apps collect textual data in the form of customer reviews, ratings, and feedback.
2. **Data Types:** Textual data includes product reviews, comments, ratings, and user feedback. It reflects customer opinions, preferences, and experiences.
3. **Importance:** Analyzing customer reviews and feedback helps businesses understand customer satisfaction, identify product or service issues, and make improvements. It's crucial for reputation management and customer-centric decision-making.

8.3.2.6 IOT DEVICES

1. **Examples:** Devices of IoT, for example, smart home appliances, sensors, and industrial equipment, generate textual data logs and reports as part of their operational data.
2. **Data Types:** Text-based logs and reports include system status updates, error messages, sensor readings, and event logs.
3. **Importance:** Analyzing IoT-generated textual data is essential for predictive maintenance, anomaly detection, and optimizing the performance of IoT ecosystems (Figure 8.3).

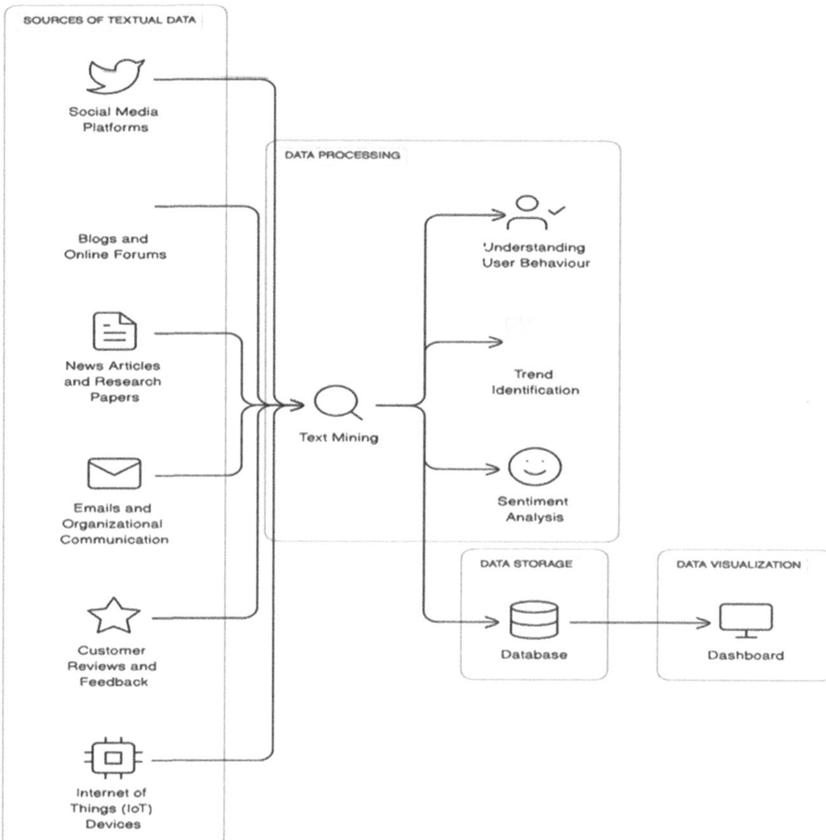

FIGURE 8.3 Sources of textual data.

8.3.3 IMPORTANCE OF UNSTRUCTURED TEXTUAL DATA

Unstructured textual data is a critical component of the Big Data landscape. Its significance lies in its ability to provide valuable insights, capture sentiments, understand customer preferences, track public sentiment, facilitate market research, unveil hidden knowledge, and serve as a cornerstone in specific domains. When effectively harnessed, unstructured textual data enriches decision-making, innovation, and competitive advantage, ensuring that organizations can derive maximum value from their Big Data investments. The importance of unstructured textual data in the context of Big Data analytics is profound and multifaceted. Here, we elaborate on the significance of unstructured textual data, highlighting its value and the various purposes it serves within the Big Data landscape:

8.3.3.1 VALUABLE INSIGHTS AND SENTIMENTS

i. Unstructured textual data contains a wealth of valuable insights, sentiments, and information that can't be easily captured through structured data alone. This data often includes opinions, reviews, comments, and feedback from users and customers.
ii. **Example:** Customer reviews on e-commerce websites can reveal product strengths and weaknesses, customer satisfaction levels, and potential areas for improvement.

8.3.3.2 UNDERSTANDING CUSTOMER PREFERENCES

i. Unstructured textual data is instrumental in understanding customer preferences and behavior. By analyzing text data from sources like social media, reviews, and customer support interactions, organizations can gain deep insights into what their customers want and how they perceive their products or services.
ii. **Example:** Analyzing social media mentions can help a company identify popular trends, customer pain points, and emerging needs.

8.3.3.3 TRACKING PUBLIC SENTIMENT

i Unstructured textual data allows organizations to track public sentiment in real-time. This is particularly valuable for businesses and governments seeking to monitor public opinion, assess reactions to policy changes, and manage public relations.
ii **Example:** Political campaigns use sentiment analysis on social media data to gauge public sentiment and adjust their strategies accordingly.

8.3.3.4 CONDUCTING MARKET RESEARCH

i Market research heavily relies on unstructured textual data to analyze consumer behavior, preferences, and trends. By studying online discussions, product reviews, and social media conversations, companies can make informed decisions about their products and marketing strategies.

ii **Example:** A company might analyze Twitter conversations to understand how customers perceive its brand compared to competitors.

8.3.3.5 HIDDEN KNOWLEDGE AND PATTERNS

i Unstructured textual data is a goldmine for discovering hidden knowledge, trends, and patterns that might not be immediately evident in structured data sources. Through text mining techniques, organizations can unveil critical insights that drive innovation and competitive advantage.

ii **Example:** By analyzing medical case notes and research articles, healthcare researchers can identify emerging treatment trends and potentially life-saving interventions.

8.3.3.6 CRUCIAL IN SPECIFIC DOMAINS

i. In domains like healthcare, legal, and customer service, unstructured textual data is indispensable. In healthcare, patient records, clinical notes, and medical literature are often in text form and are crucial for diagnosis and research.

ii. In the legal domain, contracts, court transcripts, and legal documents contain essential information that must be extracted and analyzed.

iii. In customer service, analyzing customer support interactions can help identify common issues and improve service quality.

8.3.3.7 COMPLEMENTING STRUCTURED DATA

i. Unstructured textual data complements structured data, creating a more comprehensive view of the data landscape. Combining structured data (e.g., transactions) with unstructured data (customer feedback) allows organizations to gain a holistic understanding of their operations and customers.

ii. **Example:** Summing structured sales data with unstructured customer reviews may provide a complete picture of product performance.

8.4 INFORMATION RETRIEVAL (IR) STRATEGIES

8.4.1 FUNDAMENTAL PRINCIPLES OF INFORMATION RETRIEVAL

The field of IR revolves around the essential principles that govern the retrieval of relevant information from extensive textual data collections. This paper provides an in-depth exploration of five fundamental principles that constitute the backbone of IR: relevance, query, document, indexing, and ranking.

8.4.1.1 RELEVANCE

- Relevance lies at the heart of IR. It defines the primary goal of IR, which is to deliver information that aligns with the user's query or information need.
- Understanding relevance is complex because it is inherently subjective. What one user deems relevant may differ significantly from another user's perspective. Relevance depends on various factors, including user context, intent, and the specific problem or task at hand.
- IR systems strive to capture this subjectivity by employing relevance models that take into account user feedback, query expansion, and relevance feedback mechanisms.

8.4.1.2 QUERY

- The query serves as the bridge between the user's information need and the vast corpus of documents. It is the formalized request for information, often composed of one or more keywords, phrases, or a combination of terms.
- Queries come in various forms, ranging from simple keyword queries to complex Boolean expressions or natural language sentences. The choice of query formulation depends on the user's familiarity with the system and the precision required.

- Effective query formulation is critical for achieving successful retrieval results. Researchers continue to explore techniques such as query expansion and query reformulation to enhance query quality.

8.4.1.3 DOCUMENT

- In the realm of IR, documents represent the atomic units of information. These can encompass diverse forms of textual content, including web pages, articles, books, emails, and more.
- Documents are the building blocks of the IR system's collection. Each document holds the potential to contain information that satisfies a user's query.
- Document representation and modeling are key research areas in IR, with methods like vector space models and deep learning (DL) techniques constantly evolving to capture document semantics and structure.

8.4.1.4 INDEXING

- Indexing plays a pivotal role in the efficiency of IR systems. It involves the creation of a structured data structure called an index, which facilitates rapid document retrieval.
- The index contains terms (words or phrases) extracted from the documents along with references to the documents where these terms appear. This inverted index allows for efficient access to documents containing specific terms.
- Ongoing research in indexing focuses on improving index compression, distributed indexing, and handling multilingual collections to accommodate the ever-growing volume of textual data.

8.4.1.5 RANKING

- Upon receiving a user's query, IR systems must determine the relevance of documents in the collection to that query.
- Ranking algorithms assign scores or ranks to each document based on their perceived relevance to the query. These algorithms leverage

various techniques, including word frequency, inverse document frequency, and ML models like BM25 and neural networks.
- Ranking is crucial because it streamlines the presentation of results, presenting the most relevant documents at the top of the list. This ensures that users find pertinent information quickly and effectively (Figure 8.4).

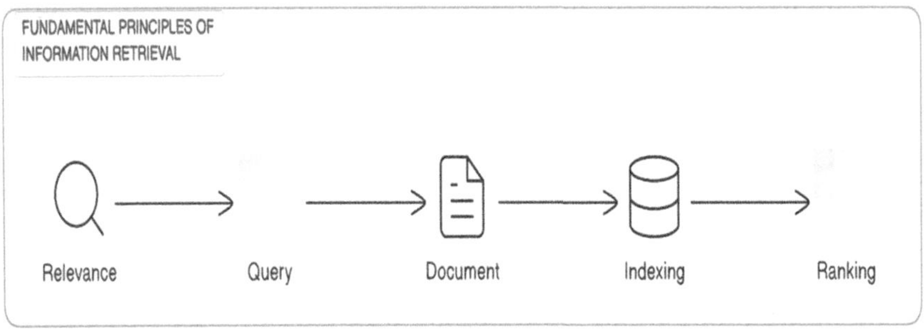

FIGURE 8.4 Principles of information retrieval.

8.4.2 COMPONENTS OF IR, INCLUDING INDEXING, QUERYING, AND RANKING ALGORITHMS

The components of IR, including indexing, querying, and ranking algorithms, work synergistically to facilitate the retrieval of relevant information from vast document collections. These components are continuously evolving as researchers seek to enhance the efficiency, effectiveness, and user-friendliness of IR systems in the ever-expanding digital information landscape.

8.4.2.1 INDEXING

1. **Definition:** Indexing is the process of creating a structured data structure called an index that efficiently organizes and stores information about the documents in a collection.
2. **Function:** Indexing is essential for speeding up the retrieval process. It allows IR systems to quickly locate documents that contain specific terms, significantly reducing the time required to search through large document collections.

3. **Process:**
   i. **Tokenization:** During indexing, documents are tokenized, meaning they are split into individual terms or tokens, typically words or phrases.
   ii. **Inverted Index:** The most common form of index used in IR is the inverted index. It consists of a mapping between terms and files in which those terms appear. Every term is associated with a list of document IDs or pointers.
   iii. **Term Frequency-Inverse Document Frequency:** In addition to simple term-document mappings, more advanced indexing methods like this are used to compute the importance of terms in documents relative to their importance in the entire collection.
   iv. **Challenges and Research:** Ongoing research in indexing focuses on improving compression techniques to reduce the size of the index, distributed indexing to handle massive document collections, and handling multilingual and multimedia content.

8.4.2.2 QUERYING

1. **Definition:** Querying is the process of formulating user queries and matching them against the indexed documents to identify those that are relevant to the query.
2. **Function:** The querying component is the interface between users and the IR system. It allows users to express their information needs and retrieve documents that fulfill those needs.
3. **Query Types:** Queries can take various forms, including:
   i. **Keyword Queries:** Simple queries composed of one or more keywords.
   ii. **Boolean Queries:** Queries that use Boolean operators (AND, OR, NOT) to combine keywords.
   iii. **Natural Language Queries:** Queries expressed in everyday language.
   iv. **Query Expansion:** Techniques like query expansion may be employed to improve query effectiveness. This involves adding synonyms or related terms to the original query to retrieve more relevant documents.

v. **Challenges and Research:** Query understanding and semantic search are areas of ongoing research. Natural language processing (NLP) techniques and semantic models aim to improve the system's ability to understand user queries and provide contextually related results.

8.4.2.3 RANKING ALGORITHMS

1. **Definition:** Ranking algorithms are algorithms that assess the relevance of documents to a given query and order them based on their perceived relevance.
2. **Function:** Ranking is crucial because it determines the order in which search results are presented to users. The goal is to present the most relevant documents at the top of the list, enhancing the user experience.
3. **Common Ranking Models:**
   i. **Term Frequency-Inverse Document Frequency (TF-IDF):** A classic ranking model that evaluates the importance of a term in a document relative to its importance in the entire collection.
   ii. **BM25:** A popular probabilistic ranking model that considers term frequency, inverse document frequency, and document length normalization.
   iii. **ML-Based Models:** Modern IR systems often employ ML algorithms, such as neural networks, to learn complex patterns of relevance from large datasets.
   iv. **Evaluation Metrics:** To assess the effectiveness of ranking algorithms, different evaluation metrics like precision, recall, F1 score, and mean average precision (MAP) are used.
   v. **Challenges and Research:** Continuous research in ranking algorithms focuses on improving the quality of relevance models, incorporating user feedback, and addressing the challenges of personalized search (Figure 8.5).

8.4.3 INFORMATION RETRIEVAL (IR) TECHNIQUES USED IN CONTEXT OF HUGE DATA

Information retrieval (IR) techniques used in the context of Big Data are specifically tailored to handle the challenges posed by the enormous volume, variety, and velocity of data that characterize the Big Data landscape. These techniques aim to extract valuable information and knowledge from vast and complex datasets. Here's a detailed explanation of IR techniques in the context of Big Data:

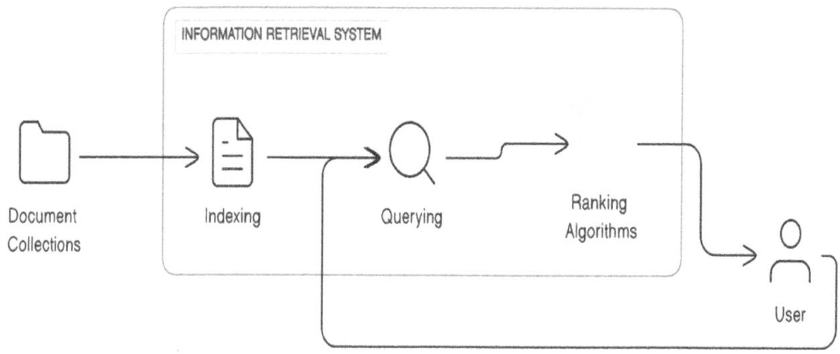

FIGURE 8.5 Information retrieval system.

8.4.3.1 DISTRIBUTED INDEXING

1. **Definition:** Distributed indexing is a technique that involves breaking down the process of creating and maintaining an index into distributed tasks that can be handled by multiple nodes or servers.
2. **Function:** In Big Data scenarios, traditional indexing techniques may not scale to handle the massive volume of documents. Distributed indexing allows for efficient storage and retrieval of data across multiple machines or clusters, distributing the computational load.
3. **Technologies:** Distributed indexing is often implemented using distributed file systems like Hadoop HDFS and distributed search engines like Apache Solr and Elasticsearch.
4. **Challenges:** Ensuring data consistency, fault tolerance, and load balancing are challenges that need to be addressed in distributed indexing systems.

8.4.3.2 SCALABLE QUERY PROCESSING

1. **Definition:** Scalable query processing involves designing algorithms and systems that can handle queries efficiently over large-scale data collections.

2. **Function:** In Big Data environments, queries need to be distributed and parallelized to process vast amounts of data in a timely manner. Techniques like MapReduce are commonly used to achieve scalability.
3. **Technologies:** Technologies such as Apache Hadoop and Apache Spark provide scalable query processing frameworks that can handle complex queries over large datasets.
4. **Challenges:** Optimizing query performance in distributed and parallel environments, dealing with data skew, and minimizing data shuffling are key challenges in scalable query processing.

8.4.3.3 DISTRIBUTED RETRIEVAL MODELS

1. **Definition:** Distributed retrieval models aim to distribute and parallelize the retrieval process, allowing for efficient searching across massive document collections.
2. **Function:** These models break down the retrieval task into smaller sub-tasks that can be executed concurrently across multiple nodes, reducing query response times and improving system throughput.
3. **Technologies:** Technologies like inverted indexes distributed across clusters, sharded databases, and distributed search engines provide the infrastructure for distributed retrieval.
4. **Challenges:** Maintaining real-time synchronization of distributed indices, load balancing, and ensuring consistent and efficient query routing are challenges in implementing distributed retrieval models.

8.4.3.4 STREAMING AND REAL-TIME INFORMATION RETRIEVAL (IR)

1. **Definition:** Streaming and real-time IR techniques focus on processing and retrieving information from data streams in real time.
2. **Function:** In Big Data contexts, data is often generated and updated rapidly. Real-time IR allows for immediate retrieval of relevant information from data streams as it arrives, making it valuable for applications like social media monitoring, financial trading, and sensor data analysis.

3. **Technologies:** Stream processing frameworks like Apache Kafka and Apache Storm, combined with real-time indexing and querying systems, enable real-time IR.
4. **Challenges:** Managing data velocity, ensuring low latency, and handling data stream variability are key challenges in real-time IR.

8.4.3.5 MACHINE LEARNING FOR RELEVANCE RANKING

1. **Definition:** ML techniques, including DL, are increasingly used to improve relevance ranking in Big Data IR systems.
2. **Function:** ML models can automatically learn complex patterns of relevance from large datasets, leading to more accurate and personalized search results.
3. **Technologies:** DL frameworks like TensorFlow and PyTorch, as well as pre-trained models like BERT and GPT-3, are applied to IR tasks to enhance ranking and retrieval.
4. **Challenges:** Annotating large training datasets, model interpretability, and handling high-dimensional data are challenges in applying ML to Big Data IR.

8.4.4 EXAMPLES OF REAL-WORLD APPLICATIONS OF INFORMATION RETRIEVAL (IR) IN HANDLING LARGE TEXTUAL DATASETS

Information retrieval (IR) plays a critical role in handling large textual datasets across various real-world applications. These applications leverage IR techniques to extract relevant information, enhance search experiences, and enable data-driven decision-making. Here are detailed examples of real-world applications of IR in handling large textual datasets:

8.4.4.1 WEB SEARCH ENGINES

1. **Description:** Web search engines like Google, Bing, and Yahoo! employ IR techniques to index and retrieve web pages from the vast and continuously evolving internet.
2. **Function:** Web crawlers index web pages, creating a massive corpus of documents. When users submit search queries, IR algorithms rank

these pages based on relevance, delivering highly relevant results within milliseconds.

3. **Challenges:** Handling the dynamic nature of the web, managing large-scale distributed indices, and combating web spam are challenges in web search engines.

8.4.4.2 E-COMMERCE PRODUCT SEARCH

1. **Description:** E-commerce platforms such as Amazon and eBay use IR to help users find products in vast online catalogs.
2. **Function:** Product information is indexed, and search queries allow users to find products by keywords, attributes, or even images. Ranking algorithms consider user preferences, reviews, and other factors to provide personalized product recommendations.
3. **Challenges:** Handling product metadata, dealing with varying product descriptions, and providing relevant recommendations are challenges in e-commerce product search.

8.4.4.3 ENTERPRISE SEARCH

1. **Description:** Enterprises employ IR solutions to search and retrieve documents, emails, and other textual data stored across their organization's repositories.
2. **Function:** IR systems crawl and index internal documents, emails, and databases. Employees can then use search queries to find relevant information quickly, improving productivity and knowledge discovery.
3. **Challenges:** Ensuring secure access to sensitive data, handling diverse data formats, and managing large-scale enterprise content are challenges in enterprise search.

8.4.4.4 INFORMATION FILTERING AND RECOMMENDATION

1. **Description:** Recommendation systems (RSs), as seen on platforms like Netflix and Spotify, utilize IR to suggest movies, music, articles, or products based on user preferences and behavior.

2. **Function:** IR techniques analyze user interactions, such as viewing history and ratings, to create user profiles. Recommender systems then use these profiles to deliver personalized content.
3. **Challenges:** Addressing the cold-start problem (recommendations for new users), avoiding filter bubbles, and maintaining recommendation quality are challenges in RSs.

8.4.4.5 LEGAL DOCUMENT RETRIEVAL

1. **Description:** Legal professionals rely on IR to search through vast collections of legal documents, case law, statutes, and regulations.
2. **Function:** Legal databases are indexed to allow lawyers and researchers to search for specific legal precedents, cases, or relevant legislation. Advanced search features, such as Boolean operators and proximity searches, aid in precision.
3. **Challenges:** Ensuring accurate and up-to-date legal information, handling complex legal terminology, and maintaining the confidentiality of legal documents are challenges in legal document retrieval.

8.4.4.6 HEALTHCARE INFORMATION RETRIEVAL

1. **Description:** Healthcare professionals use IR to access medical literature, patient records, and clinical guidelines.
2. **Function:** IR systems index medical databases, research articles, and electronic health records (EHRs). Healthcare providers can then search for medical information, treatment options, and patient histories.
3. **Challenges:** Ensuring data privacy and security, dealing with complex medical terminology, and integrating data from diverse sources are challenges in healthcare IR.

8.4.4.7 SOCIAL MEDIA SEARCH AND TREND ANALYSIS

1. **Description:** Social media platforms like Twitter and Facebook employ IR to enable users to search for posts, comments, and hashtags.
2. **Function:** Textual data from social media are indexed to support real-time search. IR techniques also facilitate trend analysis,

sentiment analysis, and user profiling for advertising and content recommendation.

3. **Challenges:** Handling the volume and velocity of social media data, managing spam and fake news, and addressing privacy concerns are challenges in social media IR.

8.5 TEXT MINING STRATEGIES

Text mining, or text analytics or NLP, is the process of extracting valuable insights and data from unstructured textual data. Its primary objective is to transform large volumes of text into structured and actionable knowledge. Text mining is crucial in handling the vast amounts of textual data generated in today's digital age, and it is employed across various industries and applications. Let's delve into the details of text mining strategies:

8.5.1 DEFINE TEXT MINING AND ITS OBJECTIVES

1. **Definition:** Text mining involves the automated analysis of textual data to discover patterns, extract useful information, and gain insights from unstructured text.
2. **Objectives:**
   i. **Extracting Valuable Information:** Text mining aims to uncover hidden insights, facts, and relationships from text data that might not be apparent through manual reading.
   ii. **Reducing Information Overload:** In the age of Big Data, there's an abundance of textual information. Text mining helps in sifting through this data to find what's relevant and valuable.
   iii. **Enhancing Decision-Making:** Text mining assists in making data-driven decisions by providing insights from text data sources.

8.5.2 STAGES OF TEXT MINING

Text Preprocessing: Involves cleaning and preparing the raw text data. Tasks include:

- Deleting special-characters, punctuation, and formatting.
- Tokenization: Splitting text into individual words or tokens.
- Deleting stop words (common words like "the" and "in").

- Stemming or lemmatization: Reducing words to their root form.
- Handling spelling and grammatical errors.

8.5.3 FEATURE EXTRACTION

In this stage, relevant features or attributes are extracted from pre-processed text for analysis. Common techniques are:

- **Bag of Words (BoW):** Representing text as a matrix of word recurrence.
- **Term Frequency-Inverse Document Frequency (TF-IDF):** Assigning weights to words based on their importance.
- **Word Embeddings:** Presenting words as dense vectors in a continuous vector space (e.g., Word2Vec or GloVe).

8.5.4 MODELING

Text mining models are applied to the processed and extracted features to perform tasks like sentiment analysis, topic modeling, and entity recognition.

8.5.5 COMMON TEXT MINING TECHNIQUES

1. **Sentiment-Analysis:** Sentiment analysis, or mining of opinion, aims to determine the sentiment or emotional tone expressed in a piece of text, whether this is positive, negative, or neutral. It's used for:
   - Customer sentiment analysis for products and services.
   - Tracking public sentiment on social media.
2. **Topic Modeling:** Topic modeling techniques such as LDA and non-negative matrix factorization (NMF) identify underlying topics in a collection of documents. Applications include:
   - Content recommendation.
   - Identifying trends in news articles or social media.
3. **Entity Recognition:** Named entity recognition (NER) identifies and classifies entities (e.g., people, organizations, locations) in text. It's used for:
   - Information extraction from news articles.
   - Identifying important entities in legal documents.

8.5.6 CASE STUDIES OR EXAMPLES OF TEXT MINING APPLICATIONS IN BIG DATA SCENARIOS

1. **Financial Analysis:** Text mining is used to analyze financial news and reports to predict stock market trends.
2. **Healthcare:** NLP and text mining are employed to extract valuable insights from EHRs for clinical decision support.
3. **Social Media Monitoring:** Companies use sentiment analysis to gauge public opinion about their products or services on social media platforms.
4. **Legal and Compliance:** Law firms and regulatory bodies use text mining to review legal documents for compliance violations and relevant case law.
5. **Customer Feedback Analysis:** Companies analyze customer reviews and feedback to identify product improvements and trends.

So, text mining is a very crucial technique for gaining insights and wisdom from large volumes of unstructured textual data. Its stages include preprocessing, feature extraction, and modeling, and it's applied to various tasks like sentiment analysis, topic modeling, and entity recognition. Today, in the era of Big Data, text mining is indispensable for making data-driven decisions and uncovering hidden patterns in textual information (Figure 8.6).

8.6 TOOLS AND TECHNOLOGIES

Tools and technologies for IR and text mining are essential for efficiently processing, analyzing, and extracting insights from textual data. Here, we'll introduce some popular software tools and libraries used in these domains and discuss their advantages and limitations for specific tasks:

8.6.1 ELASTICSEARCH

1. **Introduction:** Elasticsearch is an open-source, distributed search and analytics engine built on top of Apache Lucene. It is designed for high-speed text search, log data analysis, and full-text search capabilities.

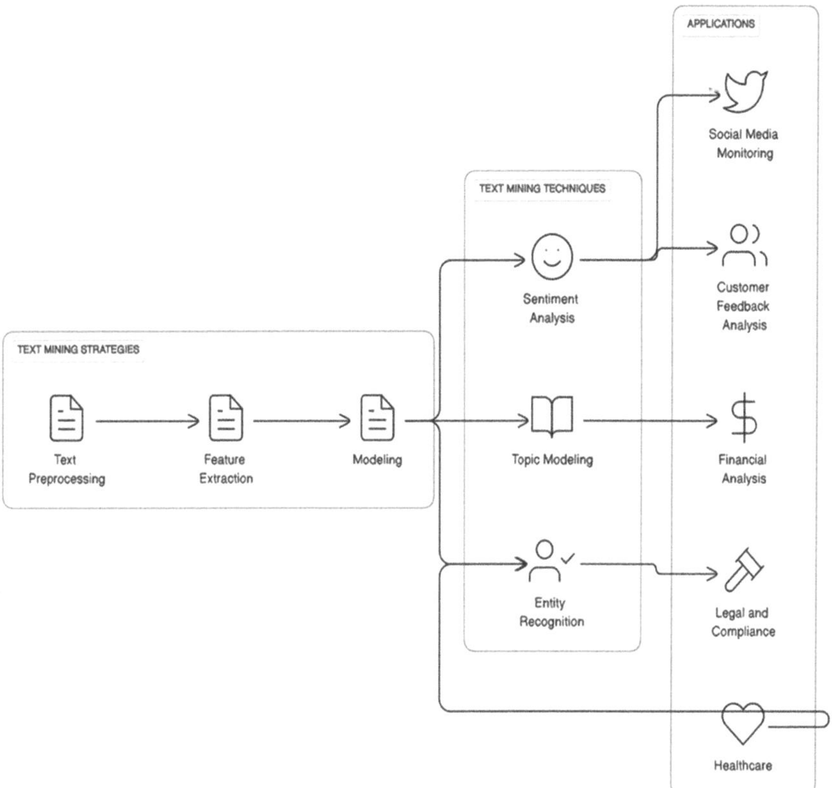

FIGURE 8.6 Text mining applications in big data scenarios.

2. **Advantages:**
   i. **Real-Time Indexing and Searching:** Elasticsearch is well-suited for applications that require fast, real-time search and retrieval of textual data.
   ii. **Scalability:** It can scale horizontally across clusters to handle large volumes of data.
   iii. **RESTful API:** Elasticsearch offers a user-friendly RESTful API for easy integration into applications.
3. **Limitations:**
   i. **Learning Curve:** Setting up and configuring Elasticsearch can be complex for beginners.
   ii. **Hardware and Resource Requirements:** Handling large datasets may require significant hardware resources.

8.6.2 APACHE-SOLR

1. **Introduction:** Apache Solr is an open-source, highly scalable search platform built on Apache Lucene. It is used for full-text search, faceted search, and distributed searching.
2. **Advantages:**
   i. **Highly Customizable:** Solr provides extensive configuration options for tailored search experiences.
   ii. **Scalability:** It can be deployed in distributed and cloud environments for handling large-scale data.
   iii. **Rich Text Processing Capabilities:** Solr includes features like tokenization, stemming, and filtering.
3. **Limitations:** Similar to Elasticsearch, Solr has a learning curve and may require significant hardware resources for large-scale deployments.

8.6.3 NATURAL LANGUAGE TOOLKIT (NLTK)

1. **Introduction:** NLTK is a Python library for NLP and text analysis. It offers a wide range of tools and resources for various NLP tasks.
2. **Advantages:**
   i. **Extensive Functionality:** NLTK includes modules for stemming, tokenization, POS tagging, and NER.
   ii. **Community Support:** It has a large and active user community, which means a wealth of resources and tutorials.
   iii. **Integration with Python:** NLTK seamlessly integrates with Python for easy data analysis and scripting.
3. **Limitations:** NLTK's performance may not be as efficient as specialized search engines like Elasticsearch or Solr for large-scale text retrieval tasks.

8.6.4 SPACY

1. **Introduction:** spaCy is another Python library for NLP, focusing on efficiency and speed. It's known for its robustness and production-ready capabilities.

2. **Advantages:**
   i. **High Performance:** spaCy is optimized for speed and memory efficiency, making it suitable for real-time applications.
   ii. **Pre-Trained Models:** It is available with pre-trained models for multiple languages and NLP tasks, including tokenization, POS tagging, and NER.
   iii. **Easy-to-Use API:** spaCy's API is designed for simplicity and ease of use.
3. **Limitations:** While spaCy is powerful for NLP tasks, it may not offer all the features of dedicated search engines like Elasticsearch or Solr for IR.

8.6.5 GENERAL ARCHITECTURE FOR TEXT ENGINEERING (GATE)

1. **Introduction:** It is an open-source software suite for text and language processing tasks. It's known for its flexibility and extensibility.
2. **Advantages:**
   i. **Modular Architecture:** GATE allows users to build custom pipelines of text processing components.
   ii. **Support for Multiple Languages:** It provides resources for a wide range of languages.
   iii. **Visualization Tools:** GATE includes visualization tools for analyzing text annotations.
3. **Limitations:**
   i. GATE's learning curve can be steep for users new to its architecture.
   ii. It may require more manual configuration compared to some other tools.

8.6.6 SPARK NATURAL LANGUAGE PROCESSING (NLP)

1. **Introduction:** Spark NLP is a library for Apache Spark, designed for scalable and distributed NLP. It's suitable for big data text processing tasks.
2. **Advantages:**
   i. **Integration with SPARK:** Spark NLP seamlessly integrates with Apache Spark for distributed text processing.

   ii. **Performance:** It's designed for high performance, making it suitable for large-scale text mining on big data.
   iii. **Pre-Trained Models:** Spark NLP includes pre-trained models for a variety of NLP tasks.

3. **Limitations:**
   i. Requires familiarity with Apache Spark for effective use.
   ii. May have a steeper learning curve compared to standalone NLP libraries like NLTK or spaCy.

The choice of tools and technologies for IR and text mining depends on specific project requirements, data scale, and the nature of the tasks. Elasticsearch and Solr are powerful search engines for full-text search, while NLTK, spaCy, GATE, and Spark NLP focus on NLP and text analysis. Each tool has its own strengths and limitations, and the selection should be based on the needs of the application and available resources.

8.7 STRATEGIES FOR EXTRACTING ACTIONABLE INSIGHTS

Strategies for extracting actionable insights from textual data involve a combination of techniques and approaches that allow organizations to derive meaningful information, make informed decisions, and gain a competitive edge. Here, we'll explore various strategies and tools commonly used for this purpose, including sentiment analysis, entity recognition, topic modeling, text classification, and the role of ML and DL:

8.7.1 SENTIMENT ANALYSIS

1. **Definition:** Sentiment analysis, also known as opinion mining, is the process of determining the emotional tone expressed in a piece of text, such as whether it is positive, negative, or neutral.
2. **Strategy:** Organizations use sentiment analysis to gauge customer opinions, identify trends, and make data-driven decisions. For instance, analyzing customer reviews can help improve products and services, and monitoring social media sentiment can inform marketing strategies.
3. **Tools:** NLP libraries like NLTK, spaCy, and pre-trained models such as VADER and TextBlob can be used for sentiment analysis.

8.7.2 ENTITY RECOGNITION

1. **Definition:** NER is a text mining technique that identifies and classifies entities in text, such as names of people, organizations, locations, dates, and more.
2. **Strategy:** Entity recognition is crucial for information extraction, as it helps in identifying key pieces of information within unstructured text. For instance, in healthcare, NER can extract patient names, diagnoses, and medications from clinical notes.
3. **Tools:** NER models are available in NLP libraries like spaCy, NLTK, and Stanford NER.

8.7.3 TOPIC MODELING

1. **Definition:** Topic modeling is a statistical technique used to discover abstract topics within a collection of documents. It identifies words that frequently co-occur and groups them into topics.
2. **Strategy:** Organizations employ topic modeling to uncover latent themes in large document collections. It is useful for content recommendation, trend analysis, and organizing content in news articles, blogs, or research papers.
3. **Tools:** LDA and NMF are popular topic modeling algorithms, and libraries like Gensim and spaCy offer implementations.

8.7.4 TEXT CLASSIFICATION

1. **Definition:** Text classification, or text categorization, is the process of assigning predefined categories or labels to text documents based on their content.
2. **Strategy:** Text classification is used for tasks like spam email filtering, sentiment analysis, news categorization, and content recommendation. It enables automation and organization of textual data.
3. **Tools:** ML algorithms such as Naive Bayes, SVM, and DL approaches like CNN and RNN can be employed for text classification.

8.7.5 MACHINE LEARNING AND DEEP LEARNING

1. **Strategy:** ML and DL techniques play a crucial role in improving the outcomes of text mining tasks. These techniques can enhance the accuracy of sentiment analysis, entity recognition, topic modeling, and text classification.
2. **Tools:** Libraries and frameworks like scikit-learn, TensorFlow, and PyTorch provide ML and DL capabilities for text mining tasks. Pre-trained models like BERT and GPT-3 have also revolutionized NLP applications.

8.7.6 UNSTRUCTURED TO STRUCTURED DATA TRANSFORMATION

1. **Strategy:** Converting unstructured text data into structured formats, such as tables or graphs, allows for easy integration with data analytics and visualization tools. Techniques like information extraction, named entity linking, and relationship extraction play a role in this transformation.
2. **Use Cases:** Transforming unstructured clinical notes into structured EHRs for healthcare analytics, or converting social media comments into structured sentiment dashboards for marketing insights (Figure 8.7).

Hence, strategies for extracting actionable insights from textual data involve a combination of techniques and tools tailored to specific tasks and objectives. Text classification, topic modeling, entity recognition, and sentiment analysis are fundamental techniques, often enhanced by ML and DL approaches. These strategies empower organizations to make data-driven decisions, gain insights from large volumes of text, and stay competitive in various industries.

8.8 CASE STUDIES AND EXAMPLES

Here are three real-world case studies that showcase the successful application of IR and text mining in big data scenarios, along with an analysis of the results and impact:

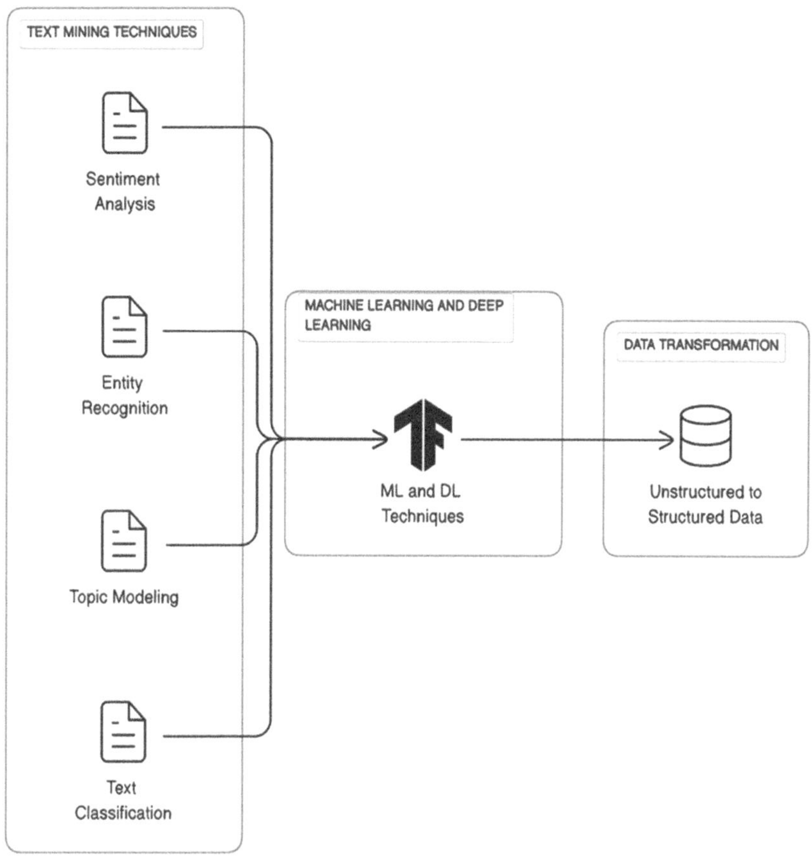

FIGURE 8.7 Extracting actionable insights in NLP and deep learning.

8.8.1 TWITTER SENTIMENT ANALYSIS FOR POLITICAL CAMPAIGNS

1. **Case Study:** During a political campaign season, a campaign team used text mining and sentiment analysis to monitor Twitter for public sentiment regarding their candidate.
2. **Application:**
   i. **Data Collection:** They collected a vast amount of Twitter data containing mentions of the candidate.
   ii. **Text Mining:** Text mining techniques were employed to preprocess the data, including tokenization, removal of stop words, and stemming.

   iii. **Sentiment Analysis:** Sentiment analysis algorithms were applied to classify tweets as positive, negative, or neutral sentiment.

3. **Results and Impact:**
   i. **Real-time Monitoring:** The campaign team gained real-time insights into public sentiment, allowing them to react quickly to emerging issues or negative sentiment.
   ii. **Targeted Campaigning:** They used the sentiment analysis results to tailor campaign messages and target swing voters more effectively.
   iii. **Increased Engagement:** By responding to tweets and addressing concerns, they improved engagement with potential voters.
   iv. **Enhanced Decision-Making:** Data-driven decisions based on sentiment analysis contributed to a successful campaign strategy.

8.8.2 HEALTHCARE DATA MINING FOR DISEASE DETECTION

1. **Case Study:** A healthcare organization used text mining to analyze EHRs for early disease detection and surveillance.
2. **Application:**
   i. **Data Collection:** They collected a vast repository of unstructured EHRs from multiple healthcare facilities.
   ii. **Text Mining:** Text mining techniques were applied to extract structured information from clinical notes, including patient symptoms, diagnoses, and treatments.
   iii. **Disease Detection:** ML models were trained on historical data to identify patterns and predict disease outbreaks.
3. **Results and Impact:**
   i. **Early Detection:** The organization achieved early detection of disease outbreaks, enabling rapid response and containment.
   ii. **Improved Patient Care:** By analyzing EHRs, they identified high-risk patients and provided timely interventions.
   iii. **Cost Savings:** Early detection and targeted interventions reduced healthcare costs associated with disease outbreaks.
   iv. **Public Health Impact:** The insights from text mining contributed to public health initiatives and improved population health.

8.8.3 CONTENT RECOMMENDATION FOR STREAMING SERVICES

1. **Case Study:** A streaming media platform leveraged text mining and recommendation algorithms to enhance content discovery for its users.
2. **Application:**
   i. **Data Collection:** The platform collected user interaction data, including search queries, viewing history, and user profiles.
   ii. **Text Mining:** Text mining was used to process textual metadata, such as movie or TV show descriptions and user reviews.
   iii. **Recommendation Engine:** ML algorithms, including collaborative filtering and content-based filtering, were used to suggest content to users based on their preferences and viewing habits.
3. **Results and Impact:**
   i. **Improved User Engagement:** The personalized content recommendations increased user engagement and time spent on the platform.
   ii. **Enhanced Retention:** Users were more likely to stay subscribed and continue using the service.
   iii. **Revenue Growth:** Increased user satisfaction and retention led to higher revenue through subscription fees and ad revenues.
   iv. **Data Monetization:** The platform monetized user data by offering targeted advertisements.
4. **Analysis:**
   i. **Data-Driven Decision-Making:** In all three case studies, IR and text mining enabled data-driven decision-making. Real-time insights, early disease detection, and content recommendations were all based on the analysis of large textual datasets.
   ii. **Enhanced Efficiency:** IR and text mining enhanced efficiency in processing large volumes of unstructured text data. They automated tasks that would be impractical or time-consuming to perform manually.
   iii. **Improved User Experience:** In the case of content recommendation and sentiment analysis, the impact was directly felt by end users. Personalized content recommendations improved user experience, while sentiment analysis allowed campaigns to connect with the public more effectively.

iv. **Cost Savings and Revenue Generation:** The applications of IR and text mining also led to cost savings (e.g., in healthcare) and revenue generation (e.g., in content recommendation).

Hence, these case studies illustrate the wide-ranging impact of IR and text mining in big data scenarios. They highlight how these techniques can provide actionable insights, improve decision-making, enhance user experiences, and have tangible benefits in various domains, from politics and healthcare to entertainment and marketing.

8.9 CHALLENGES AND FUTURE DIRECTIONS

Information retrieval and text mining in the context of Big Data are closely intertwined with advancements in technology, privacy, ethics, and data management. Addressing these challenges and embracing responsible, ethical practices will be essential to utilize the full power of Big Data analytics while preserving individual privacy and promoting fairness and transparency.

1. **Scalability:** One of the foremost challenges in Big Data text mining is the scalability of existing techniques. As data volumes continue to grow, traditional text mining approaches may struggle to process and analyze large datasets efficiently.
2. **Data Variety:** Big Data often includes diverse data types such as text, images, audio, and video. Integrating and analyzing these heterogeneous data sources remains a significant challenge.
3. **Data Quality:** Ensuring the quality of textual data is crucial. Noise, errors, and inconsistencies in the data can adversely affect the results of text mining tasks.
4. **Real-Time Processing:** In some applications, such as social media monitoring and financial trading, real-time text analysis is required. Handling high-velocity data streams while maintaining low-latency processing is challenging.
5. **Privacy and Security:** Preserving privacy while performing text mining on sensitive data, such as healthcare records or customer information, is an ongoing concern. Ensuring that personally identifiable information (PII) is not inadvertently exposed is critical.
6. **Interpretability:** As ML and DL techniques become more prevalent in text mining, ensuring the interpretability of complex models is

challenging. Users and decision-makers often require transparent insights into why a particular result was obtained.

7. **Ethical Concerns:** The ethical use of text mining and IR in Big Data is an emerging challenge. Ensuring that data analysis is unbiased, respects privacy, and adheres to ethical guidelines is essential.

8.9.1 POTENTIAL FUTURE DEVELOPMENTS AND AREAS FOR RESEARCH

1. **Advanced Machine Learning:** Developing advanced ML models, particularly DL architectures, tailored to text mining tasks will be a key focus. Research in areas like explainable AI (XAI) aims to improve the interpretability of these models.
2. **Efficient Scalability:** Research into more efficient algorithms and distributed computing techniques will help address the scalability challenge. This includes optimizing indexing methods, distributed text processing, and parallelized ML.
3. **Multimodal Analysis:** Integrating text with other data modalities like images, audio, and video will be a future research direction. Multimodal analysis will enable a more comprehensive understanding of Big Data.
4. **Real-Time Processing:** Developing real-time text mining solutions that can process data streams as they arrive will continue to be a priority. This involves improving stream processing frameworks and algorithms.
5. **Privacy-Preserving Techniques:** Research on privacy-preserving text mining techniques, such as federated learning and secure multiparty computation (MPC), will help address privacy concerns while still deriving insights from sensitive data.
6. **Ethical AI and Fairness:** Research in ethical AI and fairness will focus on ensuring that text mining techniques are used responsibly, free from biases, and aligned with ethical principles.
7. **Domain-Specific Solutions:** Tailoring text mining techniques to specific domains (e.g., healthcare, finance, legal) will be essential to address domain-specific challenges and requirements.
8. **Explainable AI (XAI):** Developing methods to make complex text mining models more interpretable and transparent will be a crucial research area to build trust in AI-driven decisions.

8.9.2 ETHICAL AND PRIVACY CONCERNS IN BIG DATA ANALYSIS

1. **Data Privacy:** Protecting individuals' privacy while analyzing textual data is paramount. Techniques like anonymization, differential privacy, and data masking should be employed to ensure data privacy.
2. **Bias and Fairness:** Mitigating biases in textual data and text mining models is essential. This includes addressing bias in training data, algorithms, and decision-making processes.
3. **Transparency and Accountability:** Ensuring transparency in data collection, analysis, and decision-making processes is crucial. Organizations should be accountable for their actions in Big Data analysis.
4. **Informed Consent:** When dealing with sensitive data, obtaining informed consent from data subjects is essential. Individuals should be aware of how their data is used and have the option to opt out.
5. **Regulatory Compliance:** Adhering to relevant data protection regulations, such as GDPR, HIPAA, or CCPA, is non-negotiable. Organizations must ensure they comply with legal and regulatory requirements.

8.10 CONCLUSION

Here, we summarize the key findings and contributions of the paper, reiterate the importance of these techniques, and provide closing remarks on the future of big data analytics:

8.10.1 KEY FINDINGS AND CONTRIBUTIONS

1. **Information Retrieval and Text Mining Significance:** The research paper has emphasized the critical significance of IR and text mining techniques in extracting actionable insights from large volumes of unstructured textual data. These techniques are essential for converting data into knowledge, enabling data-driven decision-making, and gaining a competitive edge in various domains.
2. **Applications Across Diverse Sectors:** The paper has highlighted numerous real-world applications of IR and text mining, ranging

from web search engines and healthcare IR to social media sentiment analysis and legal document retrieval. These applications demonstrate the versatility and wide-reaching impact of these techniques.

3. **Challenges and Ethical Considerations:** The paper has also discussed the challenges and ethical considerations associated with IR and text mining in big data scenarios. Scalability, data quality, privacy, and fairness are among the challenges that must be addressed. Ethical practices, transparency, and compliance with data protection regulations are crucial for responsible data analysis.

8.10.2 IMPORTANCE OF INFORMATION RETRIEVAL-TEXT MINING

Information retrieval and text mining play a pivotal role in utilizing the potential of Big Data. They transform raw textual data into structured formats with meaningful information, enabling organizations to:

1. **Gain Insights:** IR and text mining uncover hidden patterns, trends, and sentiments in textual data, providing valuable insights for decision-makers.
2. **Enhance Decision-Making:** By facilitating data-driven decision-making, these techniques contribute to better-informed strategies, whether in marketing, healthcare, finance, or other sectors.
3. **Improve User Experiences:** In applications like content recommendation and search engines, IR and text mining enhance user experiences by delivering personalized, relevant content.
4. **Ensure Compliance:** They assist in regulatory compliance by enabling the extraction of essential data from large textual datasets while preserving data privacy.
5. **Address Real-Time Needs:** IR and text mining can be applied in real-time scenarios, such as monitoring social media trends, detecting anomalies, and responding to emerging issues promptly.

8.10.3 CLOSING REMARKS ON FUTURE OF BIG-DATA ANALYTICS

Coming time of Big Data analytics is promising, driven by continual advancements in IR and text mining techniques. Here are some closing remarks on the future of Big Data analytics:

1. **Integration of Multimodal Data:** The future will witness an integration of text with other data modalities like images, audio, and video, allowing for more comprehensive and context-rich analysis.
2. **Advanced Machine Learning Models:** The development of advanced ML and DL models will enhance the accuracy and efficiency of text mining tasks, enabling more sophisticated applications.
3. **Responsible AI:** Ethical considerations, fairness, and transparency will play a central role in the future of Big Data analytics. Responsible use of information with AI will be a cornerstone of data-driven decision-making.
4. **Privacy-Preserving Techniques:** Privacy-preserving techniques will evolve to safeguard individuals' data while still allowing valuable insights to be derived from it.
5. **Efficient Scalability:** Research into scalable and efficient algorithms, as well as distributed computing methods, will enable the processing of increasingly large datasets.
6. **Interdisciplinary Collaboration:** The future will require closer collaboration between data scientists, domain experts, ethicists, and policymakers to ensure responsible and effective use of Big Data.

In conclusion, IR and text mining are at the forefront of harnessing the potential of Big Data, enabling organizations to make informed decisions, enhance user experiences, and address complex challenges. As technology and ethics continue to evolve, Big Data analytics will play an ever-increasing role in shaping our data-driven world.

KEYWORDS

- **big data analytics**
- **deep learning**
- **fake news detection**
- **information retrieval**
- **machine learning**

REFERENCES

1. Tien, J. M. (2013). Big data: Unleashing information. *Journal of Systems Science and Systems Engineering, 22*, 127–151.
2. Bizer, C., Boncz, P., Brodie, M. L., & Erling, O. (2012). The meaningful use of big data: Four perspectives—Four challenges. *ACM SIGMOD Record, 40*(4), 56–60.
3. Chen, H., Chiang, R. H., & Storey, V. C. (2012). Business intelligence and analytics: From big data to big impact. *MIS Quarterly, 36*(4), 1165–1188.
4. Bilal, M., Oyedele, L. O., Qadir, J., Munir, K., Ajayi, S. O., Akinade, O. O., & Pasha, M. (2016). Big data in the construction industry: A review of present status, opportunities, and future trends. *Advanced Engineering Informatics, 30*(3), 500–521.
5. Ibekwe-Sanjuan, F., & Geoffrey, B. (2017). Implications of big data for knowledge organization. *Knowledge Organization, 44*(3), 187–198.
6. Wu, X., Zhu, X., Wu, G. Q., & Ding, W. (2013). Data mining with big data. *IEEE Transactions on Knowledge and Data Engineering, 26*(1), 97–107.
7. Wang, H., Xu, Z., Fujita, H., & Liu, S. (2016). Towards felicitous decision making: An overview on challenges and trends of big data. *Information Sciences, 367*, 747–765.
8. Pramanik, P. K. D., Pal, S., Mukhopadhyay, M., & Singh, S. P. (2021). Big data classification: Techniques and tools. In *Applications of Big Data in Healthcare* (pp. 1–43). Academic Press.
9. Li, J., Tao, F., Cheng, Y., & Zhao, L. (2015). Big data in product lifecycle management. *The International Journal of Advanced Manufacturing Technology, 81*, 667–684.
10. Bharathi, S. V. (2017). Prioritizing and ranking the big data information security risk spectrum. *Global Journal of Flexible Systems Management, 18*, 183–201.
11. Skënduli, M. P., Biba, M., & Ceci, M. (2018). Implementing scalable machine learning algorithms for mining big data: A state-of-the-art survey. In *Big Data in Engineering Applications* (pp. 65–81).
12. Chen, C. P., & Zhang, C. Y. (2014). Data-intensive applications, challenges, techniques, and technologies: A survey on big data. *Information Sciences, 275*, 314–347.

CHAPTER 9

Reconciliation of Emotional Intelligence with Artificial Intelligence for Augmenting Organizational Effectiveness: A Conceptual Model

K. UMASHANKAR[1,*], G. RANGANATHAM[2], and H. G. CHARITRA[3]

[1]*Head of Learning and Development, Presidency University, Bengaluru, India*

[2]*Department of Commerce, School of Business Studies, Central University of Karnataka, India*

[3]*School of Liberal Arts and Sciences, Presidency University, Bengaluru, India*

**Corresponding author.*

ABSTRACT

The conceptual study focuses on reconciling Emotional Intelligence with artificial intelligence (AI). It is purported that AI has grown exponentially during the last few decades, and it is playing a crucial role in digital disruption. The intensity of growth in AI is such that it has created terror in employees' minds about being replaced by machines and technology. Besides, AI is an instrument but not an agent that can fulfill the roles by itself. The extreme dependence of humans on AI may cause irreversible damage to the natural process of cerebration and other cognitive abilities. The study cites some recent studies from cognitive neuroscience to support the argument.

Having understood the importance of emotional intelligence and AI, the study has constructed a conceptual model under 10 assumptions, and

Natural Language Processing and Artificial Intelligence: A Perspective Toward Current Trends, Challenges, and Applications. Vikas Khullar, Aryan Chaudhary, & Sunil Kumar (Eds.)

DOI: 10.1201/9781998511273-9

each assumption aims at reconciling EI with AI. The role of EI and AI in the banking industry has been reviewed extensively to understand the imperative nature of EI in the banking industry, and this cannot be replaced by AI. However, the study argues that AI can be effective when used as an instrument supporting EI but not as an agent.

9.1 INTRODUCTION

The multi-dimensional evolution of technology has paved the way for heterogeneous concerns in the human ecosystem. The development in technological clusters like biotechnology, digital technology, nanotechnology, neurotechnology, and green technology has not only offered new promises for the betterment of economic development, but it has also threatened the quality of human life. Although the embrace of artificial intelligence (AI) and machine learning (ML) has heralded a revolution in human lifestyles, it is inflicting heavy costs on the quality of human life in the long term. Indeed, the repercussions like job loss, emotional problems, and loss of quality social/personal life may be irreversible [1].

However, one of the studies implies that the exponential ontogenesis of AI can be considered a job disruptor but not a job destroyer. According to one of the fintech companies, the microfinance and banking industries were spawned with heterogeneous traditional roles with a lot of human intervention, but the advent of AI replaced those traditional job roles and expanded the scope of skill development among the employees. The quality of service has been significantly enhanced in the financial industry. As quoted by the study of the University of Warwick, AI has created more jobs in the financial industry by offering new job roles and a new scope for learning among the human capital [2].

There was significant insight into the disruptions caused by the pandemic horror and its impact on the consumer market in the banking sector. It was hypothesized that there was a transformational shift in consumer expectations from the banking industry and their digital inclination for banking services. Digital channels like phone banking, internet banking, and alternative channels like payment wallets started spawning in the business market, and the majority of consumers above 45 years of age resorted from branch banking to digital banking. The trend of digital banking imposed tremendous stress on providing secured channels, and firewall security features were advanced in most of the countries having maximum banking relationships and financial transactions. Countries like Japan, Brazil, China, Hong Kong, and Spain reported having more than 60% of consumers using digital banking, and the priority was to offer tailor-made services and enable them to self-drive their financial transactions with minimal human intervention [3].

A study by KPMG [4] inferred that the banking sector has to redefine its business models and redesign its service platforms to meet the growing demands of consumers. Digital enablement in the banking sector has become a testimonial factor for digital disruptions, and the extravagant expectations from the customers have fueled the disruptive technological advancement. The study reports some astonishing facts about consumer preferences in banking and the population included in the age group of 24 years to 44 years (Figure 9.1).

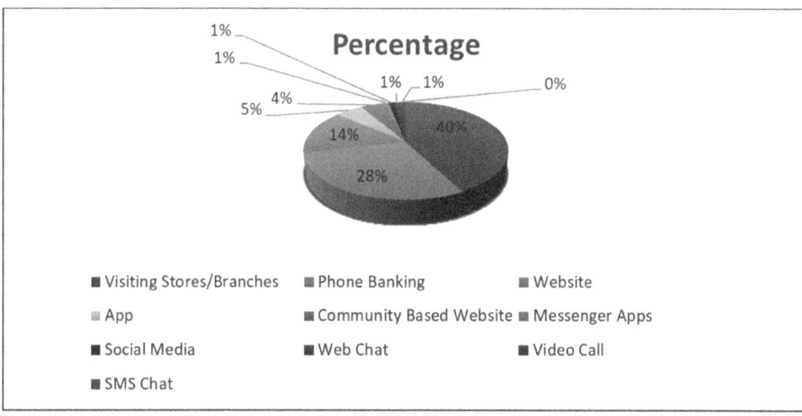

FIGURE 9.1 Consumer and the new reality.

The reports infer that consumers prefer not only advanced digital platforms for their banking transactions but also the human touch that can augment their trust in the brand. Besides, the data endorses that humans crave emotional touch, and their purchase behaviors are driven by the quality of emotional connection with their service providers. Therefore, the study reinforces the amount of influence it has on human interactions coupled with digital platforms. The outcome of the study illustrates that digital transformation in the banking industry may help increase the quality of service but never intends to replace human intelligence from the banking ecosystem [5].

The divergence in the business models of the banking sector, augmented by digitization and increasing expectations from the consumer market, has led to innovations in the world of technology. Due to tremendous pressure on technology, digital disruption has been inevitable. Driven by digital transformation, the banking sector was successful in establishing financial resilience across the globe while surpassing the economic threats posed by the pandemic era. The digital banking penetration during various financial years between 2015 and 2019 was not worth mentioning, but the pandemic era accelerated the consumption of digital products.

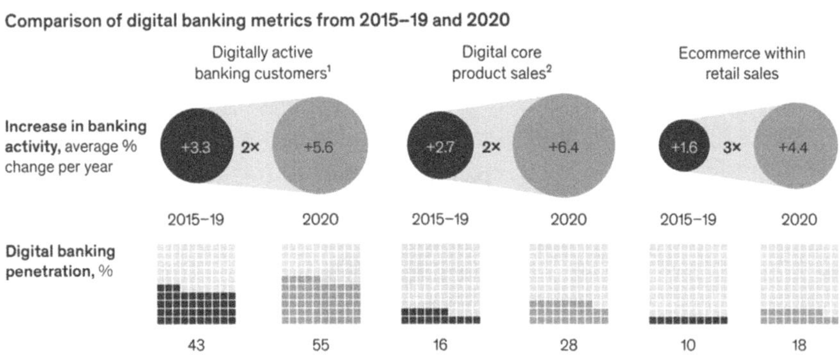

FIGURE 9.2 Building the artificial intelligence (AI) bank of the future.

Figure 9.2 substantiates the fact that the digital disruption in banking was sourced from the predicaments of the pandemic era [6].

In continuation, the report infers about the desperate reinvention of banking players to remain in the competition, and business succession planning was made divergent. Some of the best players in the banking industry have started exploring new and relevant ecosystems wherein they could position themselves to diversify their banking transactions. Further, it was presumed that digital production or digital disruption is rooted deep in the process of ideation or celebration. The process of ideation is driven by human talents, and it cannot be substituted only by AI. The process of ideation in the banking sector across the globe has led to major changes in the business models of banking giants across the globe. The report infers that human intelligence aligned with AI enables the banking players to reinvent themselves [7] (Figure 9.3).

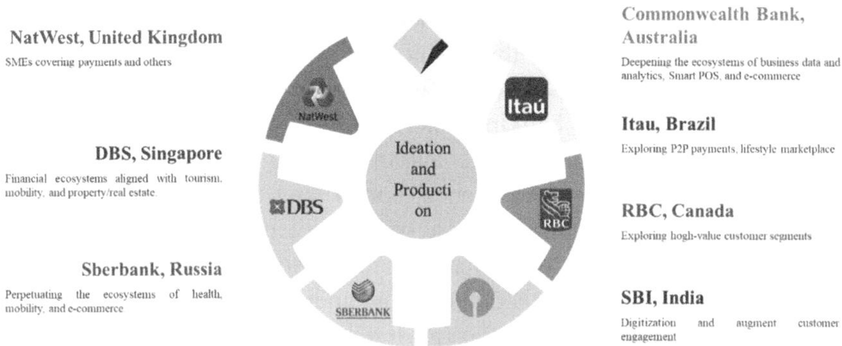

FIGURE 9.3 Global banking institutions and their innovation focus areas.

Source: Adapted from McKinsey & Company, 2021.

One of the recent technological innovations that OpenAI has made is the direction of human intellect, which has become ambiguous, and ChatGPT has created a revolution in the areas of academics and challenging human intelligence. It is observed that ChatGPT has attracted more than 100 million users within two months of its introduction, and it is expected to multiply by the middle of 2023. The banking industry is working towards the integration of ChatGPT into its business processes to achieve a competitive advantage in the banking ecosystem. It is in the process of improving the quality of the chatbot's responses through ChatGPT. However, the scholarly committee is not precise about the feasibility of integrating ChatGPT with the banking requirements [8]. It is implied that ChatGPT is meant to make human life better, but the amount of dishonesty spawned by this disruptive technology has kept the innovators' fingers crossed. One of the studies infers that ChatGPT, one of the significant technological disruptors, is based on a large language model (LLM) and has evident lapses as it comprises many biases and ethical concerns. However, technology and its new avatars cannot be terminated on the grounds of evident lapses, but it can be accompanied by human intelligence for the betterment of human life [9].

Based on the process of ideation and production, the studies in design thinking infer that ideation through human brainstorming and creating new feasible models for AI and its execution play a crucial role in creating new digital ecosystems. Besides, it is argued that human intelligence and AI are the two phases of the same coin, and they depend on each other for value addition to the business platforms. However, the current emphasis is that AI has its limitations, and it cannot eliminate human intervention from the banking interface. Although digital technology has evolved into a new era of renaissance in the macrocosm of banking and technology, human intelligence, comprising emotional quotient and social intelligence, remains the core or the nucleus of the banking sector. Besides, the banking sector is predominantly a service sector that involves regular interaction with customers. Customers crave human connection and seek human support for the resolution of their queries or concerns, considering the existing literature on digital disruptions and their influence on the frequency or the gravity of human intervention.

9.2 METHODOLOGY OF THE STUDY

9.2.1 OBJECTIVES OF THE STUDY

In continuation of the fact that AI is in the process of monopolizing the banking sector by challenging the significance of human intelligence, the present concept aims to explore the crucial role played by emotional intelligence in augmenting the banking business models. Besides, the study considers the significance of digital transformation witnessed by the banking ecosystem and bridges the gap between emotional intelligence and AI.

9.2.2 NEED FOR THE STUDY

The contemporary world of technological disruptions and intense competition among banking players in India and the globe has intensified the need to reflect on the trends of AI and the misinterpretation of its significance. It is examined that the importance of emotional intelligence has been compromised, and the very existence of human intervention has become questionable. Besides, there is a need to establish the crucial role played by emotional intelligence and human intervention during the process of customer interaction and the human efforts to augment customer relationship value. In continuation, it is observed that the significance of AI is measured based on its ability to interpret huge financial data with impeccable accuracy and support banks in reducing service costs. Accordingly, the significance of emotional intelligence has been undervalued as its visibility is veiled. Indeed, AI helps in the operational or functional process, and EI augments the sales performance of each employee, adding to organizational effectiveness.

9.2.3 SCOPE OF THE STUDY

The study comprises a conceptual approach to understanding the importance of emotional intelligence in the banking sector and its indispensable role in achieving the business goals of the banks. A conceptual model of emotional intelligence aligned to the antecedents of AI has been constructed by the study. This helps in the reconciliation of emotional intelligence and AI for organizational effectiveness. Further, the study diagnoses the existing model of emotional intelligence and its application during customer interface and explores the possibility of reconciling emotional intelligence with AI to ensure a better interface between the customer and the service provider.

9.2.4 PROBLEM STATEMENT

Emotional intelligence has a deepened relationship with the theoretical construct of neuroplasticity. Neuroplasticity is a theoretical construct of cognitive neuroscience that deals with the neural paths and the neural networks that are stimulated or formed by new human experiences. The external stimuli augment the human experience and are later stored in the synaptic structures in the brain, and each synaptic structure has its memory and a neural network. Indeed, the human brain is said to have billions of such neural networks, and each neural pathway or neural network has some amount of information that is retrieved during the information processing in the human brain. Besides, studies in cognitive neuroscience have implied that these neural pathways keep growing and expanding as humans experience new external stimuli. However, it is also inferred that the development of neural networks would slow down drastically by impairing one's intelligence if humans don't stimulate their thinking process at regular intervals [10]. The significant assumption of neuroplasticity is that the human brain must be kept triggered and stimulated constantly for sharpening cognitive precision or agility; otherwise, the cerebral ingenuity will be compromised. The same situation may be related to the burgeoning fight between emotional intelligence (human intelligence) and AI. It is evident that AI and ML have changed the phase of banking technology and enhanced operational capabilities. It is implied that disruptive technology fueled by AI has improved the quality of banking services across the globe. However, it is essential to examine the long-term consequences of excessive dependence on technology while gradually reducing human interaction and minimizing the generation of cerebral solutions to problems or concerns. Contemporary research in the areas of emotional intelligence and its counterpart, AI, has revealed that monopolizing AI at the cost of emotional intelligence may cause an irreversible catastrophe or damage to human relations and the progress of organizations.

9.3 BACKGROUND TO EMOTIONAL INTELLIGENCE MODEL AND ARTIFICIAL INTELLIGENCE

The term emotional intelligence was theorized and popularized by Daniel Goleman, a professor at Harvard University, and the theory was constructed on a four-dimensional model comprising intrapersonal awareness, intrapersonal management, social awareness, and interpersonal management. The dimensions of Goleman's emotional intelligence theory render some crucial insights into the heterogeneity of human intelligence and its application in the real world [11]. Although AI has created its virtual reality for humans and business ecosystems, it keeps evolving and changing virtual reality at sporadic intervals. However, the changing virtual realities are disruptive, and any serious conclusions cannot be drawn amid uncertainties. Contrastingly, emotional intelligence is an intrinsic and perennial component of human intelligence, and it can connect with people. Besides, humans are social animals, and their brain is not just an organ but a social organ that often craves human connection [12]. Furthermore, emotional intelligence or social intelligence enables bankers to offer an enriched social experience to customers during their face-to-face interactions.

AI has revolutionized the banking ecosystem and enabled banks to feel the pulse of consumers across the globe. AI is extensively employed in understanding consumers' buying behaviors through data mining from various online platforms and the interpretation of numerical data. Additionally, the increasing competition between banking and fintech players has heightened the need for banks' dependence on AI. It is observed that AI has been promoting seamless banking transactions and providing customer delight. The adoption of AI has strengthened the competitive advantage of banks and enabled best practices in customer engagement. AI is providing value-added services like cognitive automation, chatbots for virtual interfaces, suggestive financial decisions, and the inception of humanoids or automation of operational work. According to the study, the prediction is that AI is going to take over the affairs of banking operations, security concerns, and other repetitive functions [13].

Furthermore, AI, with its heterogeneous functions, has enabled humans to evolve into the new age of digitization and opened new horizons for humans to explore by limiting their intervention in repetitive tasks. Tasks like image recognition, ML, deciphering algorithms, audio/speech processing or recognition, linguistic processing, and deep learning (DL) are included. These functions of AI have augmented the effectiveness of organizations and drastically reduced service costs. The term DL is associated with the diagnosis of human emotions, feelings, and thought processes and offers viable banking solutions. Additionally, the algorithms seem to have no meaning in the interpretation of AI and are widely used in understanding customer buying preferences by analyzing their buying behaviors. However, it is essential to note that AI may be considered a digital support system for humans but not as a replacer or a substitute [14].

While reflecting on the background of emotional intelligence and AI in the banking ecosystem, the study examines one of the crucial technological models adopted by various banking players in India, which is customer relationship management. This model comprises four foundational dimensions of banking solutions and is powered by technological partners such as Finacle (Infosys) [15], FLEXCUBE (Oracle) [16], and others (Figure 9.4).

Based on the existing literature on emotional intelligence and AI in the banking sector, the study purports to construct a conceptual model that helps in the reconciliation of humans and technology for better organizational effectiveness. The presumed construct of the model may demystify the rumors that are creating a vacuum between humans and technology, and it is possible for both to work hand in hand without destroying each other. Besides, the conceptual model presumes attributing certain functions that are appropriate to emotional intelligence and AI. The model comprises 10 assumptions, and they are (Figure 9.5):

1. ML assists employees to initiate intrapersonal awareness (A1).
2. DL enables employees to reengineer their cognitive competencies/ deficiencies (A2).
3. Problem-solving strategies help employees to be aware of their organization/customers (A3).
4. Problem-solving solutions help in conflict resolution and managing external change (A4)
5. Knowledge representation augments empathy and organizational awareness (A5)
6. Knowledge representation sharpens influencing skills (A6).
7. Language processing nurtures empathy (A7).
8. Search Algorithms expand social awareness (A8).
9. Search Algorithms enable employees to be the change catalyst (A9).
10. DL strengthens employees' awareness of the organization and employees (A10).

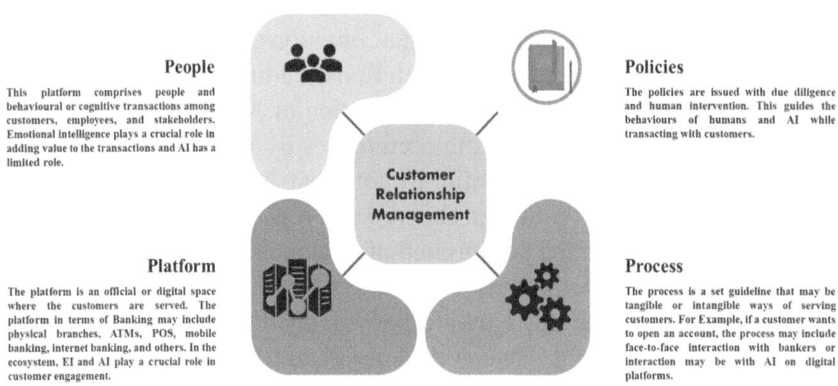

FIGURE 9.4 Customer Relationship Management.

Source: Author

9.4 MACHINE LEARNING ASSISTS EMPLOYEES TO INITIATE INTRAPERSONAL AWARENESS (A1)

The debate over emotional intelligence vs. AI has attracted global attention, and many times the debate has been misled. The debate is not about replacing each other but about building awareness of the relationship between emotional intelligence and AI. Studies have acclaimed the fact that ML through NLP helps individuals understand their emotional conditions and develop an emotion diary that may help in recognizing frequently visited emotions in one cerebral space. One of the studies implied that AI could help individuals assess their emotional conditions and predict their behavior and personality styles [17].

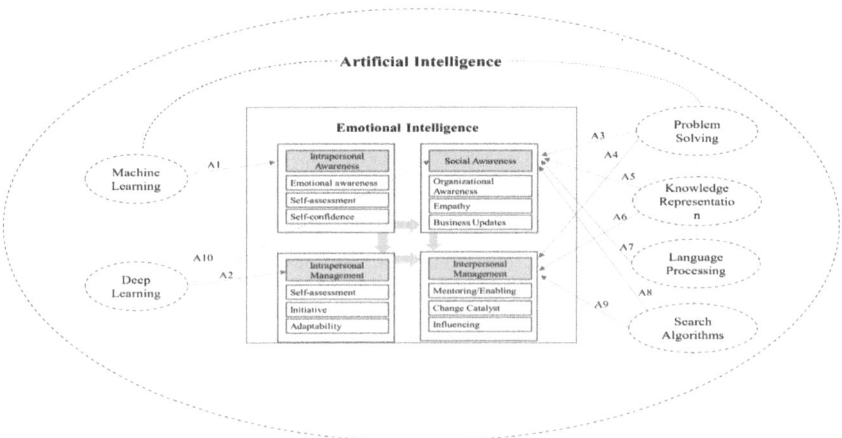

FIGURE 9.5 Conceptual model of emotional intelligence (EI) and artificial intelligence (AI) in the banking sector.

Machine learning has been a great source for introspection into self and understanding one's psychological or emotional conditions. Metaverse platforms like Meta Horizon, The Sandbox, Frostbite, and so forth, create a virtual reality and augment human information processing. Besides, this enables humans to identify their unconscious emotions or psychological conditions, contributing to better interpersonal relationships. In addition, it nurtures mindfulness about one's strengths and weaknesses as people interact on the Metaverse platforms. However, humans must be aware of virtual reality and the amount of immersion into that augmented reality, as people lose track of themselves in the landscapes of the Metaverse [18]. According to one of the studies conducted by Massachusetts institute of technology (MIT), ML can be used in a therapeutic context, and it was proven that ML is more effective than a human therapist during the virtual interaction between the client and the ML. The experimental study comprised a client who had post-traumatic symptoms, and the therapy was administered through text messages using virtual reality, and it was highly successful. In contrast, the same client was assigned to a human therapist, and the client discontinued after a couple of weeks [19].

A study aimed at understanding the feasibility of understanding human emotions by observing physiological features or changes in humans. The study captured the kinesthetic signals or physiological responses to external stimuli for further observation. It is inferred that using AI and ML can analyze subtle bodily responses and psychological signals to assess human mental conditions. Further, it is assumed that such data can help individuals understand the psychosomatic responses to certain stimuli, augmenting one's self-awareness [20].

Indeed, ML helps employees or individuals understand their preferred types of intelligence [21] and the strategies to add value to their personal development plans. However, it is implied that ML can act as a support system in the journey of self-awareness, and it cannot completely substitute for human intervention. The concept of self-awareness is deep within the cognitive dimension, and it is very complicated to penetrate the neural, psychological, and synaptic structures of the brain/mind. Hence, the role of ML becomes limited in evaluating one's emotional conditions and other cognitive concerns. Indeed, employees must initiate human efforts to understand themselves while taking some assistance from ML [22, 23].

9.4.1 DEEP LEARNING ENABLES EMPLOYEES TO REENGINEER THEIR COGNITIVE COMPETENCIES/DEFICIENCIES (A2)

The cognitive competencies may be interpreted as one's ability to redefine, reconstruct, or reinvent a set of intellectual or emotional abilities that are serviceable or feasible in the workplace. The DL modules enable prospective participants to reflect on their behavioral or cognitive deficiencies and employ feasible strategies to overcome their concerns. Besides, the DL function of AI facilitates the empowerment of one's operational, functional, and behavioral competencies. In India, the major banking players and other progressive organizations have integrated AI-powered simulation learning modules into their learning management systems (LMS). Indeed, AI has evolved a strong learning culture in organizations wherein they learn from failures and empower themselves through their successes [24]. However, an in-depth diagnosis of the prevailing trend of employing DL mechanisms to moderate a human's introspection and cognitive systematic re-engineering may generate inferential apprehensions. These apprehensions are about questioning one's cognitive ability to transform themselves and causing neurocognitive deterioration. The studies imply that the synaptic functions are substituted by the functions of DL techniques and that neurocognitive functions are naturally restrained [25].

Due to advancements in DL techniques, the tool has been widely used during human interface and has been successful in nurturing one's psychological well-being. The best examples are facial recognition, speech recognition, healthcare (the sensors that provide usable information about the health conditions of a patient), and natural language processing (NLP). One of the studies adopted chatbots for psychological intervention on an academic campus. These chatbots recorded the emotions or feelings of the subjects (experimental group) and provided real-time information about the emotional signals and their frequency. This became a breakthrough in the field of AI-intervened psychotherapy [26]. A study to alleviate women's concerns in terms of sexual or emotional abuse in personal, social, or professional landscapes invented a unique behavioral intervention technology (BIT). Since women could not openly express their emotional agony with a human therapist, the study experimented with chatbots (BITs) as therapeutic agents. These chatbots were conversational agents that alleviated the emotional or mental concerns of victimized women. Besides, these conversational agents, during the interface with the patients, were mindful of the language and emotions used by the patients. It is inferred that the experiment was highly successful with a tangible outcome [27].

9.4.2 PROBLEM-SOLVING STRATEGIES HELP EMPLOYEES TO BE AWARE OF THEIR ORGANIZATION/CUSTOMERS (A3)

Nowadays, in the banking sector, customer satisfaction has become a primary requirement for customers, and they expect more than just satisfaction. Due to excessive competition, banking players have been experimenting with the adoption of various dimensions of Extended Reality. These realities include mixed reality, augmented reality, and virtual reality [28]. The problem-solving strategies supported by AI enhance the customer experience throughout their banking journey, and the turnaround time (TAT) to resolve customer queries has been reduced drastically. Furthermore, it was proposed that AI with problem-solving features may be adopted through the cyclic process of customers' interaction with the brand. The steps include the following thresholds.

- Initiating the customer experience team
- Understanding the phases and experiences of customers through the journey
- Examining customer expectations and their changing preferences
- Diagnosing and resolving customer resentment
- Vigilant over the customers' journey

AI is indispensable in providing the expected customer experience due to its precision and accuracy in identifying the pulse of customers [29].

Artificial Intelligence comprises multiple heuristic algorithms to solve problems, and the nondeterministic polynomial (NP) problems have been addressed with minimal time. The challenges in problem-solving are well addressed by AI, and the medical field has embraced AI tools to solve some of the intricate problems associated with medicine. In particular, AI-driven medical fields like brain science, neurotransmission, and other clinical interventions have been facing tremendous development, and it is contributing to enhanced medical intervention. The study inferred that the AI-driven laboratory experiments during COVID-19 offered valuable insights into the efficacy of advanced AI [30].

In addition, AI has become indispensable in the service sector, like banking and finance. AI has been playing a crucial role in resolving the problems or queries of customers, but it lacks a deep analysis of the emotional or cognitive conditions of humans that are linked to their banking requirements. Banking or financial requirements may not be related only to wealth management, but holistic wealth management may have other dimensions of emotional concerns. The studies imply that AI cannot empathize with customer requirements, as empathy is exclusively a biological or unconscious response to human conditions. Indeed, empathy for most customers' problems cannot be programmed into AI to generate artificial responses [31]. However, there is a wide scope for research in this area to understand the intensity of empathy displayed by AI.

9.4.3 PROBLEM-SOLVING SOLUTIONS HELP IN CONFLICT RESOLUTION AND MANAGING EXTERNAL CHANGE (A4)

Artificial intelligence is enabling organizations to predict the business market in the future and propose possible changes in the business models of organizations. Besides, it assists at the individual level while translating digital algorithms to human cognition. In this context, service providers can accurately predict the changes that may take place in the future in the equity market and guide investors. Bankers or financial analysts are enabled to understand how investments will perform in the future, and the facilitator role of AI is commendable in this regard. However, contemporary studies in AI and future prediction of the equity market are trying to build accuracy in predicting future markets [32]. There are efforts to improve the empathic levels of AI, wherein they can diagnose customers or employees at the cognitive or emotional levels. The feeling AI or empathic AI is in the process of inception and can assist humans in evaluating the emotional or cognitive conditions of customers [33].

The role of AI in conflict resolution may be too limited, as conflicts are rooted in cognitive styles, behavioral patterns, or role concerns. In this regard, human intervention is crucial in identifying and resolving conflicts. Emotional Intelligence plays an imperative role in conflict resolution by recognizing the manifestation of competition in the initial stages and offering appropriate emotional or cognitive strategies to deal with the conflict. Organizational psychologists opine that conflict germinates through five stages. The first stage is the cause, and the differences in values, beliefs, or the desire for power instigate conflict. The second stage is about the way the conflict is perceived by individuals, and the third stage is about the emotions that are felt during the event of a conflict. The fourth stage is about the manifestation of conflict, and it becomes tangible in the workplace. The fifth stage reflects the repercussions of conflict. Employees with high emotional intelligence and social awareness can predict the sprouts of conflict initially and diffuse the conflict in the early stages. The involved parties may be employees and customers [34].

9.4.4 KNOWLEDGE REPRESENTATION AUGMENTS EMPATHY AND ORGANIZATIONAL AWARENESS (A5)

Knowledge representation is the combination of linguistics, psychology, and computer science. Besides, the combination of these areas to produce the representation of knowledge and create awareness about various dimensions of expertise or organization helps in augmenting one's organizational awareness and being mindful of their job behaviors and their behavioral interface with their colleagues [35].

The platforms developed by MITRE Corporation have been extensively embraced by heterogeneous organizations across the globe. Platforms like Virtual Workplace, Xpertnet, and Expert Finder have been very influential in creating awareness about the organizational culture and developing healthy workplace relationships. Besides, these platforms all enable organizations to identify emerging talents or competencies in the human capital landscape [36]. Due to the increasing efficacy of AI (knowledge representation), the field of medicine is embracing the elements of knowledge representation. One of the studies adopted in medicine used the Wittgensteinian approach for assessing the storage of data and the interpretation of data. The study implied that the tool of knowledge representation added incremental value to the awareness of the medical process and the organizational expectations [37].

Knowledge representation and logic may be directly concerned with empathy, but they are closely associated with organizational awareness, wherein the heuristics of organizational functioning and the processes are involved. The AI-driven knowledge representation and logic are based on different dimensions of knowledge programmed into AI. The five dimensions are demonstrative (facts, ideas, or objects), meta knowledge (conceptualization of various dimensions of knowledge), strategic (procedures, planning, or purpose of task), heuristics (relationship between ideas and objects, definite rules/regulations based on experiences), and formative (problem solution through previous experiences). The amount of role assistance provided by AI is exponential, and human intelligence or emotional intelligence is augmented by AI as the moderator or facilitator [38].

9.4.5 KNOWLEDGE REPRESENTATION SHARPENS INFLUENCING SKILLS (A6)

AI-driven knowledge representation may not have a direct relation to influencing skills, which means that it doesn't directly boost influencing skills but helps employees influence the purchasing decisions of customers. It is observed that employees with updated knowledge of products and services will exercise a great amount of influence. However, there is a dearth of studies on this assumption. It is suggested that the studies may focus on understanding the nuances of influencing skills concerning AI.

9.4.6 *LANGUAGE PROCESSING NURTURES EMPATHY (A7)*

The language used during behavioral transactions is significantly influenced by social or personal contexts, and these factors help experts in AI reengineer the essence of language. Accordingly, language processing has been integrated with the linguistic sentiments of the social or personal context. A study into social language processing inferred that the essence of empathy has been infused into AI language processing to elevate the persuasiveness of the language. Besides, social or personal information is consumed through mediating factors like neurophysiological elements combined with vocal-emotional processing. According to the study, language processing may help professionals identify the emotional conditions of people, which aids in empathizing with those involved in the conversation [39].

According to Cognitive Psychology, empathy has been classified into two categories: emotional and cognitive. "Emotional empathy" is the ability of a person to feel the emotions or feelings of another person, including pain. "Cognitive empathy" refers to the process of creating mental visuals or memories aligned with another person's perspectives or assumptions. The studies infer that any of these classifications of empathy are consumed or manifested through syntax, diction, and other linguistic forms. Besides, empathy, one of the major components of Emotional Intelligence, plays a significant role in defining the quality of the interface. In this context, one of the studies inferred that AI-assisted language processing can create a better linguistic structure that can add incremental value during the AI-human interface [40].

Language Processing, or NLP, is a computational model wherein programmed computers assess human emotions or feelings based on written or oral speech. This dimension of AI may facilitate customer interface with banks through chatbots and help understand customer grievances. This facility has been employed by most organizations to reduce human intervention during customer interactions. The best examples of NLP are Alexa, Google Assistant, and Siri. These virtual assistants have revolutionized human interaction with AI, and their accuracy in understanding the needs of consumers is commendable. Indeed, these chatbots have enabled banking players to reduce their service costs and enhance their business effectiveness [41]. In continuation of the crucial role played by NLP in understanding customer needs, it nurtures empathetic skills in employees during their interactions with customers. Service providers or bankers can access customer data or queries and analyze their needs before they initiate face-to-face interactions with customers.

9.4.7 SEARCH ALGORITHMS EXPAND SOCIAL AWARENESS (A8)

Organizational awareness is one of the four dimensions of emotional intelligence, and this dimension deals with factors such as awareness of the values and goals of the organization, being aware of business challenges, updates on team members, empathy, and other areas of organizational awareness. According to a study by MITRE Corporation, collaborative virtual workplace ecosystems have been gaining significance and have been widely embraced by heterogeneous organizations. These virtual workplaces help establish strong awareness among employees. One of the platforms, Expert Finder, enables employees to explore the diverse knowledge domains and skills across the organization. This allows employees to be aware of each other's competencies. In addition, these algorithms keep the talent pool utilized and contribute to organizational effectiveness [42].

Search algorithms or AI algorithms enhance social awareness or organizational awareness by providing abundant data about organizational expectations or customers' preferences. One study on Particle Swarm Optimization (a sociobiological perspective of AI) infers that solutions for heterogeneous problems are found in the global optimal, based on heuristics that provide solutions at global levels through available experiences and proven solutions [43]. This aspect of AI enables service providers to offer viable financial solutions to their customers. Search algorithms act as a repository of solutions that can be customized according to requirements. Banks have initiated the process of embracing search algorithms that could help employees reduce the turn-around time (TAT) in resolving customer grievances [44]. Although banking players have realized the crucial role played by AI in becoming aware of customer needs or market needs, full-scale implementation may take its course of time.

9.4.8 SEARCH ALGORITHMS ENABLE EMPLOYEES TO BE THE CHANGE CATALYST (A9)

The search algorithm is an oceanic source of interpretable data and enables employees to predict futuristic changes and embrace unique practices in banking and services. AI helps service providers to be change catalysts and achieve a competitive edge in the business world [45]. Emotional intelligence plays a crucial role in understanding the changing perspectives of the banking world. In terms of AI and EI association in cultivating the trend of a change catalyst, AI is one of the propellants that can help EI in the process of becoming a change catalyst in the organization.

Besides, the search algorithm has been enabling the landscapes of human resource management to reinvent their strategies or policies in the process of recruitment, talent management, and employee engagement. It is inferred that these search algorithms assist HR professionals in identifying high-potential candidates through certain keywords and building their profiles. Accordingly, the computer-generated or built profiles are recommended to the concerned HR departments. It identifies the essence of change catalysts in the profiles and helps in building a strong psychological capital of an organization [46].

9.4.9 DEEP LEARNING STRENGTHENS EMPLOYEES' AWARENESS OF THE ORGANIZATION AND EMPLOYEES (A10)

Deep learning is one of the modules of AI that helps employees reengineer their behavioral patterns and understand colleagues' and customers' behavioral or cognitive patterns during face-to-face interactions. Behavioral analysis is one of the key strengths of DL that enables service providers to understand the buying behaviors of customers. The domain of behavioral finance or behavioral economics has embraced DL techniques to understand the business market or customers [47]. DL has immensely contributed to organizational awareness and behavior, and it has provided numerous simulations to understand customer behaviors and personality types [48].

9.5 CONCLUSION

As the study examined the evolution of AI and emotional intelligence and their influence on multifarious dimensions of organizational affairs, it is inferred that AI has been mystified as a substitute for emotional intelligence. There is a dire need for demystifying the roles of AI and examining the long-term effects of AI on human intelligence and human life.

Therefore, the study made a contemplative study and proposed a conceptual model based on 10 assumptions through which the conflict between AI and EI can be resolved and embrace AI for enhancing organizational effectiveness without causing terror in the form of replacing humans with AI. The study implies that AI must not be viewed as a substitute but as an assistant to human intelligence. In addition, AI is one of the best inventions in the process of technological evolution, and it has made human life more meaningful.

Due to the disruptive nature of the consumer market and the growing demands of customers, it has become crucial for any industry to keep pace with these demands. Organizations have been compelled to redefine and restructure their business models. In order to deal with these global and local (Glocal) anomalies, the integration of human intelligence and AI is very essential. During the initial stages of AI evolution, there was an assumption that technology lacks a humanistic approach, and it was widely criticized by human activists across the globe. Accordingly, the study aimed to revisit the disruptions inflicted by AI on business and personal ecosystems. It is inferred that AI helps in creating a better human ecosystem and nurturing emotional well-being and cognitive astuteness only if it is embraced and executed strategically. The study implies that AI augments employees' intention to be aware of their emotional selves and explore hidden competencies that can add value to organizational goals (A1). In terms of managing workplace emotions and cognitive skills, the computed assisted deep-learning techniques have been widely used in cognitive neuroscience, and these techniques are administered in nurturing job skills, job attitudes, and diffusing the shortcomings that hamper work-place productivity (A2).

AI has been a tremendous foundational assistance for employees in determining the root cause of any organizational problems or challenges. Organizational strategists and HR executives have been empowered by AI to identify workplace or business problems and explore possible solutions to those problems through AI simulations (A3). Besides, the AI simulations have not only been assisting in problem identification but also in predicting the evident changes in the business market in the future. This has enabled organizations to redefine their business models or strategies and diversify their approaches to change. This has capacitated businesses to survive in the contemporary disruptive world (A4). Knowledge representation has been one of the efficacious dimensions of AI, and it is being used to infuse the organizational character into the mindsets of employees. Besides, it has been widely adopted by organizations to create brand awareness among employees and to assist in orienting new employees about the core values, mission, and vision of the organization. Additionally, this dimension of AI is also used in creating an influential personal brand for an organization (A5, A6).

In continuation of the usage of language and communication, NLP experts opine that language plays a major role in expressing empathy through employing appropriate syntactic structures, diction, and articulation. Accordingly, it is assumed that AI-assisted language processing helps employees cultivate appropriate or influential language or communication patterns during behavioral or interpersonal transactions (A7). In addition, the search algorithms accelerate the process of identifying and offering viable solutions to organizational problems or challenges. Search algorithms operate based on existing data structures and help users find multiple solutions to existing problems. Furthermore, search algorithms assist organizational strategists and HR executives in predicting possible changes in business trends and the required competency framework of human resources. Accordingly, AI enables organizations to implement the required reengineering of their business models (A8, A9). Similar to knowledge representation, the technique of DL helps in organizational awareness and inculcates a deep sense of belongingness. However, the techniques of DL are deeply rooted in cognitive neuroscience, which influences brain plasticity (A10).

KEYWORDS

- **artificial intelligence**
- **cognitive neuroscience**
- **deep learning**
- **emotional intelligence**
- **turn-around time**

REFERENCES

1. World Bank. (2018). *Development and Disruptive Technology.* World Bank Group.
2. Gaskell, A. (2022). *AI Creates Job Disruption But Not Job Destruction.* Forbes. https://www.forbes.com/sites/adigaskell/2022/01/18/ai-creates-job-disruption-but-not-job-destruction/?sh=4e4e58c93b3e (accessed on 13 March 2025).
3. Caplain, J. (2023). Digital banking has been fast-tracked. *KPMG – Responding to Consumers in the New Reality: Banking.* https://kpmg.com/xx/en/home/insights/2020/06/consumers-new-reality-banking.html (accessed on 13 March 2025).
4. KPMG (2020). *Financial Resilience in Banking: A Balancing Act.* KPMG International. https://assets.kpmg.com/content/dam/kpmgsites/xx/pdf/2020/12/financial-resilience-in-banking.pdf (accessed on 13 April 2025).
5. KPMG. (2022). *Consumer and the New Reality.* KPMG International.
6. Renny, T. (2021). *Building the AI Bank of the Future.* McKinsey & Company.
7. McKinsey & Company. (2021). *The Great Divergence: McKinsey Global Banking Annual Review 2021.* McKinsey & Company.
8. Krystal, H. (2023). *ChatGPT Sets Record for Fastest-Growing User Base–Analyst Note. Reuters.* https://www.reuters.com/technology/chatgpt-sets-record-fastest-growing-user-base-analyst-note-2023-02-01/ (accessed on 13 March 2025).
9. Eva, A. M. V., Johan, B., Willem, Z., Robert, v. R., & Claudi, L. B. (2023). ChatGPT: Five priorities for research. *Nature, 614*(7947), 224–226.
10. Costandi, M. (2016). *Neuroplasticity.* The MIT Press.
11. Goleman, D. (2004). *Working with Emotional Intelligence.* Bloomsbury Publishing.
12. Huang, M. H., & Rust, R. T. (2018). Artificial intelligence in service. *Journal of Service Research, 21*(2), 155–172.
13. Khurshid, A. (2023). *Why Banks Need Artificial Intelligence. Wipro–Business Processes.* https://www.wipro.com/business-process/why-banks-need-artificial-intelligence/ (accessed on 13 March 2025).
14. 14. Brynjolfsson, E., & McAfee, A. (2017). The business of artificial intelligence. *HBR–Big Idea Series.* https://hbr.org/2017/07/the-business-of-artificial-intelligence (accessed on 13 March 2025).
15. Infosys. (2023). *Edgeverve.* Infosys–Finacle. https://www.edgeverve.com/finacle/ (accessed on 13 March 2025).
16. Oracle. (2023). *Discover Core Banking Made for Tomorrow—Oracle FLEXCUBE Universal Banking. Oracle–Oracle Services/Banking.* https://www.oracle.com/in/industries/financial-services/banking/flexcube/core-banking-software/ (accessed on 13 March 2025).
17. Bryan, P. (2021). *90% of Your Customer Sentiment is Easily Accessible and Collecting Dust.* Complete CSM–AI. https://completecsm.ai/2021/12/16/90-of-your-customer-sentiment-is-easily-accessible-and-collecting-dust/ (accessed on 13 March 2025).
18. Sergeyeva, T., Bronin, S., Turlakova, N., & Lamnytskyi, S. (2022). Integrating educational components into the metaverse. In D. Guralnick, M. Auer, & A. Poce (Eds.), *Innovative Approaches to Technology-Enhanced Learning for the Workplace and Higher Education* (pp. 345–378). Springer, Cham.
19. Jeearchive, C., & Douglas, W. H. (2021). The therapists using AI to make therapy better. *MIT Technology Review.*

20. Balamurali, R., Brannen, L. P., Taneja, K., & Krishna, G. (2021). Detecting human emotions through physiological signals using machine learning. *Artificial Intelligence and Technologies, 806*, 587–602.
21. Gardner, H. (1993). *Multiple Intelligences: The Theory in Practice*. Basic Books.
22. Shneiderman, B. (2020a). Human-centered artificial intelligence: Reliable, safe & trustworthy. *International Journal of Human-Computer Interaction, 36*(6), 495–504.
23. Shneiderman. (2020b). Design lessons from AI's two grand goals: human emulation and useful applications. *IEEE Transactions on Technology and Society, 1*, 73–82.
24. Alison, J., Karjian, R., & Niederkorn, M. (2011). *Lean Management – New Frontiers for Financial Institutions.* McKinsey & Company.
25. Fröhlich, E. E., Aitak Farzi, R. M., Angela, J., Bernhard, W., Erwin, Z., & Holzer. (2016). Cognitive impairment by antibiotic-induced gut dysbiosis: Analysis of gut microbiota-brain communication. *Brain, Behavior, and Immunity, 56*, 140–155.
26. Junjie, Y., Zixun, C., Kelai, Z., & Chongyuan, Y. (2019). A deep learning-based chatbot for campus psychological therapy. *Artificial Intelligence (Cornell University),* 1–31.
27. Priyanshu, P., Mauajama, F. B., & Asif, E. (2023). A multi-task learning framework for politeness and emotion detection in dialogues for mental health counseling and legal aid. *Expert Systems with Applications, 224.*
28. Chaturvedi, R. (2022). Artificial intelligence-driven customer experience: Overcoming the challenges. *California Management Review.*
29. McAfee, A., & Brynjolfsson, E. (2012). Big data: The management revolution. *Harvard Business Review, 90(10), 61–67.*
30. Lushun, J., Zhe, W., & Yunqing, Q. (2021). Opportunities and challenges of artificial intelligence in the medical field: Current application, emerging problems, and problem-solving strategies. *Journal of International Medical Research, 49(3), 3000605211000157.*
31. Montemayor, C., Halpern, J., & Fairweather, A. (2022). In principle obstacles for empathic AI: Why we can't replace human empathy in healthcare. *AI & Society, 37*, 1353–1359.
32. Primest. (2020). *Artificial Intelligence is Changing Change Management. Primeast–Insights.* https://primeast.com/us/insights/leading-change/Artificial-Intelligence-is-changing-change-management#:~:text=AI%20%E2%80%93%20Creating%20More%20Efficient%20Change%20Management&text=AI%20gives%20change%20managers%20the,to%20predict%20what%20will%20happen (accessed on 13 March 2025).
33. Esmaeilzadeh, H., & Vaezi, R. (2022). Conscious empathic AI in service. *Journal of Service Research, 25*(4).
34. Nicole, J. (2010). Interventions in workplace bullying: A multilevel approach. *European Journal of Work and Organizational Psychology, 19*, 51–75.
35. Vickery, B. (1986). Knowledge representation: A brief overview. *Journal of Documentation, 42*(3).
36. Mark, M., Ray, D., & David, H. (2009). Awareness of organizational expertise. *International Journal of Human–Computer Interaction, 14(2), 199–217.*
37. Laszlo, B., & Ronald, C. (2019). The interplay of knowledge representation with various fields of artificial intelligence in medicine. *IMIA Yearbook of Medical Informatics,* 27–34.

38. Leone, N., Pfeifer, G., Faber, W., Eiter, T., Gottlob, G., Perri, S., & Scarcello, F. (2006). The DLV system for knowledge representation and reasoning. *ACM Transactions on Computational Logic, 7*(3), 499–562.
39. Daniëlle, B., Van Berkum, J. J., Marcel, C. B., Cathelijne, M. T., Miriam, K., Jan, K. B., & Peter, H. (2010). Empathy matters: ERP evidence for inter-individual differences in social language processing. *Social Cognitive and Affective Neuroscience, 7*(2), 173–183.
40. Allison, L., Lahnala, C. W., David, J., & Lucie, F. (2022). A critical reflection and forward perspective on empathy and natural language processing. *Association for Computational Linguistics,* 2139–2158.
41. IBM. (n.d.). *What is a Chatbot?* IBM–Chatbots. https://www.ibm.com/in-en/topics/chatbots#:~:text=A%20chatbot%20is%20a%20computer,to%20them%2C%20simulating%20human%20conversation (accessed on 13 March 2025).
42. Maybury, M., D'Amore, R., & House, D. (2002). Awareness of organizational expertise. *International Journal of Human-Computer Interaction, 14*(2), 199–217.
43. Tam, A. (2021). A gentle introduction to particle swarm optimization. *Machine Learning Mastery.* https://machinelearningmastery.com/a-gentle-introduction-to-particle-swarm-optimization/ (accessed on 13 March 2025).
44. Ebert, J. T., Berlinger, F., Haghighat, B., & Nagpal, R. (2022). A hybrid PSO algorithm for multi-robot target search and decision awareness. *IEEE/RSJ International Conference on Intelligent Robots and Systems (IROS),* 11520–11527.
45. Ganesh, U. (2020). *Catalyst of Change: Bringing Artificial Intelligence to the Forefront.* Financial Express.
46. Posthumus, J., Bozer, G., & Santora, J. (2016). Implicit assumptions in high potentials recruitment. *European Journal of Training and Development, 40*(6), 430–445.
47. Königstorfer, F., & Thalmann, S. (2020). Applications of artificial intelligence in commercial banks – A research agenda for behavioral finance. *Journal of Behavioral and Experimental Finance, 27.*
48. Schuller, D., & Schuller, B. W. (2018). The age of artificial emotional intelligence. *Computer, 51*(9), 38–46.

CHAPTER 10

Innovations in Future Mobility and Intelligent Virtual Assistants: Lensing Machine Learning AI-Based Chatbots for Digital Users

BHUPINDER SINGH[1,*] and CHRISTIAN KAUNERT[2]

[1]*Sharda School of Law, Sharda University, Greater Noida, Uttar Pradesh, India*

[2]*International Security, Dublin City University, Ireland*

**Corresponding author.*

ABSTRACT

The emergence of advanced technologies such as artificial intelligence (AI) and machine learning (ML) has revolutionized the way we interact with digital platforms and transformed the future of mobility. A significant development in future mobility is the integration of intelligent virtual assistants (IVAs). These assistants leverage AI algorithms and ML models to provide personalized and seamless experiences for digital users. By harnessing the power of chatbots, virtual assistants can understand user queries and provide accurate and relevant information. ML AI-based chatbots play a crucial role in enhancing the efficiency and effectiveness of future mobility solutions. These chatbots are designed to understand natural language and context, enabling them to deliver personalized assistance to users. By leveraging advanced ML algorithms, these chatbots continuously learn and improve their performance, providing users with an ever-evolving and IVA. The integration of ML AI-based chatbots in

Natural Language Processing and Artificial Intelligence: A Perspective Toward Current Trends, Challenges, and Applications. Vikas Khullar, Aryan Chaudhary, & Sunil Kumar (Eds.)

DOI: 10.1201/9781998511273-10

future mobility offers numerous benefits. Firstly, it enhances the overall user experience by providing quick and accurate responses to user queries. Users can easily interact with the chatbot through voice commands or text, eliminating the need for complex user interfaces. Secondly, these chatbots enable seamless integration with various mobility services, such as ride-sharing platforms, public transportation systems, and smart city infrastructure. Users can effortlessly book rides, plan their routes, and access real-time information about traffic conditions, all through a single interface. These chatbots leverage AI and ML technologies to provide personalized and efficient assistance to digital users. By enhancing the overall user experience, enabling seamless integration with mobility services, and promoting sustainability, these chatbots are shaping the future of mobility.

10.1 INTRODUCTION

The future mobility solutions and intelligent virtual assistants (IVAs) stand as a formidable harbinger of innovation in the era of rapid technological evolution. As urban landscapes metamorphose into interconnected hubs of smart infrastructure and digital ecosystems, the synergy between IVAs and the impending wave of futuristic transportation modes becomes increasingly pronounced. The technology, as a revolutionary paradigm in the realm of machine learning (ML), serves as the linchpin in this transformative interplay. This chapter endeavors to unravel the intricacies and ramifications of incorporating lensing ML AI-based chatbots into the digital user experience, scrutinizing their impact on the realms of mobility, convenience, and overall user satisfaction.

Machine Learning (ML) and Artificial Intelligence (AI) have emerged as transformative forces, reshaping the way users engage with technology. The remarkable application of this amalgamation is evident in the evolution of AI-based chatbots and IVAs. These entities have transcended the traditional boundaries of programmed responses, leveraging sophisticated algorithms and learning mechanisms to provide users with a more intuitive and personalized experience. ML, the backbone of this technological advancement, empowers chatbots and virtual assistants with the ability to learn from vast datasets and adapt their responses over time. This adaptive learning process enables these digital entities to understand user preferences, anticipate needs, and continuously enhance their performance. Consequently, interactions with users become more natural, efficient, and tailored to individual requirements.

The emergence of AI-based chatbots and IVAs marks a paradigm shift in user engagement. This transformative capability not only streamlines digital interactions but also fosters a more human-like communication experience, bridging the gap between man and machine. As we delve deeper into the realm of AI-driven digital assistance, we set the stage for exploring the multifaceted impact of ML-powered chatbots and IVAs. From revolutionizing customer support to redefining user interfaces, the implications of these technologies extend far beyond mere convenience, heralding a new era where digital interactions are not just functional but deeply intuitive and responsive to the ever-evolving needs of users. This paper serves as a compass navigating the uncharted territories of technological innovation, offering a panoramic view of the landscape where IVAs and future mobility intersect. Through an in-depth examination of the capabilities, challenges, and ethical considerations, it aspires to contribute valuable insights to scholars, practitioners, and enthusiasts alike, guiding them through the evolving contours of a digitally augmented future.

10.1.1 CURRENT STATE OF MOBILITY AND VIRTUAL ASSISTANTS

The contemporary landscape of mobility and virtual assistants is characterized by a confluence of transformative technologies and shifting paradigms. In mobility, traditional transportation models are giving way to an era defined by smart, connected, and autonomous systems. The deployment of internet of things (IoT) technologies and the integration of sensors and actuators into vehicles and urban infrastructure are ushering in an age where vehicles communicate with each other and with the environment in real-time. Electric and autonomous vehicles are becoming more prevalent, promising a future where transportation is not only more sustainable but also more efficient [1].

Simultaneously, the domain of virtual assistants has evolved from simple rule-based systems to sophisticated, context-aware entities powered by AI. IVAs, equipped with natural language processing (NLP) and ML capabilities, have transcended mere voice recognition to understand and respond to user queries in a conversational and personalized manner.

They have become ubiquitous in our daily lives, residing in smartphones, smart speakers, and various other digital interfaces. From providing weather updates and setting reminders to controlling smart home devices, virtual assistants have become integral facilitators of our digital experiences.

The intersection of these two domains is where the synergies and opportunities lie. As smart cities emerge and the demand for seamless, personalized experiences grows, the integration of IVAs with future mobility solutions becomes not just a convenience but a necessity. Imagine a scenario where a user, through a simple voice command, can seamlessly plan a multi-modal journey, receive real-time traffic updates, and interact with their vehicle—all facilitated by an IVA embedded in their digital ecosystem.

This convergence is not without its challenges, as the security and privacy implications of collecting and processing vast amounts of data in the context of mobility are significant concerns. Additionally, the ethical considerations surrounding the use of AI in decision-making, particularly in autonomous vehicles, demand careful scrutiny. Striking a balance between innovation and responsible deployment is crucial to harness the full potential of these technologies.

The current state of mobility and virtual assistants paints a picture of rapid evolution and integration. The synergy between these two domains holds immense promise for shaping a future where transportation is intelligent, personalized, and seamlessly connected to our digital lives. As we navigate this dynamic landscape, it is imperative to critically assess the current state, understand the challenges, and envision a future where these innovations enhance not only convenience but also the overall quality of life.

10.1.2 SIGNIFICANCE OF INTEGRATING MACHINE LEARNING (ML) ARTIFICIAL INTELLIGENCE (AI)-BASED CHATBOTS IN FUTURE MOBILITY SOLUTIONS

Machine learning AI-based chatbots in future mobility solutions mark a paradigm shift with profound implications for the way we perceive, interact with, and experience transportation. At its core, the significance of this integration lies in the enhancement of user experience and the optimization of operational efficiency. These chatbots, imbued with ML capabilities, have the capacity to learn from user behaviors, preferences, and historical data, thereby enabling a level of personalization that transcends conventional modes of transportation. Users can expect tailored recommendations, predictive analysis of travel patterns, and real-time updates that align seamlessly with their individual needs. The introduction of AI-based chatbots in future mobility contributes to the creation of intelligent and interconnected transportation ecosystems. These systems go beyond mere conveyance; they evolve into dynamic entities that adapt to changing conditions, proactively manage traffic, and optimize routes based on live data. ML algorithms empower these chatbots to analyze vast datasets, from traffic patterns to weather conditions, enabling them to make informed decisions that result in smoother and more efficient journeys. This level of adaptability and responsiveness is crucial for mitigating congestion, reducing travel times, and ultimately fostering a sustainable and resilient urban mobility landscape [2].

In addition to enhancing the efficiency of transportation, the use of AI-based chatbots in future mobility solutions also has significant implications for safety. These chatbots can contribute to the development of advanced driver-assistance systems, facilitating real-time monitoring of road conditions, identifying potential hazards, and even assisting in autonomous vehicle navigation. The predictive capabilities of ML algorithms can foresee and mitigate risks, fostering a safer transportation environment for both drivers and pedestrians.

The economic and environmental implications of this integration should not be understated. AI-based chatbots can optimize energy consumption, reduce emissions through intelligent traffic management, and contribute to a more sustainable urban infrastructure. By fostering a shift towards shared mobility models and promoting efficient use of resources, these chatbots play a crucial role in aligning future transportation solutions with broader environmental and economic sustainability goals.

The ML AI-based chatbots in future mobility solutions transcend the realm of convenience; they are a catalyst for a holistic transformation of our transportation ecosystems. From personalized user experiences and enhanced operational efficiency to improved safety and sustainability, the significance of this integration lies in its potential to redefine how we perceive and engage with the mobility of the future. As technology continues to advance, the synergy between AI-based chatbots and future mobility represents a powerful narrative of innovation, responsiveness, and progress in the realm of urban transportation [3].

10.1.3 OBJECTIVE OF CHAPTER

The primary objective of the paper titled is to comprehensively explore and analyze the intersection of future mobility solutions and IVAs, with a specific focus on chatbots enhanced by ML AI and lensing technology. The paper aims to achieve the following specific objectives:

1. **Examine the Current Landscape:** Provide an in-depth overview of the current state of future mobility solutions, including smart transportation, autonomous vehicles, and interconnected urban infrastructure. Simultaneously, assess the capabilities and prevalence of IVAs in the digital realm, emphasizing their role in daily user interactions.
2. **Explore Technological Convergence:** Investigate the synergies and opportunities arising from the integration of ML AI-based chatbots with future mobility solutions. Analyze how lensing technology enhances the conversational abilities of chatbots, making them context-aware and capable of personalized responses.
3. **Evaluate User-Centric Applications:** Highlight the practical applications of lensing ML AI-based chatbots in enhancing the digital user experience within the context of future mobility. Explore scenarios such as personalized travel recommendations, real-time traffic management, and seamless interaction with smart vehicles, showcasing how these innovations improve convenience and efficiency.
4. **Assess Implications and Challenges:** Delve into the ethical, privacy, and security considerations associated with the deployment of AI-driven chatbots in mobility contexts. Evaluate the potential societal impacts, both positive and negative, and identify areas where responsible deployment is imperative.
5. **Analyze Technological Advancements:** Investigate recent advancements and emerging trends in ML, AI, and lensing technology that have paved the way for this convergence. Provide insights into the state-of-the-art developments and their implications for the future [4].
6. **Offer Strategic Insights:** Provide strategic recommendations and insights for stakeholders in the fields of transportation, technology, and policy-making. Discuss potential strategies for responsible deployment, regulation, and collaboration to harness the full potential of these innovations.

7. **Contribute to Academic Discourse:** Add to the body of knowledge in the domains of future mobility and virtual assistants by presenting a comprehensive and interdisciplinary analysis of this emerging intersection. Serve as a resource for researchers, practitioners, and policymakers seeking to understand and leverage the evolving landscape of AI-driven innovations [5].

It is to provide a holistic examination of the integration of ML AI-based chatbots, empowered by lensing technology, into future mobility solutions. It seeks to elucidate the transformative potential, ethical considerations, and practical implications of this convergence while contributing valuable insights to the broader discourse on intelligent transportation and digital user experiences.

10.2 FUTURE MOBILITY TRENDS

The landscape of future mobility is undergoing a transformative shift, driven by a confluence of technological advancements, societal shifts, and environmental imperatives. One prominent trend that looms large on the horizon is the rise of autonomous vehicles. The development and deployment of self-driving cars and trucks represent a paradigm shift in how we perceive and engage with transportation. Companies across the globe are investing heavily in autonomous vehicle technology, promising a future where vehicles navigate seamlessly without human intervention, potentially reshaping the very fabric of urban mobility and logistics.

Connected and smart mobility is another trend that continues to gain momentum. The advent of the IoT and vehicle-to-everything (V2X) communication is fostering an ecosystem where vehicles communicate not only with each other but also with the surrounding infrastructure. This connectivity facilitates real-time data exchange, enabling smarter traffic management, improved safety through collision avoidance systems, and enhanced navigation experiences. The emergence of smart cities further amplifies this trend, as urban planning integrates technology to create more efficient, sustainable, and user-centric transportation systems [6].

The electrification of transport is a key trajectory in future mobility, fueled by the imperative to address environmental concerns and reduce reliance on fossil fuels. Electric vehicles (EVs) are becoming increasingly mainstream, with advancements in battery technology extending the range and improving the affordability of electric cars. Governments and industries worldwide are investing in charging infrastructure, offering incentives, and setting ambitious targets to accelerate the adoption of electric mobility and transition towards a more sustainable and carbon-neutral transportation ecosystem.

The concept of mobility as a service (MaaS) is gaining prominence, reflecting a shift from traditional ownership models to a more service-oriented approach. MaaS integrates various modes of transportation into a unified, user-centric platform, allowing individuals to plan, book, and pay for their journeys seamlessly across different modes such as buses, trains, ride-sharing, and even micromobility options like electric scooters or bikes. This trend aligns with the growing demand for flexible, efficient, and sustainable mobility solutions tailored to individual preferences [7].

The trend of shared mobility is reshaping how people perceive personal transportation. Ride-sharing services, car-sharing platforms, and bike-sharing programs are becoming integral components of urban transportation ecosystems. These shared mobility solutions contribute not only to reducing traffic congestion but also to optimizing resource utilization and lowering the overall environmental impact of transportation.

These trends collectively herald a new era of transportation that is not only technologically advanced but also more sustainable, efficient, and responsive to the evolving needs and expectations of users in our rapidly changing world. As these trends continue to evolve, they are poised to redefine the way we move, commute, and experience transportation in the years to come.

10.2.1 AUTONOMOUS VEHICLES, ELECTRIC VEHICLES, SHARED MOBILITY AND SMART TRANSPORTATION SYSTEMS: EMERGING TRENDS IN FUTURE MOBILITY

The future of mobility is evolving at an unprecedented pace, ushering in a wave of transformative trends that promise to reshape the way we move and interact with transportation systems. One of the most prominent trends is the advent of autonomous vehicles. The development of self-driving cars and trucks is gaining momentum, propelled by breakthroughs in AI, sensor technologies, and connectivity [8]. The prospect of vehicles navigating without direct human intervention not only holds the promise of enhanced safety but also the potential to revolutionize urban planning, reduce traffic congestion, and redefine the very concept of personal and public transportation.

The electrification of transport stands as a pivotal trend in shaping the future mobility landscape. EVs are rapidly gaining traction as advancements in battery technology drive down costs and improve performance. Governments and industries globally are investing in EV infrastructure, offering incentives, and setting ambitious targets to accelerate the transition towards sustainable and environmentally friendly transportation [9]. This shift toward electric mobility represents a critical response to the imperative of mitigating climate change and reducing dependence on fossil fuels. Another transformative trend is the rise of shared mobility. Ride-sharing services, car-sharing platforms, and bike-sharing programs have become integral components of urban transportation ecosystems. The paradigm of ownership is gradually giving way to a model where individuals seek on-demand access to vehicles and mobility services. Shared mobility not only addresses issues of urban congestion and parking but also promotes resource efficiency by maximizing the utilization of vehicles, contributing to a more sustainable and economically viable transportation future.

Concurrently, the emergence of smart transportation systems is revolutionizing the way we conceive and manage urban mobility. The IoT and advanced connectivity solutions are facilitating seamless communication between vehicles, infrastructure, and even pedestrians. Smart traffic management, real-time data analytics, and intelligent navigation systems are enhancing the efficiency of transportation networks. Smart cities are leveraging technology to create interconnected ecosystems that prioritize sustainability, accessibility, and user-centric experiences, marking a departure from traditional models of urban planning [10].

The convergence of these trends paints a holistic picture of a future where mobility is not only automated and electrified but also shared and intelligently managed. These trends are not isolated but rather interconnected components of a broader transformation toward a more sustainable, efficient, and inclusive transportation landscape. The synergies between autonomous vehicles, electric mobility, shared transportation, and smart systems have the potential to redefine our cities, reduce environmental footprints, and enhance the overall quality of life for individuals globally. As these trends continue to mature and intersect, the trajectory of future mobility is poised to reflect a harmonized vision that leverages technology to create more accessible, connected, and environmentally conscious transportation systems.

10.2.2 CHALLENGES AND OPPORTUNITIES: FUTURE OF MOBILITY

The promising trends shaping the future of mobility also come accompanied by a host of challenges and opportunities that demand careful consideration. The advent of autonomous vehicles, for instance, presents an array of technological, ethical, and regulatory challenges. The development and implementation of robust AI systems capable of navigating complex urban environments safely is a formidable task. Ensuring the security of these systems against cyber threats and addressing the ethical implications of autonomous decision-making are critical challenges [11]. Moreover, the establishment of a comprehensive legal and regulatory framework to govern autonomous vehicles is paramount to their successful integration into existing transportation ecosystems.

Electric mobility, while offering a sustainable alternative to traditional fuel-powered vehicles, faces challenges related to infrastructure and range anxiety. The need for a widespread charging infrastructure, including fast-charging stations, is crucial for the widespread adoption of EVs. Governments and industry stakeholders need to invest heavily in building this infrastructure to alleviate concerns about the practicality and convenience of EVs [12]. Additionally, advancements in battery technology are essential to address range limitations and enhance the overall performance of EVs, making them more accessible and appealing to a broader segment of the population.

The shift towards shared mobility models presents both challenges and opportunities. While ride-sharing and other shared transportation services contribute to reduced traffic congestion and resource optimization, concerns about data privacy, security, and the impact on public transportation systems must be addressed. There is a need for a thoughtful regulatory framework that balances innovation with public safety and accessibility. On the positive side, shared mobility models open up opportunities for new business models, collaborative partnerships, and enhanced urban planning that prioritizes efficient land use and reduces the environmental footprint of transportation.

Smart transportation systems, reliant on extensive data collection and communication networks, introduce challenges related to privacy and cybersecurity. The vast amounts of data generated by interconnected devices and vehicles raise concerns about how this data is managed, protected, and shared. Striking a balance between the benefits of data-driven systems and the protection of individual privacy is crucial. Furthermore, the implementation of smart transportation systems requires substantial investment in infrastructure and technology, posing financial challenges for cities and governments.

Despite these challenges, each trend also presents a multitude of opportunities. Autonomous vehicles, once perfected, have the potential to revolutionize transportation accessibility for individuals with disabilities, reduce accidents caused by human error, and optimize traffic flow in urban areas. Electric mobility contributes to the reduction of greenhouse gas emissions, lessens dependence on fossil fuels, and stimulates innovation in energy storage technologies [13]. Shared mobility models can lead to more efficient use of vehicles, decrease the overall number of vehicles on the road, and enhance urban mobility planning. Smart transportation systems offer the opportunity to create safer, more efficient, and sustainable cities through data-driven decision-making, intelligent infrastructure, and improved connectivity.

While the trends in future mobility promise to reshape our transportation systems for the better, addressing the associated challenges is essential for the responsible and effective implementation of these innovations. Navigating these challenges offers an opportunity to create a more inclusive, sustainable, and efficient mobility landscape that aligns with the evolving needs of our societies. The key lies in a collaborative and forward-thinking approach that involves stakeholders from various sectors, including government, industry, and the public, to ensure that the future of mobility is not only technologically advanced but also ethically sound, accessible, and beneficial to all.

10.3 INTELLIGENT VIRTUAL ASSISTANTS (IVAS): TRANSFORMATIVE PARADIGM

Intelligent virtual assistants (IVAs) represent a transformative paradigm in human-computer interaction, harnessing the power of AI to provide personalized and context-aware assistance. At the forefront of this technological evolution are virtual assistants like Amazon's Alexa, Apple's Siri, Google Assistant, and others, which have become ubiquitous across various digital platforms and smart devices [14]. These IVAs are designed to understand natural language, interpret user intent, and execute commands, marking a departure from traditional rule-based systems towards more sophisticated and adaptive models.

The defining feature of IVAs is their ability to learn and adapt over time. Through ML algorithms, these assistants continuously refine their understanding of user preferences, speech patterns, and contextual cues, enabling them to deliver increasingly accurate and tailored responses. This adaptive learning ensures that the virtual assistant becomes more proficient in anticipating user needs and providing proactive assistance, ultimately enhancing the overall user experience.

The applications of IVAs span a wide spectrum, from simple tasks such as setting reminders and answering factual queries to more complex interactions like natural language conversations and task automation. In the business realm, IVAs are increasingly being deployed for customer support, streamlining processes, and enhancing operational efficiency. These virtual assistants offer a scalable solution for handling routine inquiries, allowing human agents to focus on more complex and nuanced customer interactions [15].

As the technology behind IVAs continues to advance, multimodal capabilities are becoming more prevalent. Integrating voice recognition, visual cues, and even gestures, IVAs are evolving towards a more immersive and interactive user experience. This trend is particularly evident in the integration of IVAs with smart home devices, where users can seamlessly control lights, thermostats, and other connected appliances through voice commands, creating a seamless and integrated smart living environment [16].

However, the widespread adoption of IVAs also brings forth ethical considerations and challenges. Issues related to privacy, data security, and the responsible use of AI are paramount. As virtual assistants collect and process vast amounts of user data, ensuring robust privacy safeguards and transparent data policies is essential. Additionally, there is a need to address biases in AI algorithms that may inadvertently perpetuate stereotypes or discrimination, emphasizing the importance of responsible development and deployment.

Intelligent Virtual Assistants represent a groundbreaking evolution in human-machine interaction, leveraging AI and ML to create responsive, adaptive, and user-centric digital assistants. As these technologies continue to mature, the potential applications for IVAs are poised to expand further, influencing how we work, live, and engage with the digital world. While realizing the opportunities presented by IVAs, it is imperative to navigate the associated challenges mindfully, ensuring that ethical considerations and user privacy remain at the forefront of their development and deployment [17].

10.3.1 INTELLIGENT VIRTUAL ASSISTANTS (IVAS) AND THEIR ROLE IN ENHANCING USER EXPERIENCE IN FUTURE MOBILITY

IVAs are poised to play a pivotal role in redefining the user experience in the realm of future mobility. As transportation undergoes a radical transformation with the advent of autonomous vehicles, electric mobility, and smart transportation systems, IVAs emerge as a critical interface that bridges the gap between users and the evolving technologies. These virtual assistants, powered by AI and ML, are designed not only to assist users with routine tasks but also to understand their preferences, anticipate needs, and enhance overall convenience and comfort.

In the context of future mobility, IVAs have the potential to revolutionize how users interact with their vehicles and navigate through increasingly complex urban environments. Imagine a scenario where an IVA seamlessly integrates with an autonomous vehicle, offering a natural language interface for users to specify destinations, inquire about route options, or control in-vehicle settings—all through voice commands or intuitive gestures [18]. This level of interaction not only simplifies the user experience but also contributes to a safer and more focused driving environment, as users can keep their attention on the road.

IVAs can extend their capabilities beyond the vehicle itself, integrating with broader mobility ecosystems. Users could plan and manage multimodal journeys effortlessly, combining autonomous vehicles with other modes of transportation such as public transit or ride-sharing services. The IVA becomes a personalized travel companion, providing real-time updates, suggesting optimal routes, and seamlessly handling the logistics of a multi-faceted journey. This not only enhances efficiency but also contributes to a more sustainable and interconnected urban mobility landscape. In the electric mobility domain, IVAs can alleviate the concerns associated with charging infrastructure and range anxiety. Users might interact with their virtual assistant to locate nearby charging stations, plan routes based on charging availability, or even optimize energy consumption based on real-time traffic and weather conditions. By integrating predictive analytics and learning from user behaviors, IVAs can contribute to a more seamless and reassuring EV experience [19].

However, as with any technological integration, challenges arise. Ensuring the security of data exchanged between IVAs and vehicles, addressing potential biases in AI algorithms, and establishing robust privacy protocols are critical considerations [20]. Additionally, striking a balance between the benefits of IVAs and the need for a distraction-free driving experience poses a challenge that requires careful design and user interface considerations. IVAs stand as a linchpin in the future of mobility, offering a dynamic and intelligent interface that enhances user experience, safety, and convenience. As vehicles evolve into sophisticated and interconnected entities, the role of IVAs extends beyond mere assistance to become a central element in the symbiosis between users and the future of transportation. By fostering intuitive and context-aware interactions, IVAs contribute not only to the practicalities of mobility but also to a vision of transportation that is user-centric, interconnected, and seamlessly integrated into the fabric of our daily lives.

10.3.2 EVOLUTION OF INTELLIGENT VIRTUAL ASSISTANTS (IVAS) AND THEIR INTEGRATION WITH ARTIFICIAL INTELLIGENCE

The evolution of IVAs represents a remarkable journey at the intersection of AI and human-computer interaction. Initially rooted in basic rule-based systems, IVAs have undergone a profound transformation, fueled by advancements in AI and ML. Early virtual assistants could perform predefined tasks and respond to simple commands, but their capacity for understanding natural language and context was limited. With the integration of AI, particularly NLP and ML algorithms, IVAs have evolved into dynamic and adaptive entities capable of nuanced interactions [21].

The infusion of AI has enabled IVAs to comprehend the intricacies of human language, discern user intent, and adapt their responses based on context. This evolution is exemplified by the likes of Siri, Alexa, and Google Assistant, which leverage sophisticated AI algorithms to understand user queries, engage in natural language conversations, and even learn from user behavior over time. ML, a key component in this evolution, empowers IVAs to continuously improve their performance, becoming more accurate, personalized, and efficient as they process large datasets and learn from diverse user interactions.

The integration of AI technologies has propelled IVAs beyond mere voice-activated interfaces. Multimodal capabilities, encompassing voice, visual recognition, and even gesture-based interactions, have become integral. This evolution broadens the scope of IVAs, allowing them to interact seamlessly across various digital platforms and devices. Whether it's interpreting visual cues in a smart home environment or facilitating text-based interactions on a computer, IVAs have become versatile companions that adapt to the preferences and behaviors of users [22].

Artificial intelligence (AI) has elevated IVAs beyond reactive assistants to proactive enablers of user needs. Through predictive analytics and data-driven insights, IVAs anticipate user preferences, suggest relevant information, and automate routine tasks. This transformation from reactive to proactive assistance aligns with the broader trend of creating more intuitive and anticipatory user experiences in the digital realm.

As IVAs become increasingly sophisticated, ethical considerations and challenges emerge. Concerns about data privacy, security, and potential biases in AI algorithms necessitate a careful and responsible approach to their development and deployment. Striking a balance between personalization and user privacy is a critical aspect that developers and policymakers must address as IVAs continue to evolve [23]. From rudimentary rule-based systems to context-aware, adaptive companions, IVAs have come a long way. Their integration with AI not only enhances user experiences but also paves the way for a future where IVAs seamlessly navigate the complexities of human communication, contributing to a more connected, intuitive, and user-centric digital landscape.

10.4 MACHINE LEARNING ARTIFICIAL INTELLIGENCE (AI)-BASED CHATBOTS

Machine learning AI-based chatbots represent a groundbreaking convergence of AI and NLP, reshaping the landscape of human-computer interaction. Unlike traditional rule-based chatbots, these advanced systems leverage ML algorithms to autonomously learn from data, adapt to user input, and continuously improve their conversational capabilities. This evolution marks a paradigm shift in the way we engage with technology, as these chatbots transcend scripted responses and develop a dynamic understanding of language nuances, context, and user intent [24].

The key advantage of ML AI-based chatbots lies in their ability to comprehend and respond to natural language in a manner that simulates human-like conversation. Through ML models, these chatbots analyze vast datasets, learning patterns of speech, colloquial expressions, and contextual cues. This enables them to engage in more fluid and contextually relevant conversations, enhancing the user experience by providing responses that are not only accurate but also tailored to individual preferences.

The adaptability of ML AI-based chatbots is a testament to their capacity for continuous learning. As users interact with these chatbots, the systems assimilate new information, refine their understanding of user queries, and dynamically adjust their responses. This learning process ensures that the chatbots become increasingly proficient over time, addressing a diverse range of inquiries and scenarios with a level of sophistication that was once thought to be exclusive to human communication.

In customer service and support, ML AI-based chatbots are increasingly being deployed to streamline interactions, reduce response times, and enhance overall efficiency. These chatbots can handle a myriad of customer queries, ranging from product information to issue resolution, with a level of scalability that traditional methods struggle to achieve [25]. This not only improves customer satisfaction but also allows human agents to focus on more complex and nuanced tasks, fostering a symbiotic relationship between automation and human expertise.

Despite their remarkable capabilities, challenges persist in the development and deployment of ML AI-based chatbots. Ensuring ethical use, mitigating biases present in training data, and maintaining data privacy are critical considerations. Striking a balance between automation and the human touch is also pivotal, particularly in scenarios where empathy and emotional intelligence are essential components of the interaction. ML AI-based chatbots represent a transformative leap in the evolution of conversational agents. Their integration of ML not only propels them beyond rule-based predecessors but positions them as intelligent and adaptive entities capable of providing sophisticated, user-centric interactions. As these chatbots continue to evolve, their impact is poised to extend across diverse domains, from customer service to virtual assistants, contributing to a future where human-computer interaction is characterized by seamless, responsive, and contextually aware conversations.

10.5 BENEFITS AND CHALLENGES

The integration of ML AI-based chatbots in future mobility holds the promise of revolutionizing user experiences, providing enhanced personalization, and streamlining interactions within an evolving transportation landscape. These chatbots, equipped with dynamic learning capabilities, offer benefits such as improved efficiency, seamless multi-modal integration, and a heightened level of user adaptability. However, this transformative potential is accompanied by challenges, including data privacy and security concerns, ethical considerations related to biases in ML algorithms, integration complexities with existing systems, and the need to build user trust and acceptance. Striking a balance between realizing the benefits and addressing these challenges is essential for the successful deployment and widespread acceptance of chatbots in shaping the future of intelligent and connected mobility [26].

10.5.1 BENEFITS OF CHATBOTS IN FUTURE MOBILITY: IMPROVED USER EXPERIENCE, INCREASED EFFICIENCY AND ENHANCED SAFETY

Chatbots in future mobility bring forth a plethora of benefits, contributing to a more intelligent, user-centric, and efficient transportation ecosystem. Firstly, these chatbots significantly enhance the user experience by providing personalized and context-aware interactions. Users can seamlessly engage in natural language conversations, receive tailored travel recommendations, and access real-time information, leading to a more satisfying and user-friendly interaction with transportation services [27]. Moreover, chatbots contribute to increased efficiency in the mobility landscape. They can streamline tasks such as trip planning, ticket booking, and navigation, reducing the time and effort required from users. This efficiency not only enhances convenience but also promotes the optimal use of available transportation resources.

Additionally, chatbots play a pivotal role in enhancing safety within the realm of future mobility. Through proactive notifications, real-time traffic updates, and intelligent navigation assistance, these chatbots contribute to a safer and more informed travel experience. By alerting users to potential hazards, optimizing routes to avoid congested areas, and facilitating communication in emergency situations, chatbots actively contribute to the overall safety of individuals utilizing future mobility solutions [28]. The benefits of chatbots in future mobility extend beyond mere convenience, encompassing improved user experiences, increased efficiency, and enhanced safety across diverse aspects of transportation services.

Chatbots in future mobility are poised to bring about transformative benefits, reshaping the way users interact with transportation services. One significant advantage is the substantial improvement in user experience. Chatbots, powered by NLP and ML, enable more intuitive and personalized interactions. Users can effortlessly communicate with these chatbots, receiving real-time updates, personalized recommendations, and context-aware assistance, leading to a smoother and more user-centric experience. Moreover, chatbots contribute to increased efficiency within the realm of future mobility [29]. By automating routine tasks such as trip planning, ticket booking, and providing real-time traffic information, these intelligent agents streamline processes and save users valuable time. The efficiency gains extend to optimizing routes, managing transportation logistics, and facilitating seamless interactions between users and various mobility services. As a result, users can navigate through urban environments more efficiently, reducing travel times and enhancing overall convenience.

In terms of safety, chatbots play a crucial role in enhancing the security and well-being of individuals using future mobility solutions. Through proactive communication and real-time updates, chatbots can provide users with critical information about potential hazards, road conditions, and emergency situations. This real-time awareness contributes to making safer travel decisions and ensures that users are well-informed about their surroundings. Additionally, chatbots can facilitate emergency communication and assistance, enhancing the overall safety net within the dynamic landscape of future mobility. As these intelligent agents continue to evolve, their potential to positively impact how we perceive and engage with transportation services is significant, paving the way for a more connected, user-friendly, and secure future of mobility [30].

10.5.1.1 IMPROVED USER EXPERIENCE

1. **Personalized Interactions:** Chatbots offer NLP, enabling personalized and context-aware interactions, making the user experience more intuitive.

2. **Real-Time Assistance:** Users can receive real-time updates, travel recommendations, and assistance, enhancing the overall convenience and satisfaction of the travel experience.

10.5.1.2 INCREASED EFFICIENCY

1. **Automation of Routine Tasks:** Chatbots automate mundane tasks such as trip planning, ticket booking, and providing traffic information, saving users time and effort.
2. **Optimized Routes:** These intelligent agents optimize travel routes, contributing to more efficient navigation through urban environments and reducing travel times [31].

10.5.1.3 ENHANCED SAFETY

1. **Proactive Communication:** Chatbots provide proactive communication, offering real-time information about potential hazards, road conditions, and emergencies, contributing to safer decision-making.
2. **Emergency Assistance:** In emergency situations, chatbots can facilitate communication, helping users access timely assistance and ensuring a safety net within the future mobility landscape.

10.5.1.4 ACCESSIBLE INFORMATION

1. **On-Demand Information:** Users can access information on demand, such as transit schedules, service availability, and safety guidelines, fostering a more informed and empowered travel experience [32].

10.5.1.5 SEAMLESS MULTI-MODAL INTEGRATION

1. **Coordination of Different Modes:** Chatbots facilitate seamless integration of various transportation modes, allowing users to plan and coordinate journeys that involve multiple modes of transport effortlessly.

10.5.1.6 TIME AND COST SAVINGS

1. **Streamlined Processes:** By streamlining processes and reducing manual interventions, chatbots contribute to time savings and operational efficiency, resulting in cost savings for both users and service providers.

10.5.1.7 ADAPTIVE LEARNING

1. **Continuous Improvement:** Through ML, chatbots continuously learn from user interactions, adapting and improving their responses over time, ensuring an evolving and optimized user experience [33].

10.5.1.8 24/7 AVAILABILITY

1. **Round-the-Clock Support:** Chatbots provide round-the-clock availability, offering support and information at any time, contributing to the accessibility and reliability of future mobility services.

10.5.1.9 ENHANCED COMMUNICATION IN AUTONOMOUS VEHICLES

1. **In-Vehicle Assistance:** Inside autonomous vehicles, chatbots can serve as intelligent assistants, enhancing communication between passengers and the vehicle, contributing to a safer and more user-friendly in-vehicle experience.

10.5.1.10 USER EMPOWERMENT

1. **Empowering Users:** Chatbots empower users by providing them with information and tools to make informed decisions, fostering a sense of control and confidence in utilizing future mobility services.

10.5.2 CHALLENGES AND LIMITATIONS: PRIVACY CONCERNS, TECHNICAL CONSTRAINTS AND USER ACCEPTANCE

The chatbots in future mobility, while promising transformative benefits, come with their set of challenges and limitations that must be addressed for widespread adoption and seamless functionality [34].

10.5.2.1 PRIVACY CONCERNS

1. **Data Security:** Privacy concerns arise from the collection and processing of sensitive user data. Ensuring robust data security measures is crucial to safeguard user information from unauthorized access or misuse.

10.5.2.2 TECHNICAL CONSTRAINTS

1. **Integration Complexity:** Integrating chatbots into existing transportation infrastructures and technologies may pose technical challenges [35]. Ensuring compatibility and seamless integration with diverse systems requires careful planning and execution.
2. **Technical Failures:** Technical constraints such as server outages, software glitches, or connectivity issues can disrupt the functioning of chatbots, affecting the reliability of services provided.

10.5.2.3 USER ACCEPTANCE

1. **Trust Building:** Convincing users to trust and accept chatbots as reliable sources of information and assistance poses a significant challenge. Transparency about how chatbots operate and addressing concerns about data security and privacy are crucial for building user trust [36].
2. **Resistance to Automation:** Some users may resist the shift towards automated services, preferring human interactions. Overcoming this resistance requires effective communication about the advantages and reliability of chatbot-assisted services.

10.5.2.4 ETHICAL CONSIDERATIONS

1. **Bias in Algorithms:** Chatbots, powered by ML, can inadvertently perpetuate biases present in their training data. Addressing and mitigating biases in algorithms is essential to ensure fair and unbiased interactions.
2. **Ethical Decision-Making:** Ethical considerations arise when chatbots are involved in decision-making processes, especially in critical situations. Striking a balance between automated decision-making and human oversight is crucial.

10.5.2.5 LIMITED CONTEXT UNDERSTANDING

1. **Contextual Limitations:** Chatbots may struggle to fully understand the context of complex or nuanced user queries, leading to inaccuracies in responses. Improving contextual understanding is an ongoing challenge for enhancing the effectiveness of chatbot interactions [37].

10.5.2.6 COMMUNICATION GAPS

1. **Ambiguity Handling:** Understanding ambiguous queries or requests can be challenging for chatbots. Developing more sophisticated algorithms for handling ambiguity is necessary to improve communication effectiveness [38].

10.5.2.7 LACK OF EMOTIONAL INTELLIGENCE

1. **Emotional Understanding:** Chatbots currently lack the ability to understand and respond to user emotions effectively. Incorporating emotional intelligence into their design could enhance user interactions, particularly in situations requiring empathy or sensitivity.

10.5.2.8 CONTINUOUS LEARNING

1. **Learning from User Feedback:** Ensuring that chatbots can effectively learn from user feedback and adapt to evolving user needs is

a challenge. Continuous learning mechanisms must be robust and responsive to provide ongoing improvements in performance.

Addressing these challenges requires a collaborative effort from developers, policymakers, and users. Striking a balance between innovation and addressing concerns is vital for ensuring that chatbots in future mobility contribute positively to the user experience while upholding ethical standards and privacy considerations [39].

10.6 IMPLEMENTATION STRATEGIES: INSIGHTS FOR EFFECTIVELY IMPLEMENTATION MACHINE LEARNING AI-BASED CHATBOTS IN MOBILITY SOLUTIONS

Effectively implementing ML AI-based chatbots in mobility solutions requires a strategic and comprehensive approach by organizations [40]. Firstly, a thorough analysis of the existing mobility infrastructure is essential to identify areas where chatbots can add value. Understanding user needs and preferences is crucial for tailoring chatbots to meet specific demands within the mobility landscape. Strategic partnerships with reputable AI and ML providers can expedite the development process, leveraging external expertise and cutting-edge technologies.

A phased implementation strategy is advisable, beginning with a pilot program to assess the feasibility and effectiveness of chatbots in controlled environments before widespread deployment. This iterative approach allows for continuous refinement based on user feedback and evolving requirements. Customization is key; organizations must ensure that chatbots are designed to address the unique challenges and opportunities within the transportation sector, aligning the technology with organizational goals and user expectations.

Prioritizing data privacy and security is non-negotiable. Robust measures must be in place to protect user information and comply with relevant regulations, fostering user trust and confidence. Employee training is equally important, ensuring that staff members are proficient in utilizing and troubleshooting chatbot interactions to provide seamless user experiences [41].

Continuous monitoring and analytics play a pivotal role in assessing chatbot performance. Organizations should leverage data-driven insights to identify areas for improvement and optimize user engagement. Lastly, fostering a culture of innovation and adaptability is essential. The mobility landscape is dynamic, and organizations need to be responsive to emerging trends and user preferences [42]. By combining strategic planning, user-centric design, technology partnerships, and a commitment to ongoing improvement, organizations can successfully implement ML AI-based chatbots to enhance efficiency and user experience in their mobility solutions.

10.6.1 BEST PRACTICES AND POTENTIAL STRATEGIES: OVERCOMING IMPLEMENTATION BARRIERS

Implementing ML AI-based chatbots in mobility solutions is accompanied by various challenges, but organizations can navigate these barriers effectively by adopting best practices and strategic approaches. First and foremost, conducting a comprehensive assessment of the existing infrastructure and user needs is essential. Organizations should prioritize user-centric design, ensuring that chatbots are tailored to meet specific mobility requirements and enhance user experiences [43].

Strategic partnerships with reputable AI providers can accelerate the development process, leveraging external expertise and staying at the forefront of technological advancements. Employing a phased implementation strategy, starting with a pilot program, allows organizations to test and refine chatbot functionality in a controlled environment before full-scale deployment. Iterative testing based on user feedback and continuous improvement are crucial components of this approach.

Data privacy and security are paramount considerations. Implementing robust measures to safeguard user information, complying with regulations, and fostering transparency in data handling practices build trust among users [44]. Comprehensive employee training programs ensure that staff members are adept at utilizing and troubleshooting chatbot interactions, contributing to a smoother user experience.

Additionally, continuous monitoring and analytics play a key role in identifying performance gaps and areas for enhancement. Organizations should leverage data-driven insights to make informed decisions, optimize chatbot functionality, and address evolving user needs. Furthermore, fostering a culture of adaptability and innovation within the organization encourages the proactive identification of emerging trends and the rapid incorporation of technological advancements. By embracing these best practices, organizations can successfully implement ML AI-based chatbots in mobility solutions, creating a seamless and efficient user experience [45].

10.7 CONCLUSION AND FUTURE SCOPE

The implementation of ML AI-based chatbots in mobility solutions presents a transformative paradigm for the future of transportation. The fusion of AI with chatbot technology offers unparalleled opportunities to enhance user experiences, increase efficiency, and bolster safety across diverse aspects of mobility services. As organizations navigate the challenges of integration, emphasizing user-centric design, strategic partnerships, and robust data security measures will be pivotal in ensuring successful deployments.

The future scope of ML AI-based chatbots in mobility is incredibly promising. Continued advancements in NLP, ML algorithms, and multimodal interfaces will further refine chatbot capabilities, enabling more sophisticated and context-aware interactions. Integrating chatbots seamlessly into the expanding ecosystem of smart cities and connected vehicles holds the potential to create a holistic and interconnected transportation infrastructure.

As ML models become more adept at understanding user behavior and preferences, chatbots will evolve into IVAs that not only provide information and assistance but also anticipate user needs. The integration of these chatbots with emerging technologies like autonomous vehicles, electric mobility, and smart transportation systems will redefine the landscape of urban mobility, making it more responsive, efficient, and sustainable. The journey ahead involves addressing ongoing challenges, including ethical considerations, data privacy, and the need for widespread user acceptance. Organizations must stay vigilant, adapting their strategies to align with evolving technological landscapes and regulatory frameworks.

The future of mobility lies at the intersection of AI and chatbot technology, where IVAs seamlessly navigate the complexities of urban transportation, offering users a more personalized, efficient, and enjoyable travel experience. As innovations continue to unfold, the collaborative efforts of technology developers, policymakers, and users will play a crucial role in shaping a future where ML AI-based chatbots are integral to the fabric of modern and connected transportation systems.

KEYWORDS

- **artificial intelligence**
- **intelligent virtual assistants**
- **machine learning**
- **natural language processing**
- **smart cities and connected vehicles**

REFERENCES

1. Burns, M. B., & Igou, A. (2019). "Alexa, write an audit opinion": Adopting intelligent virtual assistants in accounting workplaces. *Journal of Emerging Technologies in Accounting, 16*(1), 81–92.
2. Rawassizadeh, R., Sen, T., Kim, S. J., Meurisch, C., Keshavarz, H., Mühlhäuser, M., & Pazzani, M. (2019). Manifestation of virtual assistants and robots into daily life: Vision and challenges. *CCF Transactions on Pervasive Computing and Interaction, 1*, 163–174.
3. Austerjost, J., Porr, M., Riedel, N., Geier, D., Becker, T., Scheper, T., & Beutel, S. (2018). Introducing a virtual assistant to the lab: A voice user interface for the intuitive control of laboratory instruments. *SLAS Technology: Translating Life Sciences Innovation, 23*(5), 476–482.
4. Soofastaei, A. (2021). Introductory chapter: Virtual assistants. In A. Soofastaei (Ed.), *Virtual Assistant* (pp. 1–9). London, United Kingdom: IntechOpen.
5. Singh, B. (2024). Legal dynamics lensing metaverse crafted for videogame industry and e-sports: Phenomenological exploration catalyst complexity and future. *Journal of Intellectual Property Rights Law, 7*(1), 8–14.
6. Uğurlu, Y., & Karabulut, M. (2021). Artificial intelligence supported Turkish university virtual assistant. In *2021 Innovations in Intelligent Systems and Applications Conference (ASYU)* (pp. 1–5). IEEE.
7. Weber, M., Weiss, T., Gechter, F., & Kriesten, R. (2023). Approach for improved development of advanced driver assistance systems for future smart mobility concepts. *Autonomous Intelligent Systems, 3*(1), 2.
8. Sarker, I. H., Hoque, M. M., Uddin, M. K., & Alsanoosy, T. (2021). Mobile data science and intelligent apps: Concepts, AI-based modeling and research directions. *Mobile Networks and Applications, 26*, 285–303.
9. Singh, B. (2023). Blockchain technology in renovating healthcare: Legal and future perspectives. In *Revolutionizing Healthcare Through Artificial Intelligence and Internet of Things Applications* (pp. 177–186). IGI Global.
10. Batura, A. (2019). *Integrating Virtual Assistant Technology into Omni-Channel Processes.*

11. Jones, V. K. (2018). Voice-activated change: Marketing in the age of artificial intelligence and virtual assistants. *Journal of Brand Strategy, 7*(3), 233–245.
12. Page, L. C., & Gehlbach, H. (2017). How an artificially intelligent virtual assistant helps students navigate the road to college. *AERA Open, 3*(4), 2332858417749220.
13. Kamoonpuri, S. Z., & Sengar, A. (2023). Hi, may AI help you? An analysis of the barriers impeding the implementation and use of artificial intelligence-enabled virtual assistants in retail. *Journal of Retailing and Consumer Services, 72*, 103258.
14. Singh, B. (2023). Federated learning for envision future trajectory smart transport system for climate preservation and smart green planet: Insights into global governance and SDG-9 (industry, innovation and infrastructure). *National Journal of Environmental Law, 6*(2), 6–17.
15. Perez Garcia, D. M., Saffon Lopez, S., & Donis, H. (2018). Everybody is talking about virtual assistants, but how are people really using them? In *Proceedings of the 32nd International BCS Human Computer Interaction Conference 32* (pp. 1–5).
16. Zhang, S., Meng, Z., Chen, B., Yang, X., & Zhao, X. (2021). Motivation, social emotion, and the acceptance of artificial intelligence virtual assistants—Trust-based mediating effects. *Frontiers in Psychology, 12*, 728495.
17. Sharma, A., & Singh, B. (2022). Measuring impact of e-commerce on small-scale business: A systematic review. *Journal of Corporate Governance and International Business Law, 5*(1).
18. Arora, S., Athavale, V. A., Maggu, H., & Agarwal, A. (2020). Artificial intelligence and virtual assistant—Working model. In *Mobile Radio Communications and 5G Networks: Proceedings of MRCN 2020* (pp. 163–171). Singapore: Springer Singapore.
19. Jain, S., Joshi, S., & Parashar, R. (2021). ALPHA: Personal assistant in chat box. *International Journal of Scientific Research in Computer Science Engineering and Information Technology.*
20. Ciappelloni, R., Avellini, L., Marenzoni, L., & Serenelli, F. (2020). *Smart Chatbots as Virtual Assistants in the Library.* Sanità Pubblica Veterinaria.
21. Singh, B. (2022). Understanding legal frameworks concerning transgender healthcare in the age of dynamism. *Electronic Journal of Social and Strategic Studies, 3*, 56–65.
22. Verma, M. (2023). Novel study on AI-based chatbot (ChatGPT) impacts on the traditional library management. *International Journal of Trend in Scientific Research and Development, 7*(1), 961–964.
23. Dongbo, M., Miniaoui, S., Fen, L., Althubiti, S. A., & Alsenani, T. R. (2023). Intelligent chatbot interaction system capable of sentimental analysis using hybrid machine learning algorithms. *Information Processing & Management, 60*(5), 103440.
24. Okuda, T., & Shoda, S. (2018). AI-based chatbot service for the financial industry. *Fujitsu Scientific and Technical Journal, 54*(2), 4–8.
25. Singh, B. (2022). Relevance of agriculture-nutrition linkage for human healthcare: A conceptual legal framework of implication and pathways. *Justice and Law Bulletin, 1*(1), 44–49.
26. Rawat, B., Bist, A. S., Rahardja, U., Aini, Q., & Sanjaya, Y. P. A. (2022). Recent deep learning-based NLP techniques for chatbot development: An exhaustive survey. In *2022 10th International Conference on Cyber and IT Service Management (CITSM)* (pp. 1–4). IEEE.

27. Maedche, A., Legner, C., Benlian, A., Berger, B., Gimpel, H., Hess, T., & Söllner, M. (2019). AI-based digital assistants: Opportunities, threats, and research perspectives. *Business & Information Systems Engineering, 61*, 535–544.
28. Hamdoun, S., Monteleone, R., Bookman, T., & Michael, K. (2023). AI-based and digital mental health apps: Balancing need and risk. *IEEE Technology and Society Magazine, 42*(1), 25–36.
29. Adam, M., Wessel, M., & Benlian, A. (2021). AI-based chatbots in customer service and their effects on user compliance. *Electronic Markets, 31*(2), 427–445.
30. Singh, B. (2022). COVID-19 pandemic and public healthcare: Endless downward spiral or solution via rapid legal and health services implementation with patient monitoring program. *Justice and Law Bulletin, 1*(1), 1–7.
31. Mantravadi, S., Jansson, A. D., & Møller, C. (2020). User-friendly MES interfaces: Recommendations for an AI-based chatbot assistance in Industry 4.0 shop floors. In *Asian Conference on Intelligent Information and Database Systems* (pp. 189–201). Cham: Springer International Publishing.
32. Nadarzynski, T., Puentes, V., Pawlak, I., Mendes, T., Montgomery, I., Bayley, J., & Newman, C. (2021). Barriers and facilitators to engagement with artificial intelligence (AI)-based chatbots for sexual and reproductive health advice: A qualitative analysis. *Sexual Health, 18*(5), 385–393.
33. Paliwal, S., Bharti, V., & Mishra, A. K. (2020). AI chatbots: Transforming the digital world. In *Recent Trends and Advances in Artificial Intelligence and Internet of Things* (pp. 455–482).
34. Singh, B. (2020). Global science and jurisprudential approach concerning healthcare and illness. *Indian Journal of Health and Medical Law, 3*(1), 7–13.
35. Maher, S. K., Bhable, S. G., Lahase, A. R., & Nimbhore, S. S. (2022). AI and deep learning-driven chatbots: A comprehensive analysis and application trends. In *2022 6th International Conference on Intelligent Computing and Control Systems (ICICCS)* (pp. 994–998). IEEE.
36. Liu, L., & Duffy, V. G. (2023). Exploring the future development of artificial intelligence (AI) applications in chatbots: A bibliometric analysis. *International Journal of Social Robotics, 15*(5), 703–716.
37. Singh, B. (2019). Profiling public healthcare: A comparative analysis based on the multidimensional healthcare management and legal approach. *Indian Journal of Health and Medical Law, 2*(2), 1–5.
38. Khan, H. A., & Walcott, T. H. (2023). The role of chatbots in Industry 4.0. In *2023 International Conference on Computing, Electronics & Communications Engineering (iCCECE)* (pp. 85–88). IEEE.
39. Nadarzynski, T., Miles, O., Cowie, A., & Ridge, D. (2019). Acceptability of artificial intelligence (AI)-led chatbot services in healthcare: A mixed-methods study. *Digital Health, 5*, 2055207619871808.
40. Hocutt, D., Ranade, N., & Verhulsdonck, G. (2022). Localizing content: The roles of technical & professional communicators and machine learning in personalized chatbot responses. *Technical Communication, 69*(4), 114–131.
41. Pandey, S., Sharma, S., & Wazir, S. (2022). Mental healthcare chatbot based on natural language processing and deep learning approaches: Ted the therapist. *International Journal of Information Technology, 14*(7), 3757–3766.

42. Zierau, N., Hausch, M., Bruhin, O., & Söllner, M. (2020). Towards developing trust-supporting design features for AI-based chatbots in customer service. In *ICIS* (Vol. *2020*, pp. 1–9).
43. Ayanouz, S., Abdelhakim, B. A., & Benhmed, M. (2020). A smart chatbot architecture based on NLP and machine learning for healthcare assistance. In *Proceedings of the 3rd International Conference on Networking, Information Systems & Security* (pp. 1–6).
44. Sharma, M., & Kumari, A. (2023). AI-based deep learning chatbot for career and personal mentorship. In *2023 IEEE 3rd International Conference on Technology, Engineering, Management for Societal Impact using Marketing, Entrepreneurship and Talent (TEMSMET)* (pp. 1–6). IEEE.
45. Maroengsit, W., Piyakulpinyo, T., Phonyiam, K., Pongnumkul, S., Chaovalit, P., & Theeramunkong, T. (2019). A survey on evaluation methods for chatbots. In *Proceedings of the 2019 7th International Conference on Information and Education Technology* (pp. 111–119).

CHAPTER 11

FedNLP: Secure and Efficient Federated Learning Using NLP (Architecture, Challenges, and Solutions)

EKTA BHAGGI[1,*], VARSHA[2], KOMAL[2], and MAMTA DEVI[3]

[1]*GNA University, Phagwara, Punjab, India*

[2]*School of Computer Science and Engineering, Lovely Professional University, Punjab, India*

[3]*CT Group of Institutions, Punjab, India*

**Corresponding author.*

ABSTRACT

The use of federated learning (FL), which prioritizes privacy protection, has grown significantly in recent years. FL is now a well-known paradigm for training machine learning (ML) systems that span distributed, large-scale contexts. NLP has undergone a revolution thanks to transformer-based pre-trained models, which offer better performance and generality. Federated learning is the de facto method for fine-tuning pre-trained models for downstream tasks, which frequently require private data (FedNLP). Nevertheless, our observations reveal that the massive model sizes and associated high network/computation costs make FedNLP unreasonably slow. The main building blocks for practical FedNLP are adapters, which are modest bottleneck modules introduced at different model layers. The correct configuration of adapter depth and width significantly affects training efficiency and speed. Instead of requiring data to be centralized at the training location, as is the case with traditional ML techniques, FL makes

Natural Language Processing and Artificial Intelligence: A Perspective Toward Current Trends, Challenges, and Applications. Vikas Khullar, Aryan Chaudhary, & Sunil Kumar (Eds.)

DOI: 10.1201/9781998511273-11

use of the computational power of several edge devices to train a global model jointly. Importantly, this is done while maintaining the privacy of the local data on each device. Federated learning is a distributed machine-learning technique in which several computers or servers cooperate to train a single model. Federated learning delivers the model to the data on individual devices or servers, in contrast to conventional systems where data is routed to a central server. Thus, data privacy and security are guaranteed as each device or server can participate in the model's training using its local data. The central server acts as an orchestrator in a typical FL arrangement. It gathers and combines the local models that each client has trained using their unique private data, iteratively until completion. Comparing this decentralized approach to conventional ML techniques, there are many benefits. This chapter provides information about what federated learning is, its types, why to use it, the working of federated learning, and some privacy challenges and issues along with some of the solutions in federated learning.

11.1 INTRODUCTION

With the arrival of artificial intelligence (AI) in 1956, it has had a profound effect on human beings. Diverse industries have adopted intelligent solutions as major improvements in AI technology have been made in recent years. Academic research has focused on technical issues like models, processing power, chip performance, and other technological aspects throughout the evolution of AI. These initiatives support the continued development of AI technology. Extensive training on significant amounts of real data is required to develop cognitive abilities comparable to human reasoning. The inability of AI to gain user confidence, gather sensitive data, and grow efficiently is hampered by issues such as cloud computing limits, data security flaws, and data segmentation. Therefore, the need for a practical and effective strategy to allay these worries and restore AI's vibrancy is urgent. Against this backdrop, the notion of "federated learning (FL)" has emerged. The search for reliable models that can extract insights from enormous and varied datasets meets a serious hurdle in the quickly developing fields of machine learning (ML) and AI. The complex problem of balancing data privacy considerations with desired model performance is at the center of this dilemma. Conventional approaches that entail centralizing data for training face significant challenges, from privacy concerns to real-world difficulties like isolated data repositories and communication barriers. The emergence of FL presents a revolutionary paradigm in response to these problems. This innovative method efficiently addresses the aforementioned issues by fostering collaborative and privacy-focused ML across decentralized devices and distributed data sources. A change in our ML strategy may be seen in the old idea of replacing centralizing data in one place with a decentralized and collaborative training approach in federated learning. Through the use of mobile phones, edge servers, and internet of things (IoT) devices, this paradigm uses an interactive methodology to improve and update ML models collectively. A key factor is that federated learning accomplishes this while maintaining the delicate nature of localized data, protecting the privacy of sensitive data on the device. The fundamental idea behind federated learning is sophisticated yet revolutionary. Federated learning takes the model to distributed data sources rather than transporting the data to a centralized server for training. Each device independently trains on its local data, sending updates to the models to a central server for aggregation. The global model develops through iterative communication and aggregation, including learnings from the many data landscapes across participating devices. Federated learning has enormous potential in this era of growing privacy concerns and the proliferation of data-generating gadgets. It addresses problems that affect a wide range of sectors, including industry, banking, and transportation. Federated learning stands out as a versatile solution that balances the need for both data-driven insights and privacy compliance as enterprises and researchers struggle with it. It fosters collaboration while maintaining anonymity.

To let Android smartphone users locally update models without jeopardizing their data, Google created the idea of FL in 2016. Google subsequently developed a functional FL system that was suited to particular purposes. This specifically created FL system focused on using federated average (FedAvg) algorithms on smartphones to support federated analytics. It also found use in monitoring statistics for large-scale cluster equipment without transferring raw device data to cloud servers. In 2016, Google researchers published a paper titled "Communication-Efficient Learning of Deep Networks from Decentralized Data," in which they first suggested the idea of federated learning. The paper was presented at the 2016 neural information processing systems (NIPS) conference. The necessity to solve problems caused by the centralized nature of conventional ML models led to the development of federated learning. Data is gathered from numerous sources and transmitted to a central server for model training in centralized ML. However, this method poses issues with data security, privacy, and the significant communication and bandwidth needs when working with large amounts of data dispersed among several devices. The idea of federated learning was developed as an answer to these problems. The goal was to make it possible for ML models to be jointly trained across decentralized devices while maintaining local data. Federated learning provided a technique to execute ML without sacrificing individual data privacy. It allowed devices to train models on their local data and exchange only model updates rather than raw data. This decentralized strategy was in line with the rising popularity of edge devices and the growing concern over difficulties with data privacy. Federated learning emerged as a method that may unleash the ML capabilities of mobile phones, IoT devices, and other edge devices while protecting user privacy. Fundamentally, federated learning was developed as a solution to the requirement for privacy-preserving ML that could utilize the dispersed capability of edge devices without compromising data security.

The use of FL, which prioritizes privacy protection, has grown significantly in recent years. FL is now a well-known paradigm for training ML systems that span distributed, large-scale contexts. Instead of requiring data to be centralized at the training location, as is the case with traditional ML techniques, FL makes use of the computational power of several edge devices to train a global model jointly. Importantly, this is done while maintaining the privacy of the local data on each device. Federated learning is a distributed ML technique in which several computers or servers cooperate to train a single model. Federated learning delivers the model to the data on individual devices or servers, in contrast to conventional systems where data is routed to a central server. Thus, data privacy and security are guaranteed, as each device or server can participate in the model's training using its local data. The central server acts as an orchestrator in a typical FL arrangement. It gathers and combines the local models that each client has trained using their unique private data, iteratively until completion. Comparing this decentralized approach to conventional ML techniques, there are many benefits.

Figure 11.1 shows the overall FL process as follows: First, each client downloads the initial common model from the central server; second, each client trains the model using their local data until the local model converges; and finally, each client returns the model to the central server. Third, the client encrypts the model parameters before sending them to the server. Finally, the server runs the FL algorithm on the uploaded gradients or parameters to update the global model for the next training. The client and server communicate indefinitely while looping through the preceding process until the global federated model converges.

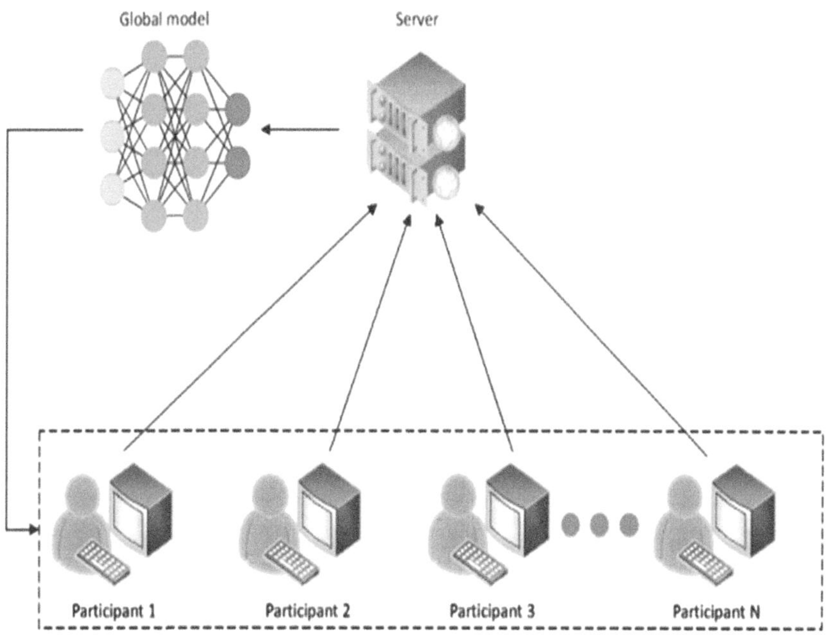

FIGURE 11.1 Federated learning.

Modern NLP models demonstrate interesting use cases on client devices, especially with the recent rise of transformers and their variants. Sentiment analysis, quality assurance, and auto-completion are some examples. It has been shown that inference with the NLP models is economical on client devices through careful engineering.

The training process of the current NLP is what makes it so successful, as Figure 11.2 shows. (i) a model is initialized on sizable text corpora during pre-training. Self-supervised and labor-intensive, the training frequently requires hundreds or even thousands of GPU days. Pre-training instructs the model on the innate structure of a language, such as word distribution; and (ii) a pre-trained model is further adjusted during the fine-tuning phase for a particular NLP job aimed at a specific domain, such as the task of classifying user emails' attitudes (a domain). Modern NLP training cannot function without fine-tuning since it is the only way the model connects the knowledge of generic language to the outputs for rich NLP tasks.

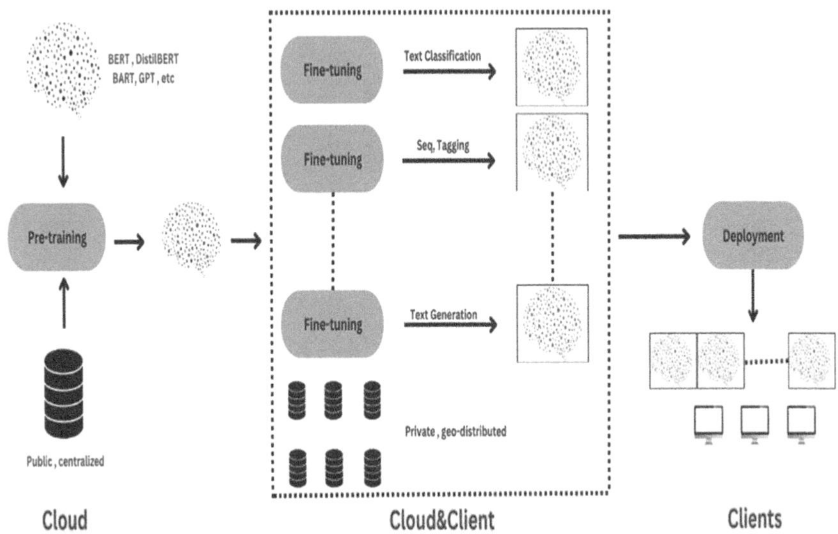

FIGURE 11.2 Architecture of natural language processing (NLP).

Different types of data are required for the two NLP training steps. While domain-specific samples like user reviews, chats, or emails are needed for fine-tuning, pre-training is usually conducted on publicly accessible text corpora like Wikipedia articles. These samples are continuously created by end-users in personal computing, dispersed across client devices, and frequently regarded as private. The de facto method for fine-tuning models using private, dispersed data is federated learning. A cloud service chooses several clients to take part in training for a certain NLP task and domain. Using its private data, a client trains a local copy of the model and updates it remotely over the cloud. The cloud provides the customers with an updated model after compiling model changes from several clients. Hundreds of thousands of rounds of training are usually conducted until the model's accuracy converges to the target level. Thus, the key NLP task in personal computing that is the subject of this study is NLP model fine-tuning in a federated system.

11.2 LITERATURE REVIEW

In Society 5.0, models and methods are utilized to predict a movie's rating. To validate the accuracy of the results, decision trees, clustering (K-means) analysis, extraction, transformation, and loading (ETL) processes are employed. For data visualization, graphs and charts are used. Society 5.0 is a human-centered society that employs technological innovation to address social issues and promote economic progress. The proposed model indicates that a movie's rating is influenced by factors such as its premise, budget, and performance. This approach can assist in elevating the economic performance of the entertainment industry and improve the predictability of movie ratings, bringing Society 5.0 closer to reality [1].

With the introduction of FL, a secure distributed approach created to address the difficulties presented by data silos in building unified models, the ML landscape has undergone a paradigm shift. FL takes full advantage of both the benefits of safe aggregation techniques and its unique decentralized training process, making it especially well-suited for a variety of practical applications that require strict privacy protocols. However, as FL moves from theoretical ideas to practical implementations, several training process bottlenecks have appeared that impact the effectiveness and performance of FL models in real-world circumstances. Therefore, researchers have focused on these issues, trying to develop efficient techniques to get around these roadblocks [2].

The ML environment has been greatly impacted by the emergence of FL, a new paradigm that solves ML problems by coordinating efforts across numerous clients under the direction of a central aggregator. This strategy guarantees the protection of personal device privacy by decentralizing the distribution of training data. This paper conducts a thorough investigation of federated learning systems with a focused attention on their use in the healthcare industry. FL's frameworks, structures, and applications are carefully examined to reveal its capabilities. The ability of FL to overcome obstacles by utilizing a shared global deep learning (DL) model managed by a central aggregator server is essential to FL's efficacy. This work engages with recent breakthroughs in FL research, taking into account the field's explosive expansion, to compile a comprehensive list of outstanding problems that call for additional research. The topic covers a variety of privacy-preserving methods, including differential privacy, homomorphic encryption, secure multiparty computation, and stochastic gradient descent. These techniques are placed within the context of FL, highlighting the importance of maintaining data confidentiality. This review also explores federated TL as well as several FL categories, including horizontal and vertical FL. Using examples from wireless communication, service recommendations, intelligent medical diagnosis systems, and healthcare, the research highlights FL's adaptability. These various fields demonstrate FL's applicability and flexibility in handling challenging issues. This study of the literature is an essential tool that offers a thorough investigation of the importance, workings, and applications of Federated Learning, with a focus on the medical field. The review helps us understand FL's transformative potential and the directions it will shape in the evolving field of ML and data-driven applications through its comprehensive analysis of frameworks, architectures, privacy methodologies, classifications, applications, challenges, and future trajectories [3].

The proliferation of intelligent services powered by AI coincides with the IoT ubiquitous infiltration into various facets of our daily lives. In the past, centralized data aggregation and processing were required for AI methods. However, this method is unworkable in many real-world application settings due to the enormous scalability of modern IoT networks and growing concerns around data privacy. FL, a distributed, cooperative AI paradigm that can support a wide range of intelligent IoT applications, has developed as a solution to these problems. This is accomplished by enabling AI training on dispersed IoT devices without the need to share raw data. This paper provides a thorough investigation of the expanding FL applications within IoT networks [4].

When doing health research, maintaining privacy is of utmost importance, necessitating the use of cutting-edge protection techniques. Studies have typically used centralized data repositories, which may have exposed the private data that underlies the analysis. However, recent developments in federated learning have opened up new possibilities by making it possible to build complex machine-learned models through distributed training. These developments make it easier to calculate research study endpoints while ensuring that personal information stays on particular devices or within healthcare systems. In this review, we examine the transformative effects of federated learning on a variety of single- and multi-site studies in the field of health research. This innovative method shows that it is possible to achieve accuracy, precision, and generalizability levels that are comparable to those of conventional centralized statistical models. Additionally, it produces consistent interpretations while considerably enhancing privacy protections and adding only a small amount of computational overhead. The transformative potential of federated learning in health research is highlighted by this literature review. It attests to the capability of federated models to safeguard privacy while retaining the quality and integrity of research by demonstrating its application across a variety of research paradigms. This work is a key contribution that opens the door for moral, efficient, and privacy-conscious health research as science and privacy collide through the lens of federated learning [5].

This study introduces FedIns, a unique FL technique that enables instance-specific inference within the FL framework to solve the issue of intra-client data heterogeneity. The suggested method uses a parameter-efficient fine-tuning technique known as "scale and shift deep features" (SSF) on a pre-trained model instead of deploying extensive instance-specific models. The method involves training an SSF pool for each client and then aggregating these SSF pools on the server side to maintain minimal communication overhead. The algorithm dynamically chooses the best-matched SSF subsets from the pool for a specific instance to enable instance-specific inference. After that, these subsets are combined to create an adaptive SSF specific to the instance, which reduces both intra-client and inter-client heterogeneity [6].

Language is used as a means of group communication, and accent refers to the distinctive way that individuals or groups pronounce words and phrases in that language. The goal of this research project is to develop a privacy-focused intervention built on federated learning. This intervention aims to give people the ability to recognize a system that can improve their English accents. The technique used entails using the AccentDB dataset to train and validate a DL system. This exact paradigm is subsequently translated into a Federated Learning Ecosystem. To determine the model's effectiveness in data-constrained contexts, the research includes data distribution scenarios, including identical and independent distributions. The findings section focuses on a thorough examination of the suggested implementations, analyzing their actual use and impact. The trained model is also added to mobile and web applications, making it easier to get English accent instruction that is specifically tailored to your needs. Investigating incremental learning made possible by federated learning is being done in parallel. The rating is based on important factors including memory, accuracy, precision, and loss. This groundbreaking approach has the potential to be incorporated into a variety of applications in the future, magnifying the improvement of effective communication skills. This research illustrates a path toward more proficient and inclusive language abilities by fusing the possibilities of federated learning with accent reduction [7].

In the context of IoT malware detection, this research program explores the revolutionary potential of federated learning. It acknowledges the security issues that are inherent in this new paradigm of learning. This framework introduces a novel method for malware detection in IoT devices that makes use of federated learning. The N-BaIoT dataset, which simulates network traffic from actual IoT devices impacted by malware, forms the basis for assessing this proposed architecture. Multi-layer perceptron and autoencoder models, as well as unsupervised and supervised federated models, are all trained and evaluated in this study. The N-BaIoT dataset shows that these algorithms can identify malware that affects both observed and unobserved IoT devices. Additionally, their performance is compared to two established approaches, one of which involves participants sharing data with a central organization for the training of global models, and the other which allows individuals to train local models using personal data. In conclusion, this thorough literature review highlights the crucial part federated learning plays in supporting IoT malware detection. The study advances the security landscape in IoT networks by tackling security issues and utilizing the potential of decentralized data training. The investigation of privacy-preserving methods and the assessment of model resilience highlight the complex interaction between technical advancement and security augmentation in the dynamic IoT environment [8].

The goal of the recommendation systems (RS) sector is to offer users precise and pertinent recommendations. The rating-based many-objective hybrid recommendation model is a novel strategy that has been presented to improve the accuracy and diversity of recommended outcomes. Accuracy, recall, diversity, novelty, and coverage are among the many variables that this model aims to simultaneously maximize. This is a considerable shift from single-objective optimization techniques, to provide customers with recommendations that are both accurate and take a larger range of preferences into account. This literature review emphasizes how RSs have advanced beyond simple precision and highlights the critical role played by hybrid multiple-objective optimization techniques. This method paves the way for a more thorough and customized recommendation experience by concurrently optimizing accuracy, recall, diversity, novelty, and coverage, which is further strengthened by cutting-edge evolutionary techniques. The algorithm's skill in producing recommendations that go beyond conventional measures is further demonstrated by the empirical validation, which ultimately results in increased user satisfaction and engagement [9].

As a collaborative and decentralized system, FL holds the promise of protecting privacy while reducing the problems caused by data silos and sensitive data. Both the field of industrial engineering and the trend of ongoing research that is fueling FL have been captured. This study launches a thorough investigation of FL, probing its history and the crucial route it travels to address the underlying problems in its development process. This project's major goal is to shed light on FL's current uses in the field of industrial engineering to prepare the ground for potential future deployment. This literature study captures federated learning's dynamic development, from its conception as a system that protects privacy to its eventual emergence as a paradigm-shifting force in academia and business. It highlights the importance of continuing research, outlines FL's evolutionary history, and emphasizes its uses. This study highlights FL's potential and the paths it is positioned to evolve in, adding to the larger conversation about FL's role in determining the future of industrial engineering [10].

The idea of FL has become increasingly well-known in recent years as a paradigm for training large-scale, distributed, and privacy-conscious ML systems. By deviating from traditional ML approaches, which call for data to be gathered at the exact site of training, FL makes use of the processing power provided by a variety of edge devices. While maintaining the privacy of each device's data, these devices work in harmony to train a common global model. The central server acts as an orchestrator in the context of a typical FL system. It manages the iterative process of gathering and aggregating local models until convergence is reached. Each client trains their local models using their private data. Although FL has many advantages over classical ML, such as its built-in protection of private data ownership, it is not without flaws. Managing the dangers posed by the centralized orchestration common to traditional FL client-server designs is one of its key concerns. These weaknesses include, among others, susceptibility to man-in-the-middle attacks and single points of failure. Decentralized FL solutions have arisen to reduce these possible hazards, enabling a collaborative environment where FL clients can communicate and work together without the need for a central server. This study provides a thorough synthesis and analysis of decentralized FL approaches now in use that have appeared in the academic literature. It also draws attention to emerging problems that have developed in this comparatively untapped field. The survey broadens its scope by highlighting attractive directions for additional study and emphasizing the changing environment of decentralized FL techniques. This survey thus not only consolidates the current level of knowledge but also offers academics looking to delve into this dynamic and expanding topic a roadmap [11].

FL is a groundbreaking privacy-focused ML technique that has the potential to completely transform the way intelligent models are trained using distributed data sources. Through the communication of local model updates throughout each iteration of the learning process, this novel approach enables the use of private data without disclosing raw data. Despite its benefits, a problem with sizable model update sizes, which contain a lot of parameters, arises, requiring lengthy communication rounds for efficient training. It is necessary to find a solution that takes these issues into account because this significant communication overhead burdens both clients and the environment. In response, FedKD, a federated learning approach created to be effective and efficient in communication, is introduced in this study. This strategy makes use of dynamic gradient compression and adaptive mutual knowledge distillation techniques to enable simplified communication without sacrificing efficacy. The effectiveness of FedKD is examined in three different situations when privacy protection is crucial. Surprisingly, it shows how communication costs could be cut by up to 94.89% while still producing competitive results on par with centralized model learning. This innovation has the potential to efficiently deploy privacy-preserving intelligent technologies in industries like healthcare and personalization, paving the way for a more efficient and sustainable future [12].

FL, which acts as a decentralized method for protecting users' private data, has become well-known as a crucial paradigm for learning, especially in light of the stricter enforcement of laws and regulations across countries.

As a result, a variety of FL frameworks have been developed to speed up the creation and use of this innovative strategy. Despite a sizable corpus of research devoted to the security and privacy of FL models and systems, a thorough investigation of the security issues inherent in FL frameworks themselves has not yet been conducted. The study's conclusion is revealed with the release of nine noteworthy discoveries, each of which has vital ramifications. Our discussion covers these results, elucidating their larger implications and revealing insights that can be helpful to both FL framework creators and security researchers. We aim to stimulate discussion and action within the FL framework landscape through our investigation, fostering a more stable and safer environment for the spread of federated learning solutions [13].

Nine key results with beneficial ramifications are revealed as the study's conclusion is unveiled. Our discussion of these discoveries explores their wider importance while revealing information that can be helpful to both FL framework developers and security researchers. By conducting this investigation, we hope to stimulate discussion and action within the FL framework ecosystem and promote a more stable and safer environment for the spread of federated learning solutions. This paper explores the subtleties of these difficulties and suggests a solution in the form of data augmentation. It focuses on how real-world IoT datasets can be used to augment data to prevent performance deterioration and improve detection accuracy in anomaly detection tasks. Challenges like the class balance problem and the device heterogeneity problem, which result from different event types across heterogeneous devices, are significant. These problems are caused by differences in the classes of events generated by participating devices. This study highlights the possibility of data augmentation as a tactic for overcoming difficulties in federated learning in the IoT environment. Data augmentation emerges as a potent method for extending the capabilities of anomaly detection in real-world IoT scenarios by successfully addressing performance deterioration and significantly enhancing detection accuracy [14].

FL for NLP has advanced significantly with the release of AutoFedNLP, making it more useful and effective. AutoFedNLP optimizes the utilization of client resources and adaptively configures model adapters to simplify the training process and drastically reduce computational overhead and convergence time. This discovery has the potential to significantly accelerate the development of numerous NLP applications.

Improved performance and versatility are two key benefits of transformer-based pre-trained models in NLP. FedNLP is a typical method for using private data to fine-tune these pre-trained models for specific tasks, which is often necessary. Nevertheless, the enormous model sizes and associated high network and computing costs have shown our measures to be extremely slow in FedNLP. We have found that a key component needed to make FedNLP more useful is called "adapters." These are small components that are incorporated into the model at different levels. Finding the ideal depth and width for these adapters is a significant task because it has a big impact on training effectiveness and speed. There is no one-size-fits-all setup; rather, the optimal option changes based on the particular NLP task at hand, the target model accuracy, and the client's resource availability. It provides AutoFedNLP, a system that adds two novel features to FedNLP to tackle this problem. During a training session, AutoFedNLP first progressively enhances the adapter setup. By training fewer and smaller adapters at the upper layers of the model, the objective is to quickly acquire fundamental knowledge, and by incorporating larger adapters in the lower layers, the model will gradually learn deeper knowledge. Second, by allocating participant devices to trial groups, AutoFedNLP continuously profiles possible adapter configurations [15–19].

Differential privacy (DP) is a commonly used technique in FL to protect model updates, as it has become an essential tool for guaranteeing bounds on privacy leaks during model training. However, every FL job with DP produces cumulative privacy leakage; hence, a unified platform must be implemented to enforce a global privacy budget for every user-owned dataset. This paper presents a novel DP-enhanced FL platform that effectively schedules many FL processes across sensitive data while considering privacy as a valued resource. The platform offers a novel data abstraction that allows fine-grained composition of privacy consumption across several FL processes by introducing the notion of "device-time blocks" for distributed data streams. Because the privacy resource is not replenishable, unlike more conventional hardware resources like CPU and RAM, the platform uses an "allocation-then-recycle" scheduling mechanism. For every incoming FL job, our method first allots an anticipated upper-bound privacy budget. Then, while training goes on, it gradually recycles the budget that isn't spent, making it accessible for more FL jobs. The platform's efficacy is demonstrated through extensive experiments. Under a restricted privacy budget, it produces up to 2.1 times more completed jobs while lowering the violation rate by up to 55.2%. This demonstrates the platform's capacity to optimize the allocation and recycling process to boost the effectiveness of DP-enhanced federated learning across various datasets while also managing privacy as a crucial and limited resource [20].

Many methods for training deep neural networks on decentralized data have been presented to strike a compromise between the necessity of protecting user privacy and the potential of mobile intelligence. Nevertheless, the complexity of neural network design is exacerbated in decentralized training, especially when customizing structures for various mobile platforms. In this study, a unique method for integrating automatic neural architecture search into decentralized training—federated neural architecture search, or federated NAS—is presented. In response to the problem of constrained on-client computing and communication capacity, the paper introduces DecNAS, a highly optimized framework created especially for effective federated NAS. DecNAS leverages the limited opportunity to re-train model candidates during the architecture search process by implementing three important optimizations: training candidates in parallel on partial clients, removing candidates that perform poorly early on, and dynamically adjusting round numbers. When compared to the most advanced NAS algorithms that use centralized data, DecNAS exhibits model accuracy that is comparable to that of large-scale datasets and standard CNN structures. In addition, it significantly lowers client expenses compared to a simple federated NAS solution by up to 200 times or more. This demonstrates how well DecNAS works to enable decentralized neural architecture search, improving the effectiveness of federated learning on mobile devices while taking privacy concerns into account [21, 22].

11.3 BASIC FORMULA FOR FEDERATED LEARNING (FL)

A novel and secure decentralized learning system called FL was created to solve the problem of remote clients contributing to a collective model without disclosing their raw data. A global model is created in this virtual paradigm to compile insights from every client that takes part. Then, each participant makes use of this global model to accomplish their local learning goals. FL's primary objective is to generate modeling outputs that closely mirror those from conventional centralized training models, which collect and handle input from several clients on a single server. All participants in a federated learning environment are expected to follow shared data policies and have the same identity.

Interestingly, what makes FL special is that model changes or gradients are sent in place of the raw data being physically transferred to a central server. Because only aggregated knowledge is shared and sensitive information stays on individual devices, this strategy protects data standards and user privacy. As a result, FL offers a solution for collaborative ML that protects privacy, making it ideal in situations where centralized data sharing or storage is unfeasible or raises privacy issues.

To begin, we define the essential concept in the context of FL. FL comprises N individuals, each contributing to collectively train a global model using their respective data P_1, P_2, …, P_N. Traditionally, all the data would be aggregated into a single dataset, indicated as total data $P = P_1$, P_2, …, P_N, which would then be used to train a model M_{sum} with a performance level of V_{sum}. FL, on the other hand, takes a different strategy. In FL, players work together to train a shared model, indicated as M_{fed}, with a performance level of V_{fed}, without revealing any private data to others. To measure the performance loss of the FL model, we introduce the non-negative parameter ε, and it is expressed as follows:

$$|V_{fed} - V_{sum}| < \varepsilon \tag{11.1}$$

The learning process in FL involves minimizing a loss function, which is computed independently on each client using a weighted aggregation method. The overarching objective of FL is to minimize the following specific objective function, denoted as:

$$\min f(w) = N\ k{=}1\ nk\ n\ Fk\ (w) \tag{11.2}$$

In this equation, N represents the total number of clients participating in the federated learning process. The term nk denotes the amount of data available on the k-th client, and Fk(w) represents the local objective function of the k-th client.

FL allows for a restricted range of performance differences between the trained global federated model and the centralized training model. Through numerous rounds of efficient federated global training, the global model's performance gradually increases and approaches that of the centralized model.

The overall FL process is as follows: First, each client downloads the initial common model from the central server; second, each client trains the model using their local data until the local model converges; and finally, each client returns the model to the central server. Third, the client encrypts the model parameters before sending them to the server. Finally, the server runs the FL algorithm on the uploaded gradients or parameters to update the global model for the next training. The client and server communicate indefinitely while looping through the preceding process until the global federated model converges.

11.4 WHY FEDERATED LEARNING

The emergence of FL as a paradigm shift in ML has raised its profile. Several primary factors drive its adoption:

1. **Privacy Preservation:** FL's main motivation is to safeguard sensitive and confidential data. Data privacy problems exist when traditional ML practices require centralizing data on a single server for model training. FL is perfect for companies and applications where data confidentiality is crucial because it enables training models without requiring the sharing of raw data.
2. **Decentralized Data:** In a world that is more and more connected, data is produced across many different devices and locations. By training models locally on devices instead of requiring data to be uploaded to a central server, FL makes use of this decentralized data. This guarantees data localization, lowering the possibility of data breaches and upholding data sovereignty laws.
3. **Scalability:** The growth of mobile devices, IoT sensors, and edge devices has produced enormous amounts of data. With FL, it is possible to combine knowledge from several sources and create sophisticated global models while preserving the locality of the data.
4. **Efficiency:** Traditional ML frequently results in high communication costs and bandwidth restrictions because it requires a significant volume of data to be delivered to a central server. By training models locally and collecting just model updates, FL lowers these communication costs, increasing productivity and easing network congestion.
5. **Collaborative Learning:** FL promotes cooperation across multiple entities, including businesses, hospitals, and individuals, all while maintaining data security. By combining knowledge from many

sources, this enables the development of more precise and reliable models.

6. **Real-time Learning:** FL enables continuous learning without the need for constant data transfer in situations where data is created quickly, such as in autonomous vehicles or real-time monitoring systems. This guarantees that models are current and useful in changing situations.
7. **Customization:** FL enables the creation of customizable models while maintaining the privacy of each user's data. This is particularly important in fields like healthcare, where customized therapies can be created without revealing private patient data.
8. **Regulation Compliance:** Innovative methods that respect user privacy are required by stricter rules like the general data protection regulation (GDPR) and health data protection laws. FL makes it possible to comply with these rules by limiting data exposure.
9. **Data Sovereignty:** FL permits model training without relocating the data in circumstances where data standards specify that certain data cannot leave a particular geographic area, ensuring compliance with data sovereignty obligations.
10. **Reduced Latency:** FL makes it possible for models to be trained on devices, which can greatly cut down on latency in applications where making decisions quickly is essential, such as industrial automation or autonomous vehicles.
11. **Distributed Data Diversity:** FL makes use of data from several sources to produce models that are more thorough and reliable. Applications with highly variable data features, like medical diagnoses across many hospitals, can benefit especially from this.
12. **Cost Savings:** Sending huge amounts of data to a central server for training comes at a high expense in terms of bandwidth and storage. The requirement for data transit is diminished by FL, which lowers the cost of data usage and storage.
13. **Greater Participation:** FL enables model training on devices with low processing power. Participation in ML processes is democratized through this.
14. **Global Model Improvement:** FL makes it possible to combine data from multiple sources and devices. Global models are produced that are more complete and precise because of this cooperative approach.

15. **Personalized Services:** FL makes it possible to build tailored models without storing sensitive personal data centrally. This helps RSs learn about user preferences while protecting the privacy of personal information.

Federated Learning is essentially a reaction to the shifting dynamics of connectivity, distributed computing, and data privacy. By enabling collaborative, privacy-preserving, and effective model training on decentralized data sources, it tackles the shortcomings of conventional ML techniques.

11.5 WORKING OF FEDERATED LEARNING (FL)

Model training is made possible through the decentralized, collaborative operation of FL, while data privacy is upheld. This is how FL operates:

1. **Initialization:** A central server initializes a global model to start the process. For training across many devices, this model acts as the starting point.
2. **Participating Devices:** Each client, or participating device, trains locally on its dataset without sharing the raw data. By using the data from the device, this local training entails updating the global model.
3. **Local Model Updates:** Using its data, the device updates the global model during local training. This procedure involves repeatedly executing training algorithms on the data from the device, such as stochastic gradient descent (the optimization procedure stochastic gradient descent is frequently used in ML applications to identify the model parameters that best match the expected and actual outputs).
4. **Model Aggregation:** After completing local training, the device sends the new model parameters (weights and biases) to the main server. The central server combines these model changes to create a more precise overall model. There are other ways to aggregate data, such as by averaging the model parameters.
5. **Iterative Process:** The steps of local training, model updates, and aggregation are repeated in an iterative process. By utilizing insights from several datasets without disclosing sensitive information, each iteration pushes the whole model closer to achieving optimal performance.

6. **Privacy Preservation:** Data privacy is protected at all times during the procedure. Devices only share model updates, which provide details about how the model should be updated, rather than sharing raw data. This shields private information from prying eyes.
7. **Model Improvement:** As the global model combines knowledge from numerous devices and data sources, it becomes more accurate and reliable with each iteration. FL's collaborative character aids in overcoming the difficulties brought on by non-I.I.D. data distribution (non-IID data distribution is common when data is generated by user devices).
8. **Convergence:** The iterative process continues until the overall model converges to a suitable performance level. Convergence is reached when additional iterations do not appreciably enhance the model's performance.
9. **Deployment:** The global model can be distributed to all involved devices for inference tasks once it has converged. The model can be used by devices to classify or forecast new data while retaining privacy.

Federated learning is appropriate for applications where data security, decentralization, and cooperation are essential because of its distributed and privacy-preserving nature. Since it allows for the collection of data from various sources without violating individual privacy, it is a method that holds promise in industries like healthcare, banking, and IoT.

With an example, Figure 11.3 demonstrates how federated learning functions. In this illustration, data from several machines is trained, and supplied to the federated server without impairing personal storage, and a global model is developed for usage in the future.

11.6 TYPES OF FEDERATED LEARNING

FL comes in various forms and modifications that can address different situations and problems. Here are a few typical examples:

1. **Horizontal Federated Learning:** With horizontal federated learning, devices with comparable features but distinct data assist in the development of a single, universal model. It is appropriate when data from different sources, such as cell phones containing data from many users, is shared across devices but has comparable features.

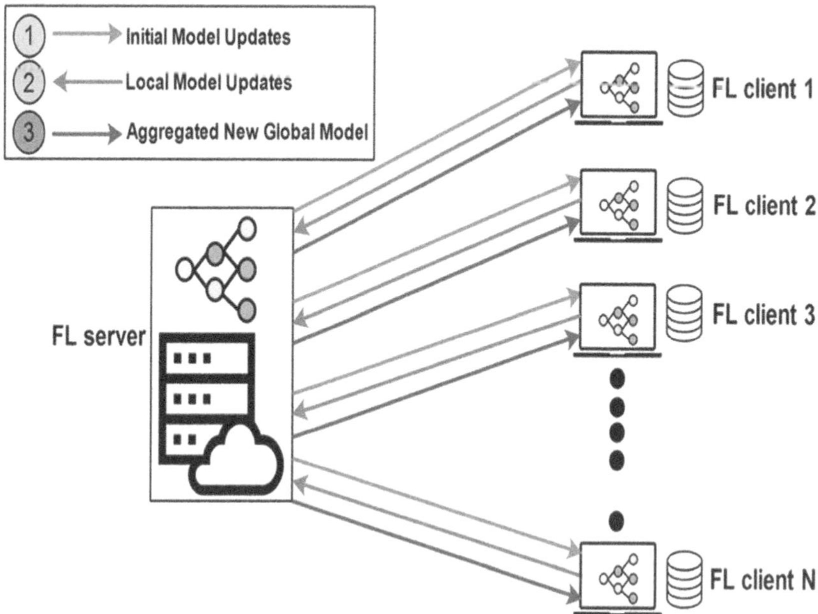

FIGURE 11.3 Working with federated learning.

2. **Vertical Federated Learning:** In vertical FL, devices that have distinct features but the same information contribute data. This can be helpful when data qualities are complementary, such as when merging medical records from several institutions with various patient attributes.
3. **Federated Transfer Learning:** This method is used when it is desired to transfer a pre-trained model from one domain to another while maintaining the decentralized nature of the data. The difficulty lies in effectively transferring information from one domain to another.
4. **Hierarchical Federated Learning:** This type uses a multi-level structure with clusters or groups of devices and aggregates model changes both within and between clusters. Scalability and hierarchical data distribution are managed.
5. **Cross-Silo Federated Learning:** In situations where data is divided into various silos (separate organizations or entities), cross-silo FL enables collaboration across these silos. It seeks to create models with data from many sources while observing privacy restrictions.

6. **Split Learning:** Split learning refers to a version in which the model is split into several segments, each of which is stored on a separate device. Due to collaborative learning still being possible, sensitive portions of the model are kept secret.
7. **Federated Meta-Learning:** The objective of this type is to create models that may change to fit new tasks or domains while requiring the least amount of data from each device. It is especially helpful when devices only have a small amount of data available yet still need to perform a variety of tasks.
8. **Asynchronous Federated Learning:** This method allows devices to update the model individually and submit updates to the central server asynchronously rather than syncing all updates at once. It accommodates devices with different connectivity.
9. **Secure Federated Learning:** To ensure that individual data is kept private in situations where data privacy is of the utmost importance, secure FL uses methods including differential privacy, secure MPC, and homomorphic encryption.
10. **Decentralized Federated Learning:** Decentralized FL is an extension of traditional federated learning that further enhances privacy and security by eliminating the need for a central server or orchestrator. In this approach, participating devices, often referred to as clients or nodes, collaborate directly with each other without relying on a central coordinator. This decentralized structure addresses some of the limitations and vulnerabilities associated with the centralized architecture.

Each kind of FL handles particular problems with data distribution, privacy, heterogeneity, and scalability, making them appropriate for a range of applications in numerous industries.

Figure 11.4 gives the view of three main types of federated learning, namely horizontal federated learning, vertical federated learning, and federated transfer learning (TL).

11.7 PROBLEMS WITH FEDERATED LEARNING PRIVACY AND SECURITY ALONG WITH SOLUTIONS

The implementation of this decentralized ML paradigm successfully depends on overcoming privacy and security issues in federated learning.

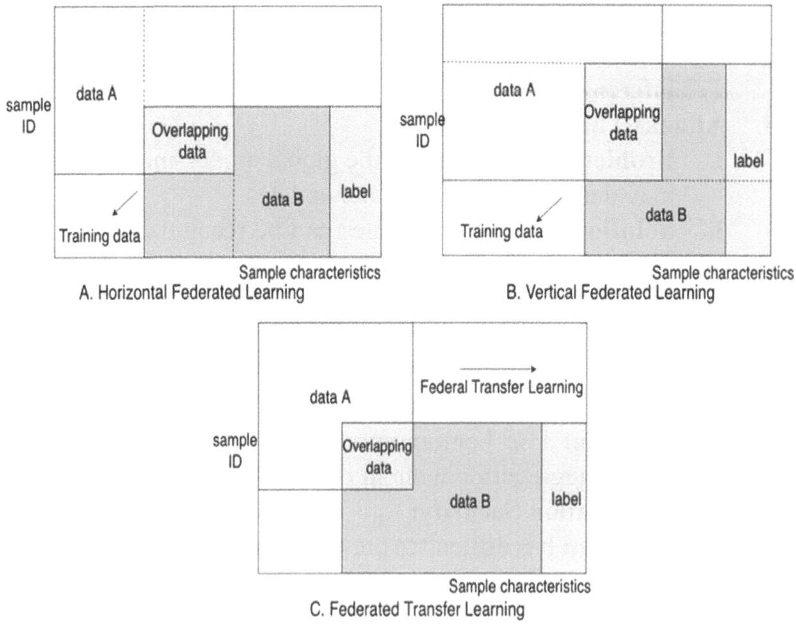

FIGURE 11.4 Types of federated learning.

While protecting the confidentiality of individual data sources, federated learning tries to make use of data from dispersed devices. The following list of significant privacy and security issues in federated learning, along with proposed remedies, includes:

1. **Data Protection:**
   i. **Problem:** Raw data is still stored on devices, lowering privacy hazards. Updates to the model could still expose private data, though.
   ii. **Solution:** To avoid inference of specific data points, techniques like differential privacy provide noise to model updates.
2. **Inversion Attacks on Models:**
   i. **Problem:** Adversaries could deduce critical training data by reverse-engineering model updates.
   ii. **Solution:** Integrate differential privacy techniques as a solution to stop attackers from identifying certain training instances.
3. **Attacks Using Membership Inference:**
   i. **Problem:** Attackers are attempting to ascertain whether a specific data point was used for training.

   ii. **Solution:** Techniques like privacy amplification and secure aggregation are used to reduce the dangers of membership inference.

4. **Attacks That Poison Data:**
   i. **Problem:** To manipulate the global model, malicious devices may deliver corrupted model updates.
   ii. **Solution:** Malicious updates can be recognized and rejected using robust aggregation techniques and anomaly detection algorithms.
5. **Securing Aggregates:**
   i. **Problem:** The difficulty is in aggregating model changes while maintaining their privacy.
   ii. **Solution:** Use homomorphic encryption or secure MPC to enable aggregation without disclosing specific updates.
6. **Communication Security:**
   i. **Problem:** It is difficult to prevent eavesdropping and tampering on communication channels between devices and the central server.
   ii. **Solution:** To maintain data confidentiality and integrity during transmission, encrypt communication using secure protocols like TLS.
7. **Watermarking a Model:**
   i. **Problem:** Authenticity and ownership of a model as it is shared among devices is a challenge.
   ii. **Solution:** The answer is to incorporate watermarks or cryptographic signatures into the model to verify its provenance and stop unwanted alterations.
8. **Fairness and Prejudice:**
   i. **Problem:** If specific groups of devices have unrepresentative data, federated learning may magnify bias.
   ii. **Solution:** To reduce bias, employ methods like federated sampling and fairness-aware aggregation procedures.
9. **Heterogeneity:**
   i. **Problem:** Devices may have different data distributions, which presents a challenge for model aggregation.
   ii. **Solution:** To tackle data heterogeneity, use federated meta-learning or adaptive aggregation techniques.

10. **Regulatory Conformity:**
    i. **Problem:** Using sensitive data while adhering to data protection laws like GDPR and HIPAA.
    ii. **Solution:** Design federated learning frameworks with privacy protections that adhere to legal standards as a solution.

A combination of cryptographic methods, developments in ML, and secure communication protocols is needed to tackle these problems. The goal of ongoing research and development is to balance model effectiveness, privacy, and security in federated learning settings.

11.8 APPLICATIONS OF FEDERATED LEARNING

With its many applications in a variety of sectors and its lack of centralized data sharing, FL enables cooperative and privacy-preserving ML. The following list of federated learning applications is accompanied by descriptions of how it is applied in each situation:

1. **Healthcare:** FL is essential to the analysis of medical data in the healthcare industry while protecting patient privacy. Collaboration between hospitals, clinics, and wearable technology is used to train models for disease prediction, individualized therapy suggestions, and medical image analysis. FL helps to comply with data privacy laws while enhancing healthcare outcomes by maintaining sensitive patient data locally.
2. **Financial Services:** FL enables market trend analysis, credit risk evaluation, and fraud detection in the financial sector. Without disclosing private transaction information, financial organizations can work together on model training. This increases the accuracy of detecting fraudulent actions while protecting data privacy.
3. **Smart Manufacturing:** FL is used in the manufacturing sector for quality assurance and proactive equipment maintenance. Production line sensors provide information to train models that anticipate equipment breakdowns and improve manufacturing procedures. FL provides instantaneous insights without disclosing confidential production data.
4. **Smart Cities:** By combining data from numerous sources, such as traffic cameras, weather sensors, and IoT devices, FL helps to

develop smart city solutions. Traffic flow, energy usage, garbage management, and other factors are all optimized using models. FL makes sure that private information, including people's movements, stays local.

5. **Telecommunications:** To optimize networks, forecast network congestion, and improve user experiences, telecommunications businesses employ FL. To train models that maximize signal strength and network stability while protecting user data, data from various devices and locations is used.
6. **Retail and E-Commerce:** FL improves supply chain management and consumer recommendations in retail and online commerce. Retailers work together to develop models for personalized product recommendations, optimized inventory management, and improved demand forecasting while maintaining consumer purchase history.
7. **Energy Management:** In smart grids, FL is used to optimize energy use. Models that forecast patterns of energy use, optimize energy distribution, and minimize energy waste are trained using data from smart meters and energy sensors, while data privacy is maintained.
8. **Agriculture:** By analyzing data from IoT devices, drones, and satellite photography, FL helps with precision agriculture. While respecting the data ownership of individual farmers, models are trained to offer insights into crop health, insect control, irrigation optimization, and yield prediction.
9. **Autonomous Vehicles:** FL is used to train autonomous vehicle simulation models. To make self-driving technology safer, data gathered from sensors and cameras on various cars is utilized to enhance object identification, road mapping, and decision-making algorithms.
10. **Education:** By examining student performance data, FL can offer individualized learning experiences. Models are taught to modify lesson plans based on student development while maintaining student privacy and abiding by rules governing educational privacy.

Each federated learning application exemplifies the adaptability of this strategy by enabling many parties to cooperatively train models while preserving data privacy and security, improving decision-making and user experiences.

Figure 11.5 shows the numerous applications of federated learning in day-to-day life.

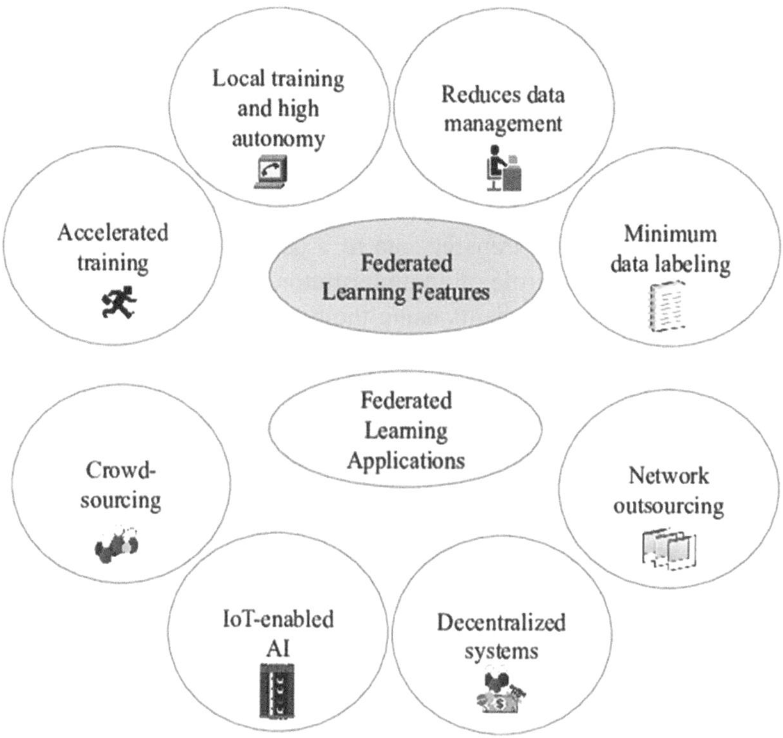

FIGURE 11.5 Applications of federated learning.

11.9 CONCLUSION

To sum up, the integration of NLP and Federated Learning exhibits significant potential in tackling the issues of data decentralization, privacy, and scalability when creating AI models. NLP is the study of human-machine interaction through natural language. It is the foundation of many applications, including sentiment analysis, language translation, and chatbots. Conversely, Federated Learning protects data privacy and permits cooperative training of ML models across decentralized data sources.

The use of FL as a privacy-preserving ML paradigm has increased significantly in recent years. FL has made a name for itself as a popular ML system training method in distributed and large-scale applications. FL uses the processing power of edge devices to cooperatively train a global model, as opposed to traditional ML techniques, which require the centralization of data for training. Data security is made possible by FL's capacity to preserve the privacy of local data on each unique device.

Through the collaboration of numerous devices or servers in the training of a single model, Federated Learning achieves its decentralized nature. This method improves data privacy and security by doing away with the requirement to transfer data to a centralized server. In FL, the central server plays the role of an orchestrator, iteratively aggregating the local models created by clients using their private data. It becomes clear that FL has many advantages compared to conventional ML methods in this novel approach.

The concept of Federated Learning, its numerous forms, the factors that led to its development, its operational structure, as well as its privacy challenges and potential remedies, have all been discussed in this chapter. FL has established itself as a game-changing technology with extensive applications across industries, thanks to its capacity to train models while preserving data privacy and security. It is anticipated that FL will play a significant role in determining the direction of ML and data-driven decision-making as the discipline develops further.

11.10 FUTURE SCOPE

The potential for FL to change many elements of data-driven technologies and businesses is enormous. Here are some crucial future application areas for federated learning:

1. **Improved Privacy Methods:** Future studies will probably concentrate on developing FL privacy-preserving methods. To further protect data while retaining model accuracy, this may entail investigating more powerful techniques for differential privacy, federated encryption, and secure aggregation.
2. **Decentralized AI Ecosystem:** It is anticipated that the idea of a decentralized AI ecosystem made possible by FL will continue to develop. There may be new frameworks for collaboration that

allow devices to work together not only to train models but also to share information, models, and insights.

3. **Dynamic and Adaptive Learning:** Future FL systems may include dynamic and adaptive learning methods that dynamically modify model updates in response to changing data characteristics and device availability. This would increase FL's capacity to adjust to changing circumstances and enhance model performance over time.
4. **Knowledge Transfer Across Areas:** Federated learning may make it possible to transfer knowledge between different areas. To speed up model convergence and cut down on data gathering, models trained in one domain may be utilized as a base for training in another.
5. **Advancements in Edge Intelligence:** As edge computing gains traction, FL's integration with edge devices may become more efficient and smoother. This might result in more advanced edge intelligence, allowing for the creation of insights and real-time decision-making.
6. **Hybrid learning methods:** By combining FL with other strategies like TL and reinforcement learning (RL), hybrid models that make use of the advantages of each method may be produced. In contexts with limited resources, this might lead to learning that is even more precise and effective.
7. **Standardization and Framework Development:** As FL becomes more widely used, the creation of standardized frameworks, protocols, and APIs may make FL's implementation simpler. As a result, it would be simpler for various platforms and devices to work together effortlessly. This would facilitate interoperability.
8. **Federated Automated Machine Learning (AutoML):** This may become an area where FL is used to automatically choose, train, and improve ML models across distributed devices, hence requiring less manual intervention.
9. **Evolving Security Measures:** Although FL improves privacy, more research needs to be done on safe protocols and techniques to defend FL systems from adversarial attacks, including Byzantine attacks and poisoning attacks.
10. **Real-Time Applications:** The combination of FL and real-time applications, such as industrial IoT and autonomous vehicles, may enable decision-making based on dynamic data streams and real-time model changes.

11. **Multi-Modal Federated Learning:** By extending FL to include more data modalities, including text, graphics, and audio, it is possible to train more diverse and thorough models using a wider range of input.
12. **Initiatives for Global Collaboration:** To assure responsible and ethical implementation, international FL standards, governance frameworks, and data-sharing protocols may be developed through collaborative initiatives between governments, businesses, and research organizations.

Fundamentally, the future of federated learning is defined by the quest for improved privacy, increased collaboration, and the creation of more advanced methods to address current issues. Federated learning is positioned to play a crucial role in influencing the ML and AI landscape as technology advances and the ethical issues around data protection grow more prevalent.

KEYWORDS

- **decentralized model training**
- **federated learning**
- **federated natural language processing**
- **machine learning**
- **privacy protection**
- **transformer-based pre-trained models**

REFERENCES

1. Ekta, & Varsha. (2024). Enhancing rating and learning through clustering in artificial intelligence. In *Artificial Intelligence and Society 5.0: Issues, Opportunities, and Challenges* (pp. 239–245). CRC Press.
2. Wen, J., & Zhang, W. (2023). A survey on federated learning: Challenges and applications. *International Journal of Machine Learning and Cybernetics, 14*(2), 513–535.
3. Bharati, S., & Prasath, V. B. (2022). Federated learning: Applications, challenges, and future directions. *International Journal of Hybrid Intelligent Systems, 18*(1, 2), 19–35.

4. Nguyen, D. C., Ding, M., & Poor, H. V. (2021). Federated learning for Internet of Things: A comprehensive survey. *IEEE Communications Surveys & Tutorials, 23*(3), 1622–1658.
5. Sadilek, A., Liu, Y., & Hernandez, J. (2021). Privacy-first health research with federated learning. *NPJ Digital Medicine, 4*(1), 132.
6. Feng, C. M., Yu, K., Liu, Y., & Zuo, W. (2023). *Towards Instance-Adaptive Inference for Federated Learning.* arXiv preprint arXiv:2308.06051.
7. Angurala, M., & Khullar, V. (2023). Federated learning-based privacy-preserved English accent training ecosystem for people with Indian language accents. *Entertainment Computing, 46*, 100572.
8. Rey, V., Sánchez, A., & Bovet, G. (2022). Federated learning for malware detection in IoT devices. *Computer Networks, 204*, 108693.
9. Cai, X., Hu, Z., & Chen, J. (2020). A many-objective optimization recommendation algorithm based on knowledge mining. *Information Sciences, 537*, 148–161.
10. Li, L., Fan, Y., Tse, M., & Lin, K. Y. (2020). A review of applications in federated learning. *Computers & Industrial Engineering, 149*, 106854.
11. Gabrielli, E., Pica, G., & Tolomei, G. (2023). *A survey on Decentralized Federated Learning.* arXiv preprint arXiv:2308.04604.
12. Wu, C., Wu, F., Liu, L., Huang, Y., & Xie, X. (2022). Communication-efficient federated learning via knowledge distillation. *Nature Communications, 13*(1), 2032.
13. Shao, W., Gao, Y., Song, F., Chen, S., & Fan, L. (2023). *An Empirical Study of Bugs in Open-Source Federated Learning Framework.* arXiv preprint arXiv:2308.05014.
14. Weinger, B., Kim, J., Sim, A., Nakashima, M., Moustafa, N., & Wu, K. J. (2022). Enhancing IoT anomaly detection performance for federated learning. *Digital Communications and Networks, 8*(3), 314–323.
15. Kairouz, P., McMahan, B., & Zhao, S. (2021). Advances and open problems in federated learning. *Foundations and Trends® in Machine Learning, 14*(1, 2), 1–210.
16. McMahan, B., Moore, E., Ramage, D., Hampson, S., & y Arcas, B. A. (2017). Communication-efficient learning of deep networks from decentralized data. In *Artificial Intelligence and Statistics* (pp. 1273–1282). PMLR.
17. Liu, Y., Kang, Y., & Yang, Q. (2020). A secure federated transfer learning framework. *IEEE Intelligent Systems, 35*(4), 70–82.
18. Kuze, N., & Murata, M. (2021). Classification of diversified web crawler accesses inspired by biological adaptation. *International Journal of Bio-Inspired Computation, 17*(3), 165–173.
19. Cai, D., Wu, Y., Wang, S., Lin, F. X., & Xu, M. (2022). *FedAdapter: Efficient Federated Learning for Modern NLP. arXiv preprint* arXiv:2205.10162.
20. Yuan, J., & Xu, M. (2023). Privacy as a resource in differentially private federated learning. In *IEEE INFOCOM 2023—IEEE Conference on Computer Communications* (pp. 1–10). IEEE.
21. Yuan, J., Xu, M., & Wang, S. (2020). *Federated Neural Architecture Search. arXiv preprint* arXiv:2002.06352.
22. Jiang, L., Svoboda, F., & Lane, N. D. (2023). *FDAPT: Federated Domain-Adaptive Pre-Training for Language Models. arXiv preprint arXiv:2307.06933.*

CHAPTER 12

LLMS-Based Human-Like Content Creation, Including Articles, Social Media Posts, and Marketing Content

VIKAS SHARMA[1,*], SUNIL KUMAR[2], TARUN KUMAR VASHISHTH[3], KEWAL KRISHAN SHARMA[3], BHUPENDRA KUMAR[3], and SACHIN CHAUDHARY[3]

[1]*Department of Computer Applications, SRM Institute of Science and Technology, Delhi NCR Campus, Ghaziabad, U.P., India*

[2]*School of Information Technology, Auro University, Gujrat, India*

[3]*School of Computer Science and Applications, IIMT University, Meerut, Uttar Pradesh, India*

**Corresponding author.*

ABSTRACT

In the ever-evolving digital landscape, content creation has become an integral aspect of communication, marketing, and information dissemination. The demand for high-quality, engaging content spans a multitude of platforms, including articles, social media posts, and marketing materials. The advent of large language models (LLMs) has ushered in a new era of content creation, enabling the generation of text that closely mimics human writing. Here we introduce the topic of "LLM-Based Human-Like Content Creation, Including Articles, Social Media Posts, and Marketing Content," providing an overview of its significance, applications, benefits, challenges, and ethical considerations. The emergence of LLMs, with exemplars such as GPT-3 and its successors, has revolutionized the way content is generated. These models, underpinned by deep learning (DL)

Natural Language Processing and Artificial Intelligence: A Perspective Toward Current Trends, Challenges, and Applications. Vikas Khullar, Aryan Chaudhary, & Sunil Kumar (Eds.)

DOI: 10.1201/9781998511273-12

techniques and neural networks, possess the remarkable capacity to understand and produce human-like text across a spectrum of styles and tones. Their operation is based on predicting the next word or phrase in a sentence, informed by the context provided in the input. As a result, they are capable of crafting coherent, contextually relevant content that often blurs the line between machine and human-generated text.

This chapter explores the multifaceted applications of LLM-based content creation, encompassing articles, social media posts, and marketing content. However, the utilization of LLMs for content creation is not without its challenges and ethical considerations. Challenges include the need for quality control to ensure generated content meets desired standards, vigilance against unintentional plagiarism, and the preservation of authenticity in the face of automation. Ethical considerations encompass the responsible use of LLMs to avoid the spread of misinformation, hate speech, or copyright violations. To maximize the benefits of LLM-based content creation while addressing these challenges, several best practices are recommended. These include implementing quality assurance measures, fine-tuning LLMs to reflect specific styles, using LLMs as a complementary tool in content planning, adhering to ethical guidelines, and maintaining human oversight in the content creation process.

12.1 INTRODUCTION

In the digital age, content creation has transcended its traditional boundaries, becoming a cornerstone of communication, marketing, and information dissemination. The demand for high-quality, engaging content spans a multitude of platforms, from informative articles to captivating social media posts and persuasive marketing materials. At the nexus of this evolving landscape lie large language models (LLMs), such as GPT-3 and its successors, which have ushered in a new era of content creation. These models, grounded in DL and neural networks, offer the remarkable capability to generate text that closely mimics human writing. This book chapter embarks on an exploration of the paradigm-shifting phenomenon of LLM-Based Human-Like Content Creation, traversing its applications, benefits, challenges, and the ethical considerations that accompany this technological innovation. LLMs operate by predicting the next word or phrase in a sentence, drawing upon the contextual cues provided in the input. This predictive ability empowers them to craft coherent, contextually relevant content that often blurs the line between machine and human-generated text.

The impact of LLMs in reshaping the content creation landscape is profound, revolutionizing how content is generated, disseminated, and consumed.

Here, the multifaceted applications of LLM-based content creation encompass the realms of articles, social media posts, and marketing content. It delves into the advantages and challenges presented by LLMs, elucidating their potential to enhance efficiency, ensure consistency, and enable customization. Simultaneously, it examines the ethical considerations that must underpin the responsible use of LLMs, safeguarding the authenticity and ethical standards inherent to content creation (Figure 12.1).

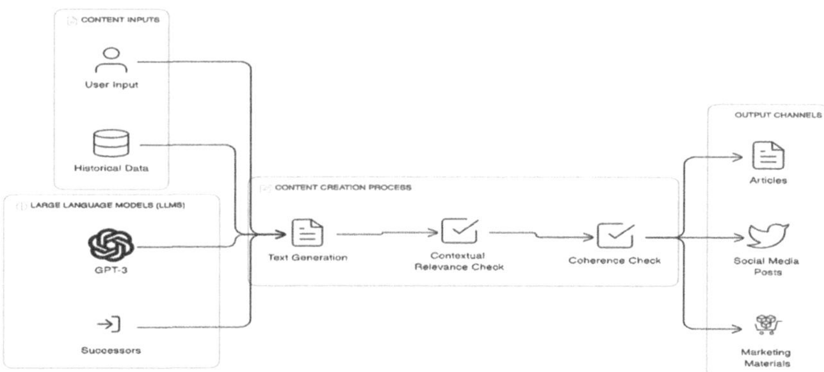

FIGURE 12.1 Large language model (LLM) based content creation system.

In sum, LLM-based human-like content creation represents a transformative force in the ever-evolving world of digital content. It opens new vistas of efficiency and effectiveness while posing profound questions about the future of content creation, the preservation of authenticity, and the ethical boundaries essential in navigating this innovative terrain. This chapter seeks to illuminate this dynamic landscape, offering valuable insights, guidance, and a deeper understanding of the opportunities and challenges presented by LLMs in shaping the future of content creation.

12.1.1 TRANSCENDING TRADITIONAL BOUNDARIES

The phrase "Transcending Traditional Boundaries" is an assertion that content creation, which was once confined to specific roles and mediums, has evolved and expanded its influence in the digital age. Here's a detailed explanation of this concept (Figure 12.2):

1. **Evolution of Content Creation:** In the past, content creation primarily encompassed traditional forms like books, newspapers, and television. It had specific roles and purposes, often limited to disseminating information or entertaining audiences. However, in the digital age, this paradigm has shifted significantly.
2. **Central Role in Modern Communication:** The digital age has witnessed content creation becoming a central and fundamental element in the broader spectrum of communication, marketing, and information dissemination. Content has transitioned from being just a part of these processes to being a cornerstone.
3. **Communication:** Content now plays a pivotal role in communication. It is not limited to one-way communication, but it also serves as a medium for interactive communication. For example, blogs, social media, and online forums allow individuals and organizations to communicate and engage with their audiences.
4. **Marketing:** Marketing has been radically transformed by the digital age. Content creation is now at the heart of marketing strategies. Content marketing, social media marketing, and digital advertising all rely heavily on creating engaging and persuasive content to connect with consumers and promote products or services.
5. **Information Dissemination:** The digital era has democratized the dissemination of information. Anyone can create and distribute content, and the internet acts as a vast platform for sharing knowledge, news, and insights. Content creation has become an essential tool for not only spreading information but also shaping public opinion.
6. **Multimedia and Multichannel:** Content is no longer confined to text-based articles or TV shows. It spans a multitude of formats, including text, images, audio, video, and interactive elements. Furthermore, content is distributed across a wide range of platforms and channels, from websites and social media to podcasts and streaming services.
7. **Audience Engagement:** Content is not just about broadcasting information; it's about engaging audiences. Content creators are increasingly focused on creating content that resonates with their target audience, encourages interaction, and builds communities.

8. **Accessibility and Ubiquity:** Digital content is accessible virtually anywhere, anytime. The ubiquity of smartphones, tablets, and computers means that content is at the fingertips of people across the globe. This accessibility has amplified the role of content in modern communication and information dissemination.

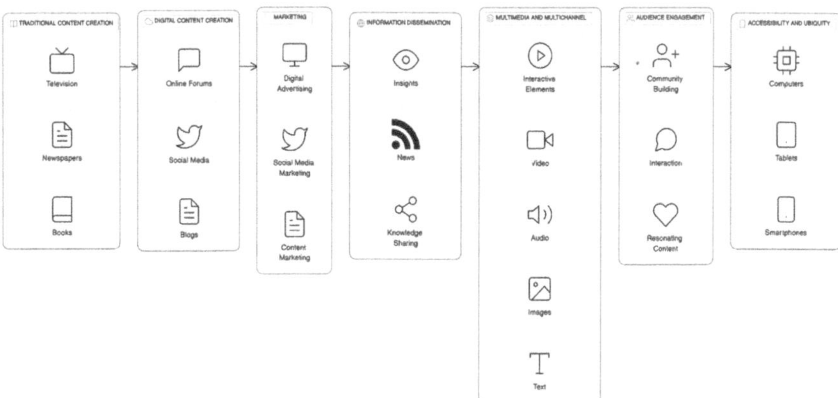

FIGURE 12.2 Transcending traditional boundaries in content creation.

The phrase "transcending traditional boundaries" encapsulates the idea that content creation has moved beyond its historical limitations and has taken on a central, multifaceted role in contemporary digital society. Content is no longer just a tool for communication; it has become a dynamic force that shapes how we interact, market, and share information in the modern world.

12.1.2 DIVERSE DEMAND FOR HIGH-QUALITY CONTENT

The concept "Diverse Demand for High-Quality Content" underscores the increasing need for engaging and top-tier content in the digital age (Figure 12.3).

1. **Extensive Demand:** In the digital age, there is a pervasive and substantial demand for high-quality content. This demand extends across various industries, sectors, and individuals, highlighting the universal value of compelling and well-crafted content.

2. **Wide Range of Platforms:** The demand for high-quality content is not limited to a single or specific platform. Instead, it spans a wide array of platforms, including but not limited to websites, social media, blogs, video-sharing platforms, podcasts, and email newsletters. Each platform requires content that resonates with its specific audience and format.
3. **Various Content Formats:** High-quality content encompasses diverse formats to cater to different preferences and needs. This includes:
   i. **Informative Articles:** These are long-form or in-depth articles that provide valuable information or insights on specific topics. They are often used for educational or informational purposes.
   ii. **Compelling Social Media Posts:** Short and engaging content designed to captivate the audience's attention and encourage interaction on social media platforms like Facebook, Twitter, and Instagram.
   iii. **Persuasive Marketing Materials:** Content designed to promote products, services, or brands, often used in advertising, email marketing, and promotional materials.
4. **Engaging Audiences:** The demand for high-quality content is driven by the necessity to capture and engage audiences effectively. It's not enough for content to merely exist; it must be engaging and valuable to keep the audience's attention and encourage them to take action, whether that's making a purchase, sharing the content, or interacting with it.
5. **Adaptation to Multiple Channels:** Content creators need to adapt to multiple channels and platforms. Each platform has its unique characteristics and audience expectations. For instance, the content that performs well on a blog might not be suitable for a short social media post or a video on a platform like YouTube. The demand for high-quality content means creating content that is not only valuable but also tailored to the medium in which it is presented.
6. **Content Quality and Brand Image:** The quality of content is closely associated with the image and reputation of a brand or individual. High-quality content not only engages audiences but also reflects positively on the creator or brand, establishing authority and trust.
7. **Competitive Landscape:** In the digital realm, competition is fierce. Brands and content creators are vying for the attention of

the same audience. To stand out and gain a competitive edge, they need content that is not only relevant but also of superior quality.

8. **User Expectations:** The modern digital consumer has high expectations for content. They demand content that is not only informative but also entertaining, visually appealing, and easily shareable. Meeting these expectations is crucial to maintaining an online presence and staying relevant.

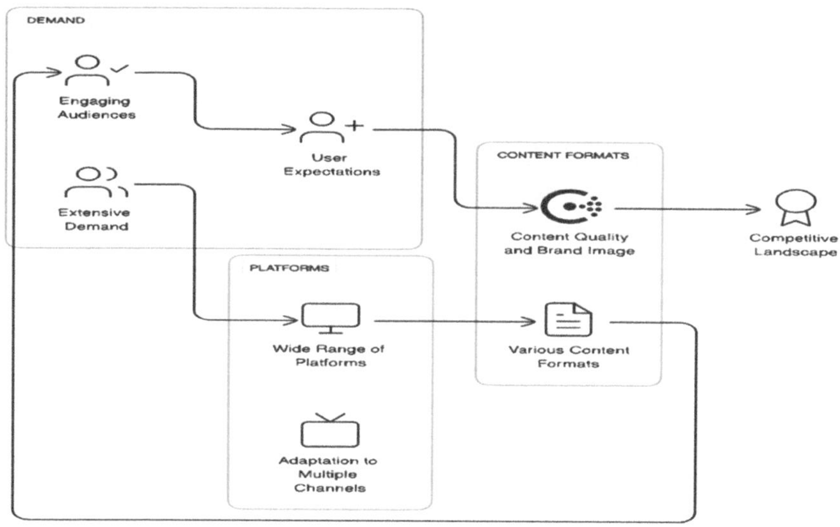

FIGURE 12.3 Demanding high quality content material.

The diverse demand for high-quality content in the digital age reflects the evolving nature of communication and marketing. It underscores the need for content creators to produce engaging, valuable, and platform-appropriate content that resonates with audiences across various channels. Meeting this demand is essential for maintaining a strong online presence and achieving communication and marketing objectives.

12.1.3 EMERGENCE OF LARGE LANGUAGE MODELS (LLMS)

Large language models, represented by GPT-3 and similar successors, are at the forefront of this content evolution. These models are underpinned by DL and neural networks, equipping them with the exceptional capability to generate text that closely resembles human writing. LLMs have ushered in a new era of content creation, where machines produce text that closely mimics human authorship (Figure 12.4).

1. **Large Language Models (LLMs):** LLMs are advanced artificial intelligence (AI) systems designed to understand and generate human-like text. These models, such as GPT-3, are trained on vast amounts of textual data and use DL techniques to process and produce text.
2. **GPT-3 and Successors:** GPT-3, which stands for generative pre-trained transformer 3, is one of the most prominent examples of LLMs. It represents the third iteration of the GPT series and is known for its exceptional language processing capabilities. "Successors" in this context refers to models that have built upon the foundations of GPT-3, incorporating improvements and refinements.
3. **Deep Learning and Neural Networks:** LLMs are underpinned by DL and neural networks. DL involves training neural networks with massive datasets to learn patterns and relationships within the data. Neural networks are computational structures inspired by the human brain, allowing machines to process and understand complex data, including natural language.
4. **Exceptional Text Generation Capabilities:** LLMs like GPT-3 possess exceptional text generation capabilities. They can understand and generate text that closely resembles human writing, both in terms of grammar and context. This ability is achieved through the DL process, which enables them to learn the nuances of language and produce coherent, contextually relevant text.
5. **Ushering in a New Era of Content Creation:** LLMs have introduced a paradigm shift in content creation. They have ushered in a new era where machines can generate text that is often nearly indistinguishable from content written by humans. This has significant implications for various industries and sectors that rely on content creation.
6. **Mimicking Human Authorship:** LLMs excel at mimicking human authorship. They can adapt to different writing styles, tones, and formats, making it challenging for readers to discern whether a human or machine generated the content. This adaptability is a key factor in their success.

7. **Versatility and Applications:** LLMs find applications in a wide range of fields. They can be used to generate articles, create social media posts, draft marketing content, assist in content planning, generate creative writing, and provide natural language interfaces for software and chatbots. Their versatility is a major advantage.
8. **Efficiency and Scalability:** LLMs significantly enhance the efficiency and scalability of content creation. They can generate text at a rapid pace and handle large volumes of content generation tasks, which can be challenging for human authors.
9. **Content Customization:** LLMs can be fine-tuned to generate content that aligns with specific requirements, styles, and brand identities. This level of customization is invaluable for organizations and individuals looking to maintain a consistent voice in their content.

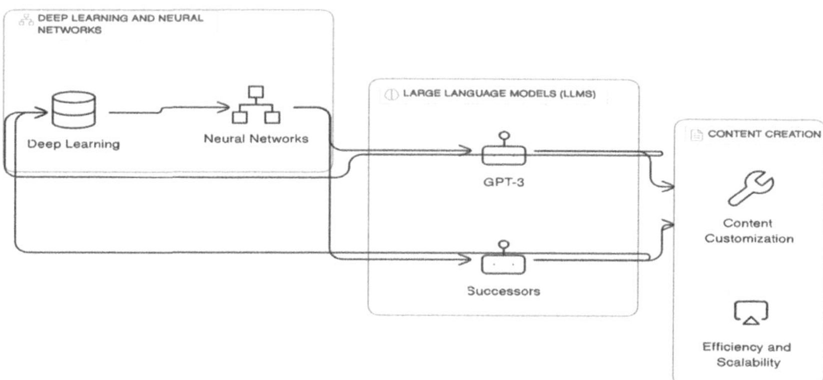

FIGURE 12.4 Large language model architecture.

Large language models like GPT-3 and their successors represent a major advancement in AI-driven content creation. These models, built on DL and neural networks, have the remarkable ability to generate text that closely resembles human writing. They have redefined the landscape of content creation, introducing efficiency, scalability, and customization, and have set the stage for a new era where machines can produce content that often rivals human-authored content in quality and authenticity.

12.1.4 PREDICTIVE ABILITIES OF LARGE LANGUAGE MODELS (LLMS)

Large language models operate by predicting the next word or phrase in a sentence, utilizing the context provided in the input. This predictive capacity empowers them to produce coherent and contextually relevant content, often indistinguishable from content authored by humans.

1. **Operation of LLMs:** LLMs, like GPT-3, operate on the principle of predicting the next word or phrase in a sentence. They do this by utilizing the context provided in the input. This predictive mechanism is at the core of their text generation process.
2. **Contextual Understanding:** LLMs are trained on vast datasets that include diverse and extensive text from the internet. During this training, they learn to understand language and context. They acquire an understanding of grammar, syntax, semantics, and the relationships between words.
3. **Predicting the Next Word or Phrase:** When LLMs generate text, they analyze the input text or prompt given to them. Based on this input, they predict what the next word or phrase in the sentence should be. This prediction is informed by the context provided in the input.
4. **Coherent and Contextually Relevant Content:** The predictive ability of LLMs empowers them to produce content that is not only coherent but also contextually relevant. This means that the generated text logically follows the given input and maintains consistency in tone and style.
5. **Human-Like Text Generation:** LLMs are often trained on a diverse range of texts, which exposes them to various writing styles and domains. This extensive training allows them to produce text that closely resembles human writing. They can adapt to different writing styles and subject matters, making it challenging for readers to distinguish between machine-generated and human-authored content.
6. **Adaptation to Input Context:** LLMs can adapt their predictions based on the context they receive. For example, if the input instructs the model to write a formal research paper, it will generate text that aligns with the academic style and tone expected in such a context. If the input requires a casual and conversational tone, the LLM will adjust accordingly.

7. **Reducing Ambiguity:** LLMs excel in handling ambiguity in language. They can disambiguate sentences by predicting the most contextually appropriate word or phrase, reducing the likelihood of confusion or misunderstandings in the generated text.

The predictive abilities of LLMs are fundamental to their text generation process. These models use contextual cues to predict and generate the next word or phrase in a sentence, resulting in coherent, contextually relevant content that often closely resembles human-authored text.

12.1.5 PROFOUND IMPACT OF LARGE LANGUAGE MODELS (LLMS)

The impact of LLMs on content creation is profound. They revolutionize how content is generated, distributed, and consumed. The ability of LLMs to mimic human writing styles and produce contextually relevant content has far-reaching implications across various industries and sectors (Figure 12.5).

1. **Revolutionizing Content Creation:** LLMs have a profound impact on content creation by fundamentally changing how content is generated, distributed, and consumed. They introduce a new era in which machines play a central role in content creation, supplementing or even replacing traditional human authors in various contexts.
2. **Content Generation:** LLMs revolutionize content generation by automating the process. They can produce a wide range of content, from articles and blog posts to marketing materials, product descriptions, and more. This automation significantly speeds up content creation, making it more efficient and cost-effective.
3. **Content Distribution:** LLMs impact content distribution by providing content creators with the ability to generate vast amounts of content that can be tailored to different platforms and audiences. This mass production capability enhances a brand's presence and reach across various online channels, including websites, social media, and email newsletters.
4. **Content Consumption:** Content generated by LLMs is often indistinguishable from human-authored content. This means that readers and consumers may not be aware that they are engaging with machine-generated text. The seamless integration

of LLM-generated content into the digital landscape affects how consumers perceive and interact with the content.

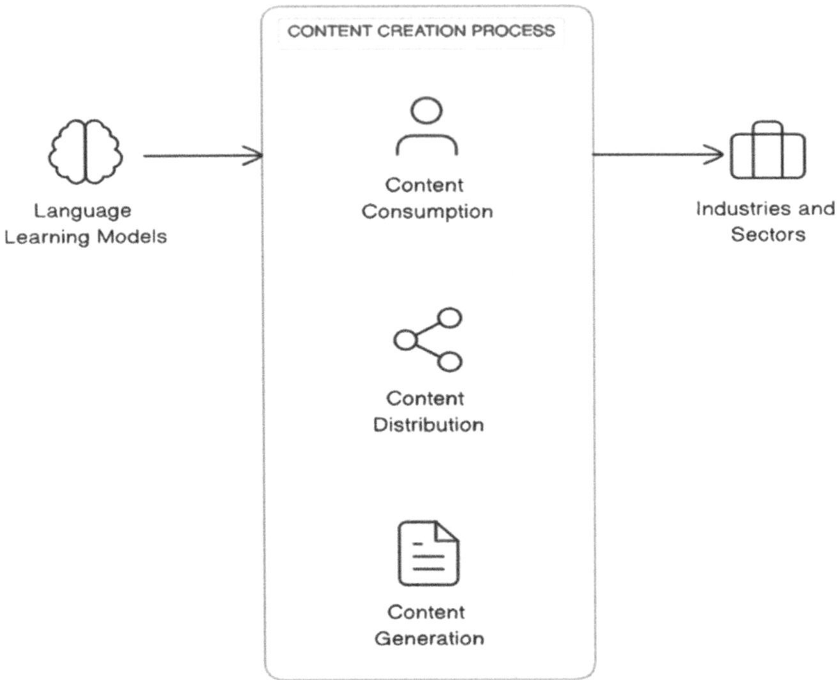

FIGURE 12.5 Language learning model in content creation.

5. **Mimicking Human Writing Style:** LLMs excel at mimicking human writing styles, including tone, voice, and even niche-specific writing styles. This mimicry allows organizations to maintain a consistent brand voice and ensures that the content resonates with the target audience.
6. **Contextual Relevance:** LLMs generate content that is contextually relevant. They analyze the input and understand the context, which enables them to produce content that aligns with the given context. This is critical for maintaining engagement and providing valuable information to the audience.

The profound impact of LLMs on content creation is marked by their ability to revolutionize the entire content lifecycle, from creation to distribution and consumption. Their capacity to mimic human writing style and produce contextually relevant content has disrupted traditional content creation practices, opening new possibilities and challenges across diverse industries and sectors.

12.2 LARGE LANGUAGE MODELS (LLM)-BASED CONTENT CREATION: AN OVERVIEW

LLMs are a type of AI model that has gained significant attention in recent years due to their remarkable ability to generate human-like text. These models operate by predicting the next word or phrase in a sentence based on the context provided in the input. In this overview, we'll delve into the details of how LLMs work and explore their applications in content creation.

12.2.1 PREDICTIVE NATURE OF LARGE LANGUAGE MODELS (LLMS)

The predictive nature of LLMs is a fundamental aspect of how these models work. LLMs are a type of AI model that excels in understanding and generating human-like text. Here, we will explore in detail how this predictive capability is achieved and why it is crucial for generating coherent and contextually relevant content (Figure 12.6):

1. **Training on Vast Text Data:** LLMs, such as GPT-3, are trained on enormous datasets that include text from the internet, books, articles, and other sources. This training process involves processing and analyzing billions of words and sentences. By training on such extensive and diverse data, LLMs learn the intricacies of language, including patterns, grammar rules, sentence structures, and the nuances of human communication.
2. **Learning Language Patterns:** During training, LLMs analyze the vast text corpus to identify statistical patterns and relationships between words. They learn which words are commonly used together, understand synonyms and antonyms, and grasp the meanings of words in various contexts. For example, they learn that "hot" and "temperature" are related in the context of weather, but "hot" and "spicy" are related in the context of food.

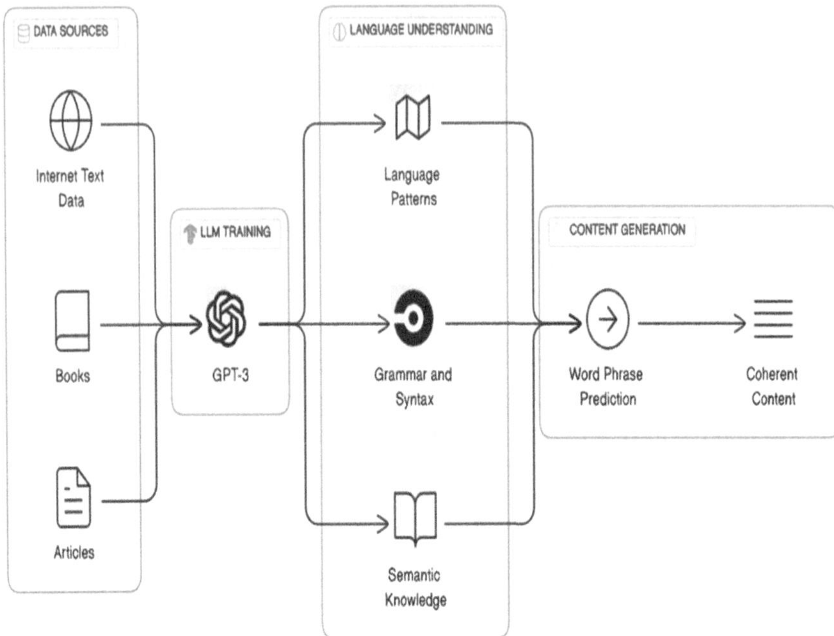

FIGURE 12.6 Predictive nature of large language models.

3. **Grammar and Syntax Understanding:** LLMs also acquire an understanding of grammar and syntax. They learn the rules governing sentence structure, such as subject-verb agreement, tense, and word order. This understanding enables LLMs to generate text that adheres to proper grammar, making the output coherent and natural-sounding.
4. **Semantic Knowledge:** In addition to grammar and syntax, LLMs develop a sense of semantics, meaning that they understand the meaning of words and how different words relate to one another. They can differentiate between homonyms (words with the same spelling but different meanings) and polysemous words (words with multiple meanings based on context). This semantic knowledge allows LLMs to generate text that is contextually relevant.
5. **Predicting the Next Word or Phrase:** When you input a partial sentence or a prompt into an LLM, the model uses its training to predict the most likely next word or phrase based on the context provided. It does this by considering the words that precede the

input and analyzing the statistical probabilities of various words or phrases occurring next. LLMs assign probabilities to each possible word or phrase, and the word or phrase with the highest probability becomes the model's prediction.

6. **Generating Coherent Content:** Because LLMs predict the next word or phrase based on the input's context, the generated content flows coherently. It follows the logical progression of a conversation or narrative, making it appear indistinguishable from human writing. LLMs can generate paragraphs, essays, or even entire articles that maintain consistency and relevance throughout the text.

12.2.2 CONTEXTUAL UNDERSTANDING

One contextual understanding is a crucial feature of LLMs that sets them apart from traditional language models. LLMs can comprehend the broader context of a given text or conversation, which enables them to generate text that is contextually appropriate and logically consistent. Here, we will explain in detail how LLMs achieve this contextual understanding (Figure 12.7):

1. **Contextual Learning:** LLMs are trained on vast amounts of text data, which encompasses a wide array of topics and subjects. During this training process, the models are exposed to a multitude of sentences and paragraphs that provide context for various words and phrases. This extensive exposure allows LLMs to learn how words and phrases are used in different contexts.
2. **Word Embeddings:** LLMs represent words in a high-dimensional vector space, where words with similar meanings or contextual usage are positioned close to each other. This technique, known as word embeddings, allows LLMs to capture semantic relationships between words. For example, words like "sunscreen," "beach," "towel," and "snacks" would have related vector representations because they often co-occur in similar contexts.
3. **Recurrent and Transformer Architectures:** LLMs are built on neural network architectures that are designed to process sequential data and capture context. RNNs and Transformer architectures are commonly used in LLMs. Transformers, in particular, have become

popular due to their ability to model long-range dependencies and context in a text.

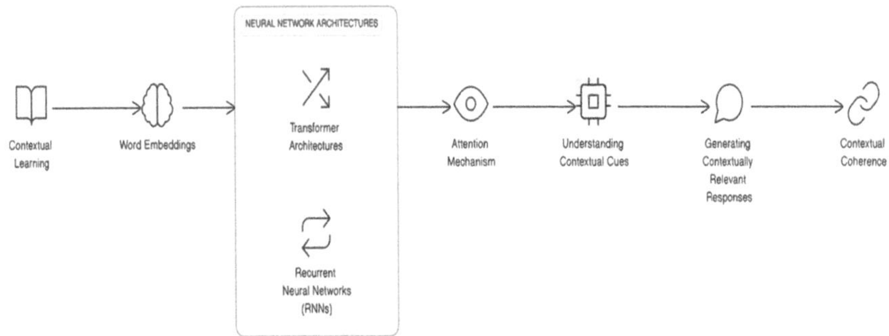

FIGURE 12.7 Large language learning model in contextual understanding.

4. **Attention Mechanism:** Transformers, which underpin many LLMs, use an attention mechanism that allows them to weigh the importance of different words in a sentence when generating the next word. This means the model can focus more on words that are contextually relevant to the input and reduce the influence of less relevant words.
5. **Understanding Contextual Cues:** When you provide an LLM with a sentence like "The weather is so nice, I think I'll go to the beach," and then ask, "What should I bring with me?" the model is not merely considering the words in isolation. Instead, it understands the entire context: the favorable weather and the intention to go to the beach. It recognizes that this context requires advice on what items to bring for a trip to the beach.
6. **Generating Contextually Relevant Responses:** Because of their contextual understanding, LLMs can generate responses that align with the input's context. In the example given, the LLM recognizes the context of a beach trip and suggests appropriate items like sunscreen, a towel, and snacks. This ability to generate contextually relevant responses makes LLMs highly valuable in natural language understanding (NLU) and generation tasks.
7. **Contextual Coherence:** Beyond individual responses, LLMs can also maintain contextual coherence over extended conversations. They remember the context from previous interactions and generate responses that make sense within the ongoing dialogue.

This is essential for chatbots, virtual assistants, and other applications that involve multi-turn conversations.

12.2.3 VERSATILITY OF LARGE LANGUAGE MODELS (LLMS)

LLMs are incredibly versatile and have applications across a wide range of content domains. Some common use cases include:

1. **Content Generation:** LLMs can generate written content for various purposes, such as articles, blog posts, product descriptions, and more.
2. **Translation:** LLMs can translate text from one language to another while maintaining the original context and meaning.
3. **Chatbots:** LLMs can power chatbots and virtual assistants, enabling natural language interactions between humans and machines.
4. **Text Summarization:** LLMs can summarize long texts, making them useful for quickly digesting large volumes of information.
5. **Question Answering:** LLMs can answer questions by providing relevant information from their trained knowledge base.
6. **Creative Writing:** LLMs can assist creative writers by suggesting ideas, generating storylines, or even co-authoring stories.
7. **Coding Assistance:** LLMs can help programmers by generating code based on descriptions or solving coding problems.

12.2.4 ETHICAL CONSIDERATIONS

The ethical considerations are paramount when it comes to the use of LLMs in content creation and other applications. While LLMs offer a range of powerful and beneficial capabilities, they also pose several ethical concerns that need to be addressed. Let's delve into these ethical considerations in detail (Figure 12.8):

1. **Misinformation and Fake News:** One of the most significant ethical concerns is the potential for LLMs to generate misinformation and fake news. Given their ability to create text that appears convincingly human-written, malicious actors can misuse LLMs to generate false and misleading information. This poses a significant threat to public trust, as well as social and political stability.

2. **Impersonation:** LLMs can be used to impersonate individuals or entities, which can have damaging consequences. For instance, an LLM could be used to craft fake emails, social media posts, or other content that impersonates a public figure, leading to reputational damage, privacy violations, or even manipulation of public opinion.
3. **Bias and Discrimination:** LLMs often inherit biases present in the training data, which can lead to biased or discriminatory content generation. This is particularly problematic when the generated content perpetuates stereotypes, promotes hate speech, or discriminates against certain groups. Addressing and mitigating these biases is an ongoing challenge.
4. **Privacy Concerns:** Content generated by LLMs may inadvertently contain sensitive or private information. The automated generation of such content can lead to privacy breaches and unintended disclosures, posing ethical and legal challenges.
5. **Consent and Authenticity:** When LLMs generate text on behalf of individuals or organizations, it can raise questions about consent and authenticity. For example, generating text on behalf of an individual without their consent may infringe on their rights and lead to issues of trustworthiness.
6. **Accountability and Responsibility:** Determining responsibility and accountability for content generated by LLMs can be complex. When a machine generates content, who should be held responsible for its accuracy, legality, or ethical considerations? This issue remains a subject of ongoing debate and legal discussion.
7. **Mitigation and Safeguards:** To address these ethical concerns, it is crucial to implement safeguards and guidelines for the responsible use of LLMs. Some potential safeguards and mitigation strategies include:

   i. **Content Verification:** Developing tools and methods to verify the authenticity and accuracy of content generated by LLMs.
   ii. **Transparency:** Encouraging transparency in the use of LLMs and disclosing when content is generated by AI.
   iii. **Bias Mitigation:** Developing techniques to reduce bias in LLM-generated content and promote fairness.
   iv. **User Authentication:** Implementing methods to verify the identity and consent of users when generating content on their behalf.

v. **Legislation and Regulation:** Governments and organizations are considering regulations and policies to address the ethical use of AI, including LLMs.
vi. **Ethical Training Data:** Ensuring that the data used for training LLMs is free from biases and ethical concerns.
vii. **Oversight and Review:** Establishing mechanisms for third-party oversight and audits of LLM use to ensure compliance with ethical guidelines.

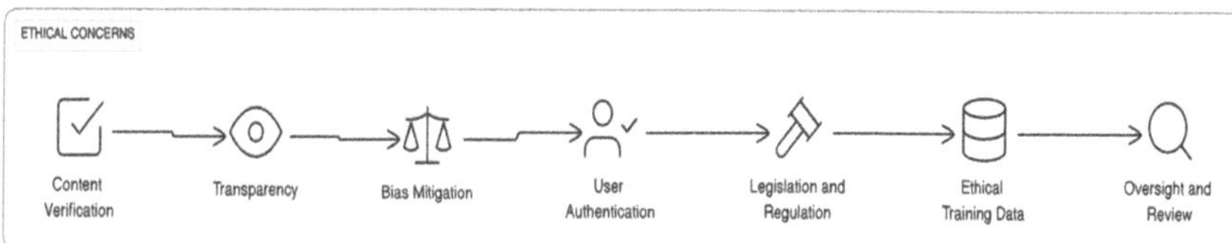

FIGURE 12.8 Integration of safeguards in large language model (LLM).

12.2.5 TRAINING AND DEVELOPMENT

Developing training and developing LLMs is a complex and resource-intensive process that underlies their impressive capabilities in content generation. This involves several stages, including initial training on large datasets and fine-tuning for specific tasks. Companies like OpenAI have indeed invested substantial resources in creating and maintaining these models. Let us explore the training and development of LLMs in detail (Figure 12.9):

1. **Data Collection:** The first step in creating an LLM involves collecting vast amounts of text data from diverse sources. This data typically includes books, articles, websites, social media posts, and more. The quality and diversity of the data are crucial to ensure that the model learns the nuances of language and can handle a wide range of topics and writing styles.
2. **Preprocessing:** The collected data undergoes preprocessing to clean and format it for training. This process may involve removing irrelevant information, standardizing text formats, and segmenting the data into sentences and paragraphs.
3. **Initial Training:** The heart of LLM development is the initial training of the model. Companies like OpenAI use powerful hardware, including graphics processing units (GPUs) and, more

recently, application-specific integrated circuits (ASICs) designed for AI workloads, to train their models efficiently. The training is done using neural network architectures, such as Transformers, which have proven to be highly effective for NLU and generation.

4. **Fine-Tuning:** After the initial training, LLMs are fine-tuned for specific tasks or domains. This fine-tuning process narrows down the model's capabilities and adapts it to particular applications. For example, an LLM may be fine-tuned for chatbot interactions, code generation, language translation, or any other specific task. Fine-tuning involves training the model on a narrower dataset that is tailored to the target domain or task.
5. **Iterative Development:** LLM development is often an iterative process. Researchers and engineers continually refine and improve the model's performance. This may involve adjusting hyperparameters, fine-tuning strategies, or even scaling up the model by increasing its size and capacity.
6. **Hardware and Computational Resources:** Developing and training LLMs demand significant computational resources. Training large models on extensive datasets requires access to specialized hardware, large-scale data centers, and energy resources. Companies invest heavily in these resources to push the boundaries of LLM capabilities.
7. **Maintenance and Updates:** Once an LLM is deployed, it requires ongoing maintenance and updates. This includes regular model retraining to ensure it remains up-to-date and continues to perform well as language and data evolve.

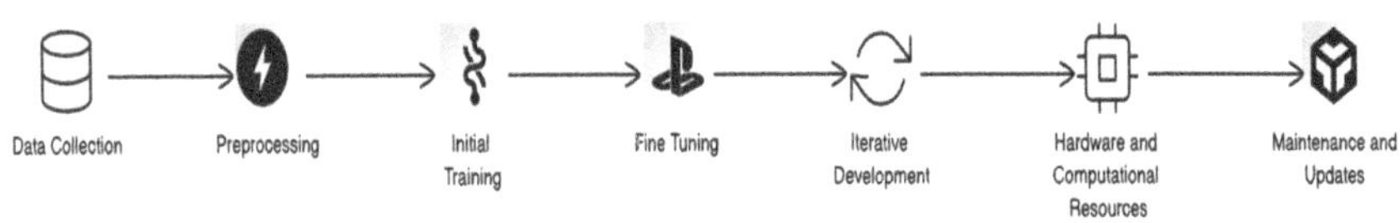

FIGURE 12.9 Model and training in large language model (LLM).

12.3 APPLICATIONS OF LARGE LANGUAGE MODEL (LLM)-BASED CONTENT CREATION

LLMs have revolutionized content creation by offering a wide range of applications across various domains. In this detailed explanation, we will explore three significant applications of LLM-based content creation:

12.3.1 ARTICLE GENERATION

LLMs play a crucial role in automating the creation of informative articles. This application extends to various forms of content, such as blog posts, news articles, technical documentation, and research papers. Here's how LLMs are applied in article generation:

1. **Streamlined Content Production:** LLMs can generate articles on a wide array of topics, making it possible to streamline content production. Content creators and organizations can use these models to quickly generate high-quality articles, saving time and effort.
2. **Constant Fresh Content:** LLMs enable the consistent production of fresh, engaging articles. This is particularly valuable for news websites and blogs where maintaining an up-to-date and dynamic content stream is essential.
3. **Customization:** LLMs can be fine-tuned to adopt specific tones, styles, or subject matter expertise. This ensures that the generated articles align with the brand's identity or the desired content style.
4. **Research Assistance:** LLMs can assist in generating research papers or summaries. This can be helpful for academics, researchers, and students looking to create concise and informative papers.

12.3.2 SOCIAL MEDIA POSTS

In the world of social media, where staying connected and engaging with audiences is crucial, LLMs offer unique advantages:

1. **Consistency:** LLMs can generate social media content consistently and at a high volume. This is essential for maintaining an active online presence, which is particularly important for brands, influencers, and individuals seeking to grow their social media following.
2. **Engagement:** LLMs can craft posts that are engaging, informative, and resonate with the target audience. This personalization can help build meaningful connections with followers and enhance online interactions.
3. **Brand Voice:** LLMs can be trained to mimic a brand's unique voice, ensuring that social media content aligns with the brand's identity and messaging.

4. **Time Efficiency:** LLMs can significantly reduce the time required to create and schedule social media posts. This is advantageous for social media managers and marketers who need to maintain a consistent posting schedule.

12.3.3 MARKETING CONTENT

Marketing heavily relies on persuasive and creative content to capture the attention of potential customers. LLMs can play a pivotal role in generating various types of marketing content:

1. **Marketing Copy:** LLMs can generate compelling marketing copy for products or services. This includes creating product descriptions, highlighting features, and crafting persuasive calls to action.
2. **Ad Campaigns:** LLMs can assist in generating ad campaign content, helping advertisers create attention-grabbing and effective advertisements.
3. **Promotional Material:** Whether it's for email marketing, flyers, brochures, or other promotional materials, LLMs can provide content that resonates with specific target demographics.
4. **Brand Consistency:** LLMs can be fine-tuned to match a brand's identity, ensuring that all marketing content remains consistent with the brand's messaging and style.

12.4 LITERATURE REVIEW

Adamopoulou and Moussiades's [1] paper on chatbots provides a comprehensive historical overview and discusses the technology, but it lacks a critical analysis of the challenges and limitations of current chatbot applications.

Alruqi and Alzahrani's [2] paper on an Arabic chatbot provides a specific focus on extractive question-answering and language transformers, but it could have benefited from a broader discussion on the generalizability of their approach to other languages and contexts.

Ashfaque's [3] paper provides an analysis of trends in chatbot design and development, but it could have benefited from a more critical evaluation of the effectiveness and limitations of these trends in practical applications.

Bilquise, Ibrahim, and Shaalan's [4] paper on emotionally intelligent chatbots provides a systematic literature review, but it could have delved deeper into the practical challenges and potential solutions for implementing emotional intelligence in chatbots.

Cherednichenko et al.'s [5] preprint on collaborative business intelligence virtual assistants provides a reference model, but it could have benefited from a more comprehensive evaluation of the model's performance and applicability in diverse business contexts.

Eltahir et al.'s [6] paper provides an overview of chatbot structure and source algorithms, but it could have delved deeper into the practical challenges and potential solutions for implementing these algorithms in real-world chatbot applications.

Huynh-The et al.'s [7] paper provides a survey on AI for the metaverse, but it could have benefited from a more critical evaluation of the challenges and limitations of applying AI in this emerging field.

Iić, Nakano, and Hajnal's [8] paper on designing dialogue systems presents an interesting approach with a mean, grumpy, sarcastic chatbot, but it could have benefited from a more comprehensive evaluation of the user experience and the chatbot's effectiveness in different contexts.

Jadhav and Thorat's [9] paper on designing conversational agent systems provides a good starting point, but it could have delved deeper into the practical challenges and potential solutions for implementing these systems in diverse contexts. The discussion seems to lack a comprehensive evaluation of the effectiveness of different design approaches.

Karyotaki, Drigas, and Skianis's [10] paper on chatbots as cognitive, educational, advisory, and coaching systems provides a broad overview, but it could have benefited from a more detailed analysis of the effectiveness and limitations of chatbots in these specific roles. The discussion seems to lack a comprehensive evaluation of real-world case studies.

Lahiri, Kandimalla, and Jeysekar's [11] paper on a hybrid multipurpose voice assistant provides an interesting approach, but it could have benefited from a more comprehensive evaluation of the system's performance and user experience in diverse real-world scenarios. The discussion seems to lack a critical analysis of the challenges and potential solutions for implementing such a system.

Luo et al.'s [12] paper provides a critical review of chatbot designs and applications, but it could have delved deeper into the practical challenges and potential solutions for implementing these designs in diverse contexts.

The discussion seems to lack a comprehensive evaluation of the effectiveness of different design approaches.

Maher, Kayte, and Nimbhore's [13] paper on chatbots and their techniques using AI provides a good overview, but it could have benefited from a more critical evaluation of the effectiveness and limitations of these techniques in practical applications. The discussion seems to lack a comprehensive evaluation of real-world case studies.

Nikhath et al.'s [14] paper on an intelligent college inquiry bot provides a specific use case, but it could have benefited from a more comprehensive evaluation of the bot's performance and user experience in diverse scenarios. The discussion seems to lack a critical analysis of the challenges and potential solutions for implementing such a system.

Nirala, Singh, and Purani's [15] paper provides a survey on using AI chatbots for customer and public administration services, but it could have delved deeper into the practical challenges and potential solutions for implementing these services. The discussion seems to lack a comprehensive evaluation of the effectiveness of chatbots in these specific roles.

Nuruzzaman and Hussain's [16] paper provides a survey on chatbot implementation in the customer service industry using deep neural networks, but it could have benefited from a more critical evaluation of the effectiveness and limitations of these implementations. The discussion seems to lack a comprehensive evaluation of real-world case studies.

Nuruzzaman and Hussain's [17] paper on IntelliBot, a dialogue-based chatbot for the insurance industry, provides a specific use case, but it could have benefited from a more comprehensive evaluation of the bot's performance and user experience in diverse scenarios. The discussion seems to lack a critical analysis of the challenges and potential solutions for implementing such a system.

Singh and Beniwal [18] paper provides a survey on near-human conversational agents, but it could have delved deeper into the practical challenges and potential solutions for achieving near-human conversation. The discussion seems to lack a comprehensive evaluation of the effectiveness of different techniques in achieving this goal.

Zafar et al. [19] present a comprehensive review and propose solution architecture for building trust in conversational AI systems. They focus on incorporating explainability and privacy awareness using LLMs and knowledge graphs. The paper offers valuable insights into addressing the challenges associated with trust in AI-driven conversations.

Zhang et al. [20] study "It's a Fair Game," or Is It? raises critical questions about the disclosure risks and benefits in the use of LLM-based conversational agents, yet leaves room for further exploration.

12.5 THE BENEFITS OF LARGE LANGUAGE MODEL (LLM)-BASED CONTENT CREATION

LLMs have become powerful tools for content creation across a wide range of applications [21–23]. Their benefits extend to various aspects of content production, making them valuable for individuals and organizations alike. Here, we will explain in detail the benefits of LLM-based content creation:

12.5.1 EFFICIENCY

Large language models excel in efficiency when it comes to content creation. This efficiency manifests in several ways:

1. **Speed:** LLMs can produce content at an impressive pace. They can generate articles, social media posts, or marketing material in a fraction of the time it would take a human content creator.
2. **Volume:** LLMs can generate substantial volumes of content. This is particularly useful for applications like news websites, blogs, and e-commerce platforms where a constant stream of fresh content is essential.
3. **Resource Savings:** By automating content creation, LLMs reduce the human resources required for producing content. This can result in significant cost savings for businesses.

12.5.2 CONSISTENCY

LLMs maintain a consistent tone and style in their generated content. This is especially valuable for maintaining brand identity and individual preferences. Here's how consistency is beneficial:

1. **Brand Alignment:** LLMs can be trained to generate content that aligns with a brand's unique voice and messaging. This ensures

that all content remains on-brand, whether it's articles, social media posts, or marketing material.

2. **Tone and Style:** Content creators can rely on LLMs to maintain a consistent tone and style throughout their content, which is important for conveying a particular image or message.
3. **Quality Assurance:** The consistency in LLM-generated content means that the quality of the output is predictable and reliable, reducing the need for extensive manual editing and quality control.

12.5.3 SCALABILITY

LLMs offer scalability that is difficult to achieve through traditional content creation methods. Here's how LLMs enable scalability:

1. **Output Flexibility:** LLMs can be adapted to produce the desired volume of content, whether it's hundreds of articles, thousands of social media posts, or personalized marketing material.
2. **Rapid Scaling:** Businesses and content creators can rapidly scale their content production efforts to meet growing demands or capitalize on emerging opportunities in the market.
3. **Multi-Purpose Use:** LLMs can generate content for various purposes, from informative articles to social media posts, allowing for a wide range of scalable content production applications.

12.5.4 CUSTOMIZATION

LLMs are highly versatile and can be fine-tuned to generate content in specific styles, tones, or languages. This customization brings several benefits (Figure 12.10):

1. **Content Variety:** LLMs can generate content in different styles, catering to diverse audiences. For example, the same model can produce formal academic content or casual, conversational blog posts.
2. **Multilingual Capabilities:** LLMs can be adapted to generate content in multiple languages, making them suitable for businesses with a global audience.

3. **Adaptation to Specific Needs:** Content creators can fine-tune LLMs to generate content tailored to specific needs, such as technical documentation, creative writing, or industry-specific material.

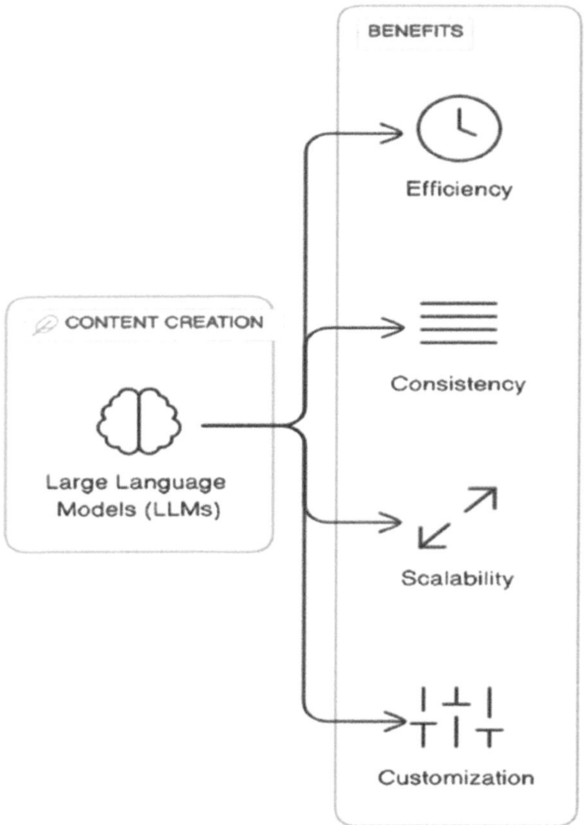

FIGURE 12.10 Large language model (LLM) based content creation benefits.

12.6 CHALLENGES AND ETHICAL CONSIDERATIONS

While LLMs offer numerous advantages for content creation, they also present several challenges and ethical considerations that require careful attention. Here, we will delve into these challenges and ethical concerns in detail:

12.6.1 QUALITY CONTROL

The automated nature of LLMs means that not all generated content will meet the desired quality standards. There are several aspects to consider:

1. **Content Review:** It's essential to implement quality control mechanisms to review and edit content generated by LLMs. Human oversight is crucial to ensure that the content is factually accurate, coherent, and free from grammatical errors.
2. **Tone and Style:** LLMs may not always perfectly capture the intended tone or style of a brand or individual. Content may require manual adjustments to align with the desired voice.
3. **Credibility:** Ensuring that the content generated is reliable and credible is vital, especially when it comes to news articles or educational materials. LLMs may inadvertently generate content that lacks factual accuracy.

12.6.2 PLAGIARISM

LLMs have access to vast amounts of online text, and this can lead to unintentional plagiarism if not used responsibly. It's important to address this concern:

1. **Originality:** Users must take precautions to ensure that the content generated by LLMs is original. Proper citations and references should be included when necessary to avoid plagiarism.
2. **Attribution:** Care should be taken to attribute ideas, facts, and phrases to their original sources, especially in academic, research, or journalistic contexts.

12.6.3 ETHICAL USE

The responsible use of LLMs is of utmost importance to mitigate potential harms. Several ethical considerations come into play:

1. **Misinformation:** LLMs should not be used to intentionally spread false or misleading information, especially in the context of news and public information.

2. **Hate Speech:** Generating content that promotes hate speech, discrimination, or incites violence is ethically unacceptable and may violate legal standards.
3. **Copyright and Privacy:** Respect for copyright laws and privacy rights is crucial. Content generated should not infringe on copyrights or invade individuals' privacy.
4. **Transparency:** It's essential to be transparent about the use of LLMs in content generation. Users should disclose when content is AI-generated, especially in cases where it might be perceived as human-authored.

12.6.4 LOSS OF AUTHENTICITY

Over-reliance on LLMs may lead to a loss of authentic human expression in content. Striking a balance between automation and human creativity is crucial for maintaining authenticity and engagement (Figure 12.11):

1. **Human Touch:** While LLMs can assist in content generation, the importance of human input, creativity, and authentic storytelling cannot be overstated. The human touch in content is what connects with audiences on a deeper level.
2. **Editorial Judgment:** Editors and content creators should play a role in curating and shaping content to ensure it reflects the values and objectives of the content creator or brand.

12.7 BEST PRACTICES FOR LARGE LANGUAGE MODEL (LLM)-BASED CONTENT CREATION

To harness the benefits of LLMs for content creation while addressing challenges and ethical concerns, it is essential to follow best practices that promote responsible and effective use. Here are detailed explanations of these best practices:

12.7.1 QUALITY ASSURANCE

1. **Robust Content Review:** Implement a thorough content review process to evaluate the quality and accuracy of content generated

by LLMs. Human editors should review and, if necessary, edit the output to ensure it meets the desired standards. This includes checking for grammatical errors, factual accuracy, and coherence.

2. **Fact-Checking:** Ensure that the information presented in the generated content is accurate and reliable. LLMs may occasionally produce inaccurate or outdated information, so fact-checking is essential, especially in contexts like news articles or educational material.

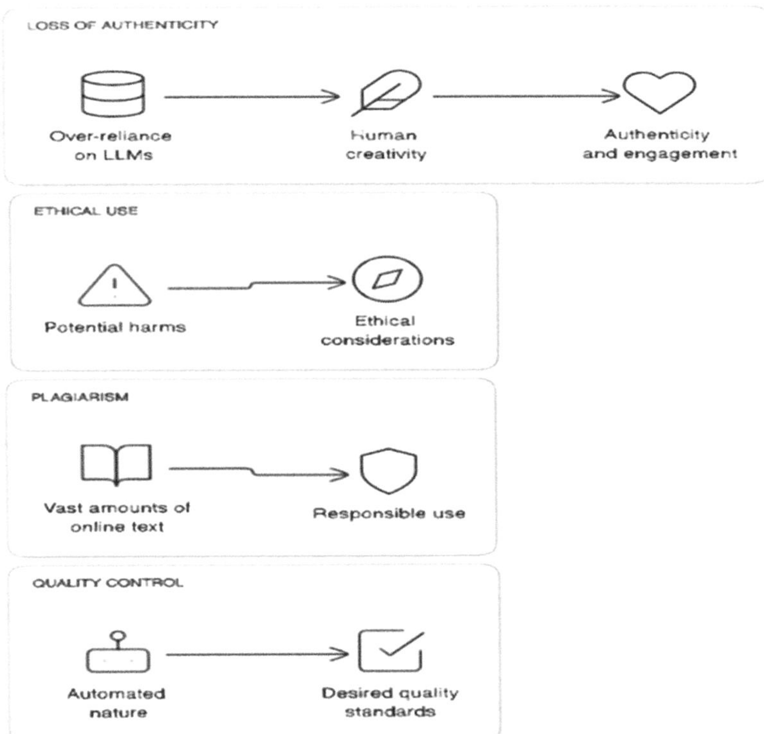

FIGURE 12.11 Large language models (LLMs) challenges and ethical concerns.

12.7.2 CUSTOMIZATION

1. **Fine-Tuning:** Customize LLMs to reflect the specific voice, style, and tone of the brand or individual they represent. Fine-tuning allows LLMs to generate content that aligns with the brand's

identity, messaging, and content style. This personalization is vital for maintaining consistency and authenticity.

2. **Language and Tone:** Adjust the language and tone settings of the LLM to match the target audience and context. LLMs can be fine-tuned to be formal, informal, technical, creative, or any style that suits the content's purpose.

12.7.3 CONTENT PLANNING

1. **Strategic Integration:** Use LLMs as a tool to assist in content creation, not as a replacement for human input. Develop a content strategy and plan that incorporates LLM-generated content within a broader content creation framework. This ensures that the generated content aligns with the content goals and objectives.
2. **Content Guidelines:** Establish guidelines and criteria for LLM-generated content. Define the types of content where LLMs are most suitable and specify the role of human editors in the review process.

12.7.4 ETHICAL GUIDELINES

1. **Fact-Checking and Plagiarism Prevention:** Adhere to strict fact-checking processes to verify the accuracy of LLM-generated content. Avoid unintentional plagiarism by ensuring originality and properly citing sources when necessary.
2. **Respect Copyright and Privacy:** Ensure that the generated content respects copyright laws and does not infringe on the privacy rights of individuals. Use copyrighted material only when appropriate permissions or fair use conditions apply.
3. **Avoiding Misinformation:** Follow strict ethical guidelines to prevent the generation or propagation of false or misleading information. Refrain from using LLMs to create content that promotes hate speech, discrimination, or incites violence.

12.7.5 HUMAN OVERSIGHT

1. **Authenticity and Ethics:** Maintain human oversight in the content creation process to uphold authenticity and ethical standards.

Human editors can apply their judgment, creativity, and expertise to ensure that the content aligns with the values and objectives of the content creator or brand.

2. **Final Review:** Include a final review stage in the content creation process to confirm that the generated content meets quality, authenticity, and ethical standards.

12.8 CASE STUDIES

Let's explore some case studies that demonstrate the successful application of LLMs in human-like content creation, including articles, social media posts, and marketing content.

1. **Case Study 1: The Washington Post's "Heliograf"**
   i. **Application:** Automated News Article Generation
      The Washington Post, a prominent newspaper, developed an LLM-based tool called "Heliograf" to generate news articles. Heliograf was initially used to cover the 2016 Summer Olympics and subsequently expanded its capabilities.
      - **Key Points:**
        - Heliograf generates news articles on various topics, including sports, politics, and local news.
        - It uses data feeds and templates to produce articles quickly.
        - The tool allows journalists to focus on more in-depth reporting while automating routine news coverage.
      - **Results:** Heliograf has successfully generated thousands of articles for The Washington Post, reducing the workload of journalists for repetitive, data-driven stories and allowing them to focus on investigative reporting and in-depth analysis.
2. **Case Study 2: ChatGPT's Social Media Engagement**
   i. **Application:** Social Media Post Generation
      A hypothetical marketing agency, "Digital Connections," uses OpenAI's ChatGPT to create engaging social media content for its clients.
      - **Key Points:**
        - Digital Connections utilizes ChatGPT to generate social media posts, tweets, and Instagram captions.

  - o The agency customizes ChatGPT's output to match the style and tone of each client's brand.
  - o ChatGPT provides a consistent and engaging online presence for various clients.
- **Results:** Digital Connections has successfully improved its clients' social media engagement and followership by maintaining a regular posting schedule and delivering content that resonates with their target audiences. The agency's clients have reported increased customer engagement and brand visibility.

3. **Case Study 3: E-commerce Marketing with GPT-3**
   i. **Application:** Marketing Content Generation
   An e-commerce company, "ShopSmart," utilizes GPT-3 to create product descriptions, email marketing campaigns, and ad copy for its online store.
   - **Key Points:**
     - o GPT-3 generates persuasive and creative product descriptions for ShopSmart's extensive product catalog.
     - o It assists in creating personalized email marketing campaigns based on customer behaviour and preferences.
     - o GPT-3 produces ad copy that resonates with specific target demographics.
   - **Results:** ShopSmart has witnessed a significant boost in its online sales and customer engagement. GPT-3's ability to generate compelling marketing content tailored to individual customers has improved click-through rates and conversion rates, leading to higher revenue for the e-commerce platform.

12.9 FUTURE DIRECTIONS

The field of LLM-based content creation continues to evolve rapidly, and several exciting future directions are emerging. These directions are poised to further enhance the capabilities and applications of LLMs in generating human-like content, including articles, social media posts, and marketing material (Figure 12.12):

1. **Enhanced Personalization:** Future developments in LLMs will focus on even more advanced personalization. These models

will be able to analyze user data, preferences, and behaviors to generate highly individualized content. This will result in social media posts, marketing campaigns, and product recommendations that are perfectly tailored to each user.

2. **Multimodal Content Creation:** LLMs have primarily focused on text-based content creation, but the future holds the promise of generating multimodal content. This means LLMs will be able to create content that includes not only text but also images, videos, and audio, resulting in more engaging and diverse content for various platforms.
3. **Real-time and Dynamic Content:** Future LLMs will be better equipped to generate real-time and dynamic content, such as live commentary on events, sports games, or breaking news. They will adapt to changing contexts and provide up-to-the-minute updates and insights.
4. **Content Collaboration:** LLMs will increasingly facilitate collaboration between humans and AI. Content creators will work alongside LLMs, using them as creative tools rather than replacements. The future may bring about platforms that seamlessly integrate human and AI-generated content in collaborative workflows.
5. **Content Translation and Localization:** LLMs will play a significant role in bridging language barriers and localizing content for global audiences. Improved language models will be able to generate content in multiple languages while preserving cultural nuances and context.
6. **Ethical and Bias Mitigation:** Future advancements will place a strong emphasis on addressing ethical concerns and mitigating biases in LLM-generated content. Efforts will focus on developing tools and techniques to reduce biased outputs and ensure that content adheres to responsible AI principles.
7. **Content Verification and Credibility:** LLMs will be equipped with built-in mechanisms for content verification and credibility assessment. These models will be able to fact-check and cross-reference information to ensure the accuracy and reliability of generated content.
8. **Augmented Content Creation:** LLMs will serve as tools for content creators rather than replacements. They will provide suggestions, recommendations, and automated drafts to assist human writers,

streamlining the content creation process while maintaining the author's unique voice.

9. **Regulatory Compliance:** As the use of LLMs in content creation expands, regulatory frameworks will emerge to address issues related to fake news, misinformation, and privacy. Future LLM development will incorporate features to facilitate compliance with these regulations.
10. **User Feedback Integration:** LLMs will increasingly integrate user feedback loops to improve content quality and relevance. Feedback mechanisms will play a vital role in training and fine-tuning models for specific applications and industries.

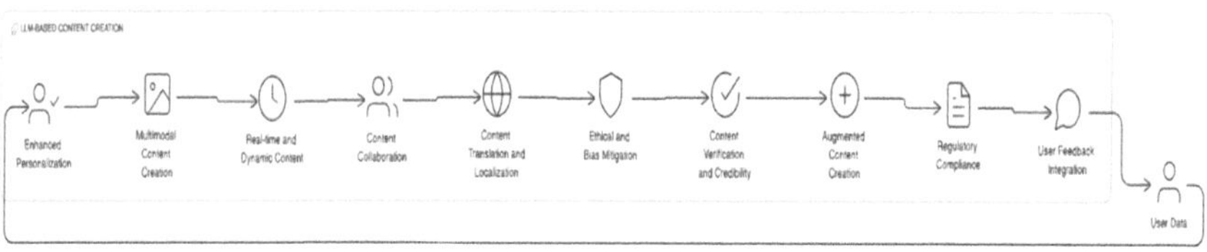

FIGURE 12.12 Future aspects in large language model based on content creation.

12.10 CONCLUSION

It is indeed a powerful tool with the potential to revolutionize content creation. It can automate the generation of articles, social media posts, marketing content, and much more, saving time and resources while maintaining a consistent and coherent style. Yet, this technology must be used responsibly and ethically. To harness the potential of LLMs in content creation, it is crucial to adhere to ethical standards. This means avoiding the spread of misinformation, hate speech, or any content that may infringe on copyright or privacy. Quality control is also essential to ensure the generated content is accurate, well-written, and coherent.

Furthermore, maintaining the authenticity and quality of content is paramount. While LLMs can be powerful tools, they are not a substitute for human creativity and judgment. Striking a balance between automation and human input is essential. Human oversight, editing, and input are vital for upholding the integrity of the content, ensuring it aligns with the brand's identity or the intended message. As technology continues to evolve, it is essential to adapt and utilize these tools judiciously. LLMs are advancing rapidly, and their potential applications are continually expanding. However, with this advancement, the responsibility to use them ethically and effectively also increases. Content creators and organizations must remain adaptable and considerate of the ethical and quality implications of LLM-based content creation.

In conclusion, LLM-based human-like content creation is a powerful tool that can transform content production but should be approached with responsibility and ethical consideration. Balancing automation with human creativity is the key to harnessing the full potential of LLMs while preserving content integrity. As technology progresses, adapting to the evolving landscape of content creation is crucial for the benefit of both content creators and consumers.

KEYWORDS

- **AI-generated text**
- **ethical AI Use**
- **human-AI collaboration**
- **LLM-based content creation**
- **marketing content AI**
- **social media automation**

REFERENCES

1. Adamopoulou, E., & Moussiades, L. (2020). Chatbots: History, technology, and applications. *Machine Learning with Applications, 2*, 100006.
2. Alruqi, T. N., & Alzahrani, S. M. (2023). Evaluation of an Arabic chatbot based on extractive question-answering transfer learning and language transformers. *AI, 4*(3), 667–691.
3. Ashfaque, M. W. (2022). Analysis of different trends in chatbot designing and development: A review. *ECS Transactions, 107*(1), 7215.
4. Bilquise, G., Ibrahim, S., & Shaalan, K. (2022). Emotionally intelligent chatbots: A systematic literature review. *Human Behavior and Emerging Technologies, 2022*.
5. Cherednichenko, O., Muhammad, F., Darmont, J., & Favre, C. (2023). *A Reference Model for Collaborative Business Intelligence Virtual Assistants.* arXiv preprint arXiv:2304.10556.

6. Eltahir, A. M., Abdulla, H., Platos, J., & Snasel, V. (2022). An overview of chatbot structure and source algorithms. In *2022 26th International Conference on Circuits, Systems, Communications and Computers (CSCC)* (pp. 91–98). IEEE.
7. Huynh-The, T., Pham, Q. V., Pham, X. Q., Nguyen, T. T., Han, Z., & Kim, D. S. (2023). Artificial intelligence for the metaverse: A survey. *Engineering Applications of Artificial Intelligence, 117*, 105581.
8. Ilić, S., Nakano, R., & Hajnal, I. (2019). *Designing Dialogue Systems: A Mean, Grumpy, Sarcastic Chatbot in the Browser*. arXiv preprint arXiv:1909.09531.
9. Jadhav, K. P., & Thorat, S. A. (2020). Towards designing conversational agent systems. In *Computing in Engineering and Technology: Proceedings of ICCET 2019* (pp. 533–542). Springer Singapore.
10. Karyotaki, M., Drigas, A., & Skianis, C. (2022). Chatbots as cognitive, educational, advisory & coaching systems. *Technium Social Sciences Journal, 30*, 109–126.
11. Lahiri, D., Kandimalla, P. C. P., & Jeysekar, A. (2023). Hybrid multi-purpose voice assistant. In *2023 2nd International Conference on Applied Artificial Intelligence and Computing (ICAAIC)* (pp. 816–822). IEEE.
12. Luo, B., Lau, R. Y., Li, C., & Si, Y. W. (2022). A critical review of state-of-the-art chatbot designs and applications. *Wiley Interdisciplinary Reviews: Data Mining and Knowledge Discovery, 12*(1), e1434.
13. Maher, S., Kayte, S., & Nimbhore, S. (2020). Chatbots & its techniques using AI: A review. *International Journal for Research in Applied Science and Engineering Technology, 8*(12), 503–508.
14. Nikhath, A. K., Rab, M. A., Bharadwaja, N. V., Reddy, L. G., Saicharan, K., & Reddy, C. V. M. (2022). An intelligent college enquiry bot using NLP and deep learning-based techniques. In *2022 International Conference for Advancement in Technology (ICONAT)* (pp. 1–6). IEEE.
15. Nirala, K. K., Singh, N. K., & Purani, V. S. (2022). A survey on providing customer and public administration-based services using AI: Chatbot. *Multimedia Tools and Applications, 81*(16), 22215–22246.
16. Nuruzzaman, M., & Hussain, O. K. (2018). A survey on chatbot implementation in the customer service industry through deep neural networks. In *2018 IEEE 15th International Conference on e-Business Engineering (ICEBE)* (pp. 54–61). IEEE.
17. Nuruzzaman, M., & Hussain, O. K. (2020). IntelliBot: A dialogue-based chatbot for the insurance industry. *Knowledge-Based Systems, 196*, 105810.
18. Singh, S., & Beniwal, H. (2022). A survey on near-human conversational agents. *Journal of King Saud University-Computer and Information Sciences, 34*(10), 8852–8866.
19. Zafar, A., Parthasarathy, V. B., Van, C. L., Shahid, S., & Shahid, A. (2023). *Building Trust in Conversational AI: A Comprehensive Review and Solution Architecture for Explainable, Privacy-Aware Systems Using LLMs and Knowledge Graphs*. arXiv preprint arXiv:2308.13534.
20. Zhang, Z., Jia, M., Yao, B., Das, S., Lerner, A., Wang, D., & Li, T. (2023). *"It's a Fair Game," or is It? Examining How Users Navigate Disclosure Risks and Benefits When Using LLM-Based Conversational Agents*. arXiv preprint arXiv:2309.11653.
21. Saxena, A. K., Asbe, C., & Vashishth, T. K. (2023). Leveraging a novel machine learning approach to forecast income and immigration dynamics. *Multidisciplinary Science Journal, 5*, 2023ss0202.

22. Vashishth, T. K., Kumar, B., Sharma, V., Chaudhary, S., Kumar, S., & Sharma, K. K. (2023). The evolution of AI and its transformative effects on computing: A comparative analysis. In B. Mishra (Ed.), *Intelligent Engineering Applications and Applied Sciences for Sustainability* (pp. 425–442). IGI Global.
23. Verma, P. K., Sharma, V., Kumar, P., Sharma, S., Chaudhary, S., & Preety, P. (2023). IoT-enabled real-time appearance system using AI camera and deep learning for student tracking. *International Journal on Recent and Innovation Trends in Computing and Communication, 11*(6s), 249–254.

CHAPTER 13

LLMS-Powered Chatbots and Virtual Assistants for Interactive and Human-Like Interactions

VIKAS SHARMA[1,*], SUNIL KUMAR[2], TARUN KUMAR VASHISHTH[3], KEWAL KRISHAN SHARMA[3], BHUPENDRA KUMAR[3], and SACHIN CHAUDHARY[3]

[1]*Department of Computer Applications, SRM Institute of Science and Technology, Delhi NCR Campus, Ghaziabad, U.P., India*

[2]*School of Information Technology, Auro University, Gujrat, India*

[3]*School of Computer Science and Applications, IIMT University, Meerut, Uttar Pradesh, India*

**Corresponding author.*

ABSTRACT

In the era of rapid technological advancement, chatbots and virtual assistants have emerged as indispensable tools for enhancing user experiences in various domains, including customer service, healthcare, and education. This chapter explores the application of Long-Short Memory networks (LLMs) to empower chatbots and virtual assistants, enabling them to engage in interactive and human-like interactions with users. The development and integration of chatbots and virtual assistants into digital platforms have witnessed significant growth in recent years. These AI-driven entities have proven their utility in simplifying tasks, providing timely information, and facilitating seamless communication. However, their ability to engage in meaningful and natural conversations remains a critical challenge. LLMs, a type of recurrent neural network (RNN), are leveraged to overcome this

Natural Language Processing and Artificial Intelligence: A Perspective Toward Current Trends, Challenges, and Applications. Vikas Khullar, Aryan Chaudhary, & Sunil Kumar (Eds.)

DOI: 10.1201/9781998511273-13

challenge by endowing these conversational agents with enhanced memory and understanding capabilities. LLMs-based chatbots and virtual assistants possess the capacity to store and retrieve context and user history, enabling them to maintain coherent and contextually relevant conversations. By maintaining a memory of the ongoing dialogue, these agents can provide more personalized and engaging interactions, leading to higher user satisfaction and a more human-like conversational experience. Moreover, LLMs networks excel in understanding and generating text that closely resembles human language, resulting in smoother and more natural exchanges.

This chapter delves into the technical aspects of LLMS integration into chatbot and virtual assistant architectures. It discusses the training process, data preparation, and fine-tuning required to optimize LLMS-powered conversational agents. The challenges of handling vast and diverse datasets and mitigating biases are also explored, emphasizing the importance of ethical considerations in chatbot and virtual assistant development. Furthermore, the practical implications and use cases of LLMS-powered chatbots and virtual assistants are examined. Industries such as e-commerce, healthcare, and customer support have seen remarkable improvements in user engagement and efficiency with the adoption of these technologies. Case studies and real-world examples illustrate how LLMS-powered agents are transforming user experiences by providing intelligent responses and a more human-like touch in their interactions. LLMS-powered chatbots and virtual assistants are at the forefront of revolutionizing user interactions in the digital age. Their ability to offer more personalized, coherent, and human-like conversations sets the stage for improved customer satisfaction, enhanced accessibility, and efficient service delivery. The application of LLMS technology in chatbots and virtual assistants opens exciting possibilities for businesses and organizations seeking to harness the potential of AI-driven conversational agents for a wide range of applications, ultimately shaping the future of human-computer interactions.

13.1 INTRODUCTION

The introduction to the chapter on "long-short memory (LLMS) powered chatbots and virtual assistants for interactive and human-like interactions" sets the stage for the discussion by presenting the context and background of the topic. This section provides readers with an overview of the key factors that have led to the development and adoption of chatbots and virtual assistants powered by LLMS networks. Here, we delve into the rise of these conversational agents, the significance of natural language processing (NLP), the emergence of large language models (LLMs), and the promise of achieving interactive and human-like interactions.

13.1.1 THE RISE OF CHATBOTS AND VIRTUAL ASSISTANTS

While chatbots and virtual assistants have steadily grown in prominence over recent years, their potential remains largely untapped. Capabilities that once seemed futuristic are now commonplace—these AI systems can instantly respond to queries around the clock from a diverse array of industries. However, more humanistic qualities continue to elude such technologies. For customers seeking support, recommendations flow mechanically from scripted exchanges lacking nuanced understanding. Businesses leverage automation for efficiency over empathy, prioritizing resolution metrics above all else. Across platforms, a uniform blandness pervades interactions; responses fail to captivate with flair or surprise in their construction. Creativity is stamped out in favor of consistency in these systems. Still, their value persists as digital dependence expands. Where humans cannot always be present, chatbots fill gaps with ease. Continued advances may one day imbue such agents with traits more reminiscent of us—versatility, subtlety, and complexities of thought and speech that genuinely connect.

For now, we utilize their capabilities judiciously while awaiting the fuller realization of their potential to understand and engage as only people can. This rise can be attributed to several key factors:

1. **Efficient and Automated Customer Service:** Routine customer queries continue to be efficiently addressed through automated chatbots, as businesses have come to appreciate their ability to swiftly handle common inquiries and liberate agents for more nuanced matters. These AI conversational agents excel at repetitive duties like answering frequently posed questions, thereby allowing people to focus on personalized exchanges requiring complexity. Within organizations, too, virtual assistants have taken on habitual responsibilities, freeing up time for intricate, customized interactions essential to excellent service overall. Meanwhile, advanced bots are progressively becoming more versatile in their mixture of brief and intricate replies befitting a human touch.

2. **Desire for Improved User Experiences:** User experience is paramount for any digital platform, determining whether it fulfills its purpose or frustrates its users. Chatbots have tremendous potential to improve the experience through individualized, immediate assistance. They can simplify intricate processes, guiding users effortlessly to desired outcomes. A virtual agent interacts in a manner reminiscent of casual conversation, answering questions clearly and concisely while maintaining a friendly, approachable demeanor. Complex inquiries can be broken down into sensible steps and clarified plainly. Lengthy explanations might occasionally expand on key details while brevity suffices for more straightforward requests. By creating a personalized interface adapted to each person's needs, these AI helpers deliver convenience with calming courtesy. Services and businesses harnessing such intuitive design seamlessly strengthen relationships with satisfied, loyal clients.
3. **Increased Reliance on Digital Communication Channels:** The modern world is characterized by a growing reliance on digital communication channels. People use various digital platforms, including websites, mobile apps, and social media, to connect with businesses, seek information, and make transactions. Chatbots and virtual assistants have seamlessly integrated into these digital channels, becoming available to users 24/7. This continuous availability means that users can obtain assistance or information at their convenience, even outside of regular business hours. The ability to interact with these AI-driven entities through familiar digital channels aligns with the preferences and habits of modern consumers, further fueling their adoption.
4. **Fast-Paced Digital World:** In today's fast-paced digital landscape, where instant gratification and efficiency are valued, chatbots and virtual assistants offer valuable solutions. Users expect quick responses and immediate access to information. Chatbots can provide instant answers, directions, or solutions, meeting these expectations. Their ability to handle multiple requests simultaneously without delays or fatigue ensures that users can have their needs met promptly. This makes chatbots indispensable in scenarios where swift responses are essential, such as e-commerce, customer support, and information retrieval (IR).

Hence, the rise of chatbots and virtual assistants can be attributed to their ability to address the increasing demand for efficient and convenient digital interactions. They have become vital tools in various sectors, catering to the need for automated customer service, improved user experiences, and seamless integration into the digital communication channels of the modern world. As technology continues to advance, these AI-driven conversational agents are expected to play an even more significant role in shaping the way businesses and organizations interact with their customers and users (Figure 13.1).

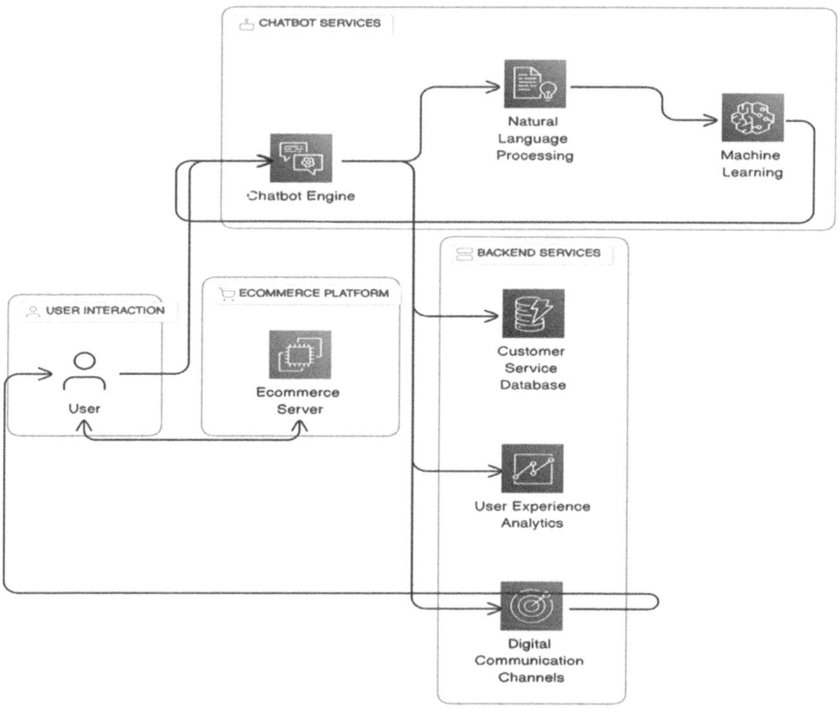

FIGURE 13.1 Chatbots and virtual assistants architecture.

13.1.2 THE POWER OF NATURAL LANGUAGE PROCESSING

NLP is a foundational technology that underpins the capabilities of chatbots and virtual assistants. NLP enables machines to understand, interpret, and generate human language, allowing for meaningful interactions between users and AI-powered entities. With the advancements in NLP, chatbots and virtual assistants have evolved from simple rule-based systems to sophisticated, context-aware agents capable of processing and responding to user queries in a more natural and human-like manner. NLP forms the basis for these conversational agents to bridge the communication gap between humans and machines.

1. **Foundation of Chatbot and Virtual Assistant Capabilities:** NLP is the core technology that underpins the capabilities of chatbots and virtual assistants. It equips these AI-powered entities with the ability to comprehend, interpret, and generate human language. This means they can process text and speech input, making sense of the language's nuances, grammar, syntax, and semantics. NLP allows chatbots and virtual assistants to not only recognize the words in a user's input but also understand the intent and context behind those words. This foundational capability forms the basis for meaningful interactions between users and AI-powered conversational agents.
2. **Evolution of Chatbots and Virtual Assistants:** Before the wide-spread adoption of NLP, chatbots and virtual assistants were often limited to rule-based systems. These early systems could only respond to specific predetermined commands or patterns. They lacked the ability to adapt to different ways users might phrase their queries or requests, making interactions feel rigid and frustrating for users. However, with the advancements in NLP, chatbots and virtual assistants have undergone a remarkable transformation. They have evolved into sophisticated, context-aware agents capable of understanding and responding to user queries in a more natural and human-like manner. NLP allows them to process a wide range of inputs, detect user intent, and provide relevant, contextually appropriate responses. This evolution has significantly improved user satisfaction and engagement.
3. **Bridging the Communication Gap:** NLP plays a crucial role in bridging the communication gap between humans and machines. It enables chatbots and virtual assistants to speak the language of humans, making interactions more intuitive and user-friendly. Users can communicate with these AI-driven entities in their own words, without needing to adapt to a specific set of commands or rigid syntax. This not only enhances the user experience but also

expands the range of applications for chatbots and virtual assistants. They can now be used in various domains, from customer service and healthcare to education and entertainment, as they can understand and adapt to the specific language and terminology of each domain.

Natural language processing is a fundamental technology that empowers chatbots and virtual assistants to operate effectively in a human language environment. It allows these conversational agents to process, interpret, and generate text and speech, enabling them to understand user intent and context. The advancements in NLP have propelled chatbots and virtual assistants from being rule-based and limited in their capabilities to becoming highly versatile, context-aware, and capable of delivering more natural and human-like interactions. As NLP continues to advance, it promises to further enhance the capabilities and applications of these AI-driven entities, making them indispensable in numerous industries and sectors (Figure 13.2).

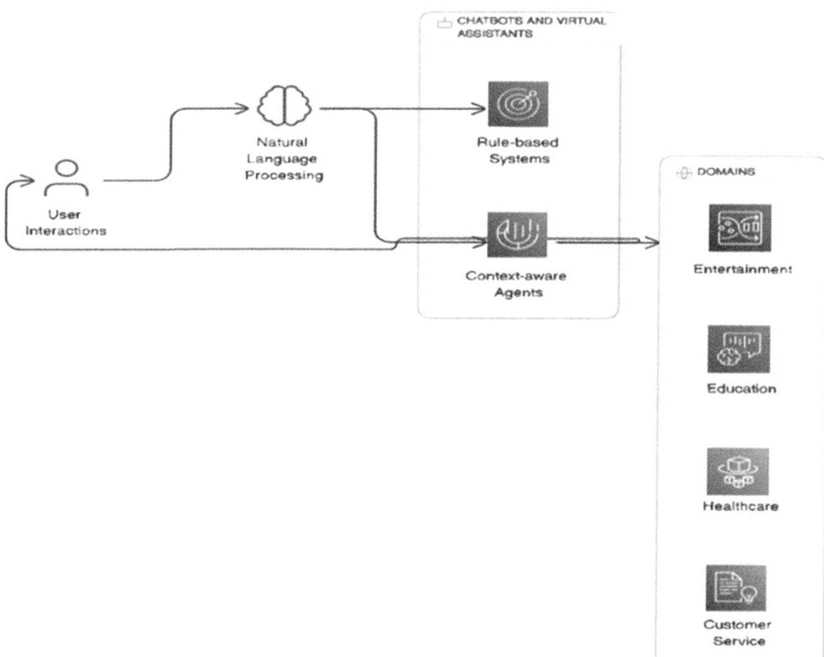

FIGURE 13.2 Natural language processing in chatbots and virtual assistants.

13.1.3 THE EMERGENCE OF LARGE LANGUAGE MODELS

The emergence of LLMs has been a transformative development in the field of artificial intelligence (AI). These models, such as GPT-3, can process vast amounts of text data and generate coherent and contextually relevant responses. They have opened new possibilities for enhancing the capabilities of chatbots and virtual assistants. LLMs like LLMS have the capacity to understand and generate text that closely resembles human language, enabling more engaging and sophisticated interactions with users (Figure 13.3).

1. **Transformational Impact on AI:** LLMs have had a transformational impact on the field of AI. These models are built upon deep neural networks, boasting millions or even billions of parameters that enable them to understand and generate human language. They leverage the vast amount of text data available on the internet to learn the intricacies of language, including grammar, syntax, and semantics. This extensive knowledge empowers them to provide sophisticated responses to user queries. GPT-3, for instance, is known for its ability to generate human-like text, making it a valuable tool for a wide range of NLP tasks.
2. **Processing Vast Text Data:** LLMs are trained on massive datasets that encompass diverse text sources, from books and articles to websites and social media. They ingest and process this enormous volume of text, learning patterns and structures within the language. This training allows them to have a broad understanding of various topics and domains, making them versatile in responding to a wide array of user queries. The ability to process vast amounts of text data is a key differentiator, as it enables these models to provide contextually relevant responses across different subjects.
3. **Enhancing Chatbots and Virtual Assistants:** LLMs, often referred to as LLMS, have significantly enhanced the capabilities of chatbots and virtual assistants. They serve as the "brains" behind these conversational agents, allowing them to understand user queries, context, and intent more effectively. By incorporating LLMS technology, chatbots and virtual assistants can provide responses that closely resemble human language. This enables more engaging and sophisticated interactions with users, leading to higher user satisfaction. Chatbots and virtual assistants equipped with LLMS can

provide nuanced, informative, and contextually appropriate answers, making them more valuable in a wide range of applications.

4. **Human-Like Text Generation:** One of the most notable features of LLMs is their capacity to generate text that closely resembles human language. This human-like text generation is a significant advancement in AI, as it enables chatbots and virtual assistants to communicate in a more natural and relatable manner. Users are more likely to engage in extended conversations and feel a sense of connection when interacting with AI-driven entities that can generate text in a human-like fashion. This not only improves the overall user experience but also has applications in content generation, copywriting, and more.

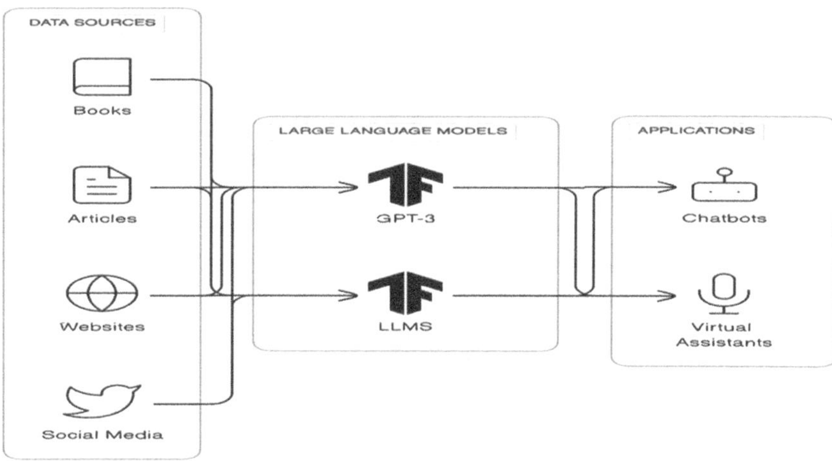

FIGURE 13.3 Large language models architecture.

Therefore, the emergence of LLMs, such as GPT-3 and similar LLMs, has brought about a paradigm shift in the field of AI. These models are adept at processing vast text data and generating contextually relevant, human-like responses. By integrating LLMs into chatbots and virtual assistants, these AI-driven entities can provide more engaging, nuanced, and sophisticated interactions with users. This development represents a significant step forward in making human-computer interactions more seamless and natural, with wide-ranging implications for industries and applications across the digital landscape.

13.1.4 THE PROMISE OF INTERACTIVE AND HUMAN-LIKE INTERACTIONS

The promise of interactive and human-like interactions is at the heart of the discussion in this chapter. By leveraging LLMS technology, chatbots and virtual assistants are poised to offer users a conversational experience that is not only informative but also engaging and natural. LLMS networks provide these AI agents with enhanced memory and understanding capabilities, enabling them to remember context, maintain coherent dialogues, and provide more personalized responses. As a result, users can expect interactions that mimic human conversation, creating a more satisfying and efficient user experience (Figure 13.4).

1. **Informative, Engaging, and Natural Interactions:** Chatbots and virtual assistants have traditionally been employed for their ability to provide information and assistance. However, the promise of interactive and human-like interactions goes beyond mere information delivery. It involves engaging users in a dynamic dialogue that is both informative and enjoyable. The aim is to create interactions that are as close to human conversations as possible, where users feel comfortable and engaged. This approach can transform chatbots and virtual assistants from functional tools into companions that users want to interact with regularly.
2. **Enhanced Memory and Understanding:** LLMS networks play a crucial role in realizing the promise of interactive and human-like interactions. They endow chatbots and virtual assistants with improved memory and understanding capabilities. Unlike earlier rule-based systems that lacked context awareness, LLMS-powered agents can remember the ongoing conversation, understand user intent, and recognize the context in which queries are made. This enables them to provide responses that are not isolated but are contextually relevant, thereby creating a more coherent and engaging conversational experience.
3. **Coherent Dialogues:** Coherence in dialogues is essential for an interaction to feel natural and human-like. LLMS-powered chatbots and virtual assistants maintain coherence by recalling prior exchanges and incorporating them into their responses. This

prevents the conversation from feeling disjointed and allows users to ask follow-up questions or provide partial information, much like they would in a human conversation. The result is a fluid and coherent dialogue that mirrors real-life interactions.

4. **Personalized Responses:** Another aspect of achieving human-like interactions is providing personalized responses. LLMS networks can analyze user input in the context of the conversation, making it possible to tailor responses to individual users' needs and preferences. This personalization leads to more meaningful and relevant interactions, increasing user satisfaction and engagement.
5. **Satisfying and Efficient User Experience:** The culmination of these capabilities is a user experience that is both satisfying and efficient. Users feel understood, engaged, and well-supported, making their interactions with chatbots and virtual assistants highly efficient. They are more likely to have their needs met promptly and accurately. The result is a user experience that not only saves time but also leaves a positive impression, which can be instrumental in building brand loyalty and customer satisfaction.

Hence, the promise of interactive and human-like interactions, made possible through LLMs technology, represents a significant step forward in the development of chatbots and virtual assistants. These AI agents can offer informative, engaging, and natural conversations, enhancing the overall user experience. By providing enhanced memory, understanding, coherence, and personalization, they transform the interaction into a satisfying and efficient experience that closely resembles human conversation. This capability opens doors to a wide range of applications in various industries, from customer service and healthcare to education and entertainment.

13.2 UNDERSTANDING LARGE LANGUAGE MODELS (LLMS)

13.2.1 LARGE LANGUAGE MODELS (LLMS) UNVEILED

This section provides an in-depth exploration of LLMs and their significance in the context of chatbots and virtual assistants. It covers the following key subtopics:

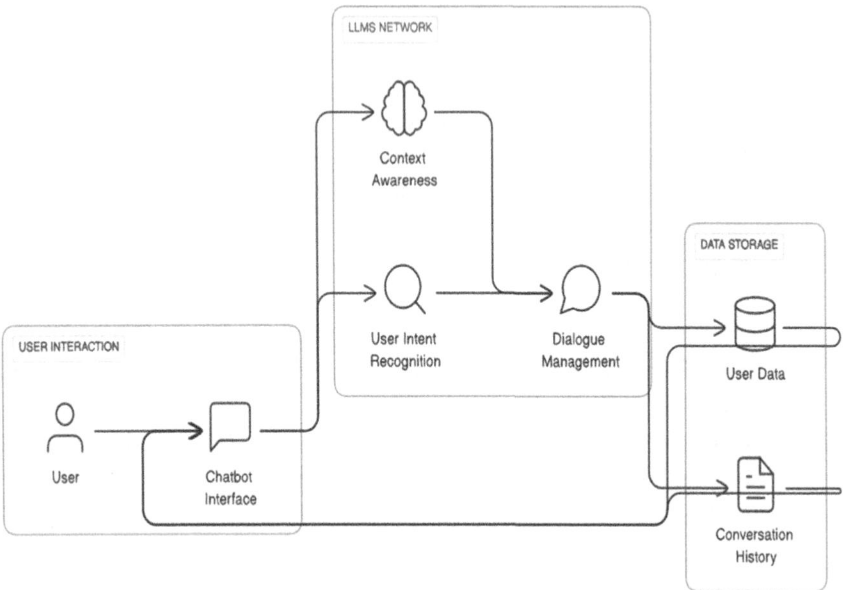

FIGURE 13.4 Cloud system architecture.

13.2.1.1 DEFINING LARGE LANGUAGE MODELS

LLMs, often abbreviated as LLMs, are a class of AI models designed to understand and generate human language at an advanced level. These models are characterized by their vast neural network architectures, consisting of millions or even billions of parameters. LLMs are not only capable of processing text but also of generating coherent and contextually relevant language, making them invaluable for natural language understanding (NLU) and generation tasks. The definition of LLMs is fundamental to comprehending their role in enhancing chatbots and virtual assistants. By grasping the essence of what makes LLMs unique, it becomes evident how they empower conversational agents to engage in more natural and sophisticated interactions with users.

1. **Advanced Language Comprehension and Generation:** LLMs are a class of AI models designed to operate at an advanced level of language understanding and generation. Unlike conventional language models, LLMs can not only comprehend human language but also generate coherent and contextually relevant text, often to a

degree that it is indistinguishable from human-generated content. This advanced language processing ability distinguishes LLMs from earlier AI language models.

2. **Vast Neural Network Architectures:** One of the defining features of LLMs is their enormous neural network architectures. These models are characterized by having millions or even billions of parameters, which are the components that the model uses to learn from data. This scale is in sharp contrast to earlier language models, which had far fewer parameters. The vast neural network architecture allows LLMs to capture and represent intricate linguistic patterns, nuances, and context in a way that was previously unattainable.
3. **Natural Language Understanding and Generation:** LLMs excel at both understanding and generating human language. Their understanding capabilities enable them to comprehend the semantics, syntax, and context of text data. This comprehension allows them to accurately interpret user queries, extract meaning from textual input, and identify user intent. Additionally, LLMs have the capability to generate text that is not only grammatically correct but also contextually coherent and contextually relevant. This means they can provide responses that align with the context of the conversation, making interactions more natural and engaging.
4. **Enhancing Chatbots and Virtual Assistants:** The definition of LLMs is foundational in recognizing their potential to enhance chatbots and virtual assistants. By incorporating LLMs into these conversational agents, they gain the ability to understand user queries more deeply, generate responses that are contextually relevant and coherent, and provide a more natural and human-like conversational experience. This empowers chatbots and virtual assistants to offer users a higher level of interaction that goes beyond mere IR, resulting in more engaging and satisfying user experiences.

LLMs represent a significant advancement in AI technology, particularly in the realm of NLP. Their advanced language understanding and generation capabilities, combined with their vast neural network architectures, make them indispensable for improving chatbots and virtual assistants. By leveraging LLMs, these conversational agents can provide more sophisticated, context-aware, and engaging interactions, bringing us closer to human-like conversations in the digital world. Understanding the fundamental characteristics of LLMs is pivotal to appreciating their role in revolutionizing human-computer interactions.

13.2.1.2 A BRIEF HISTORY

This part of the chapter provides a historical perspective on the development and evolution of LLMs. It traces the origins of LLM technology, highlighting key milestones and breakthroughs in the field of AI. LLMs have undergone significant advancements, with each iteration building on the successes and lessons learned from its predecessors. The historical context is crucial for understanding the rapid progress of LLMs and their increasing relevance in applications like chatbots and virtual assistants. It shows that these models are the result of continuous research and development, and their growth in scale and capability is a testament to the relentless pursuit of better NLP (Figure 13.5).

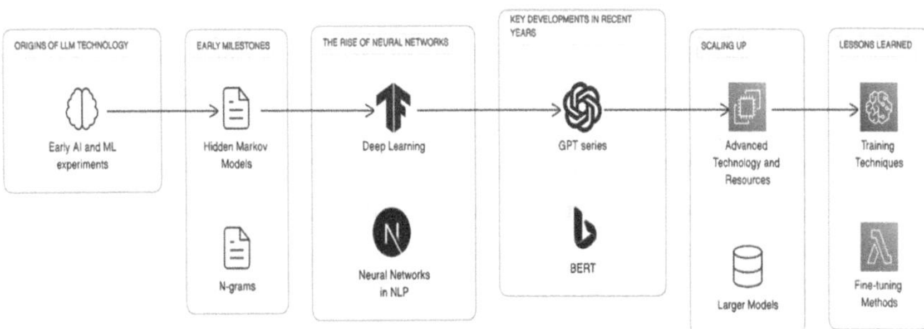

FIGURE 13.5 Brief history of large language model (LLM).

1. **Origins of LLM Technology:** The history of LLMs can be traced back to the early days of AI and machine learning (ML). Researchers began experimenting with language models that could predict the next word in a sentence based on the words that came before it. These early language models laid the foundation for more complex and capable models.
2. **Early Milestones:** In the 20th century, various language modeling techniques, such as Hidden Markov Models (HMMs)

and n-grams, were developed to process and generate text. While these models had their limitations, they represented important milestones in the journey toward advanced language understanding and generation.

3. **The Rise of Neural Networks:** The true breakthrough in the development of LLMs came with the resurgence of neural networks in the late 20th century and the early 21st century. Deep learning (DL) and the application of neural networks to NLP tasks breathed new life into language modeling. Researchers started to experiment with more complex neural network architectures that could capture the intricacies of language, syntax, and semantics.
4. **Key Developments in Recent Years:** The last decade has witnessed remarkable advancements in LLM technology. Notable models include the GPT series, bidirectional encoder representations from transformers (BERT), and more. GPT-3 represents a watershed moment in LLM development, with its 175 billion parameters and the ability to generate highly coherent and contextually relevant text.
5. **Scaling Up:** As technology and computational resources advanced, researchers began to scale up the size of LLMs. They realized that larger models, with more parameters, could capture more of the subtleties in human language. This scaling led to the creation of models with billions of parameters, such as GPT-3, which pushed the boundaries of what was previously possible.
6. **Lessons Learned:** Along this journey, researchers have learned valuable lessons about the training, architecture, and fine-tuning of LLMs. They have improved techniques for training models on massive text datasets and have devised methods for fine-tuning models for specific tasks and domains.

Hence, the historical journey of LLMs from early language models to the current era of billion-parameter models showcases the evolution of NLP in AI. This journey is pivotal for understanding the exponential growth of LLMs and their crucial role in advancing the capabilities of chatbots and virtual assistants. It demonstrates how research, innovation, and the quest for better language understanding have led to the development of LLMs that are poised to transform the way humans interact with machines.

13.2.1.3 GPT-3.5: A MARVEL OF LARGE LANGUAGE MODEL (LLM) TECHNOLOGY

Generative pre-trained transformer 3.5 (GPT-3.5) is introduced as a prime example of a state-of-the-art LLM. This subsection delves into the features and capabilities of GPT-3.5, emphasizing its remarkable ability to generate human-like text and its capacity to handle complex NLU tasks. GPT-3.5 serves as a model for the advancements achieved in LLM technology, setting a high bar for the potential applications of such models in chatbots and virtual assistants. It showcases the impressive language generation capabilities and contextual understanding that can be harnessed to improve user interactions.

1. **Exceptional Language Generation:** GPT-3.5 is renowned for its exceptional language generation capabilities. It has the capacity to produce text that closely resembles human writing. When given a prompt or context, GPT-3.5 can generate coherent and contextually relevant responses. This ability extends beyond mere grammatical correctness; it encompasses the generation of text that reflects an understanding of context and an apt use of language, making it a marvel in human-like language generation.
2. **Complex Natural Language Understanding:** Beyond its language generation prowess, GPT-3.5 is highly skilled in NLU. It can comprehend and interpret complex textual inputs, discerning the nuances of language, context, and user intent. This understanding is essential for providing meaningful and contextually appropriate responses. GPT-3.5's ability to grasp intricate language structures and semantics sets it apart as a state-of-the-art LLM.
3. **Vast Scale and Parameters:** GPT-3.5 is characterized by its substantial scale, featuring a vast number of parameters. While the specific number may vary, GPT-3.5, like its predecessors, boasts billions of parameters. This immense scale contributes to its language processing proficiency, enabling it to capture fine-grained details and nuances in the language it encounters.
4. **Versatility in Applications:** GPT-3.5's versatility extends to a wide range of applications. It can be fine-tuned for specific tasks and domains, making it adaptable for various purposes. This adaptability is a testament to its capabilities, as it can excel in different

contexts, from content generation and language translation to chatbot interactions and more.

5. **Setting a High Bar for LLMs:** GPT-3.5 stands as a model for the advancements achieved in LLM technology. Its capabilities represent a high bar for the potential applications of such models in chatbots and virtual assistants. By demonstrating the capacity to generate human-like text and understand complex natural language, GPT-3.5 showcases the possibilities for improving user interactions and enhancing the capabilities of AI-driven conversational agents.
6. **Driving Innovation:** GPT-3.5 has spurred innovation in the field of AI and NLP. Its capabilities have motivated researchers and developers to explore new avenues and applications for LLMs. This innovation has the potential to revolutionize how humans interact with machines and use AI-driven technologies.

Therefore, GPT-3.5 represents a marvel in LLM technology, exemplifying exceptional language generation and complex NLU. Its vast scale, versatility, and the benchmark it sets for LLMs make it a pivotal development in AI. GPT-3.5's capabilities have the potential to redefine the way chatbots and virtual assistants operate, promising more natural, engaging, and efficient interactions with users. Its impact on the field of AI serves as a testament to the ever-evolving and promising landscape of LLMs (Figure 13.6).

13.2.1.4 THE ARCHITECTURE AND TRAINING OF MISUNDERSTANDING

This section provides insight into the neural network architecture employed in LLMs, explaining how the vast number of parameters allows them to capture language patterns and semantics. Additionally, the training process of LLMs is discussed. It involves exposing the model to extensive text data from the internet, enabling it to learn the nuances of language, including grammar, vocabulary, and context. This training phase is instrumental in shaping the model's language generation and understanding abilities.

Knowledge of LLM architecture and training is critical for comprehending how these models can enhance chatbots and virtual assistants. It illustrates how LLMs can be integrated into conversational agents to endow them with advanced NLP capabilities, resulting in more interactive and human-like interactions.

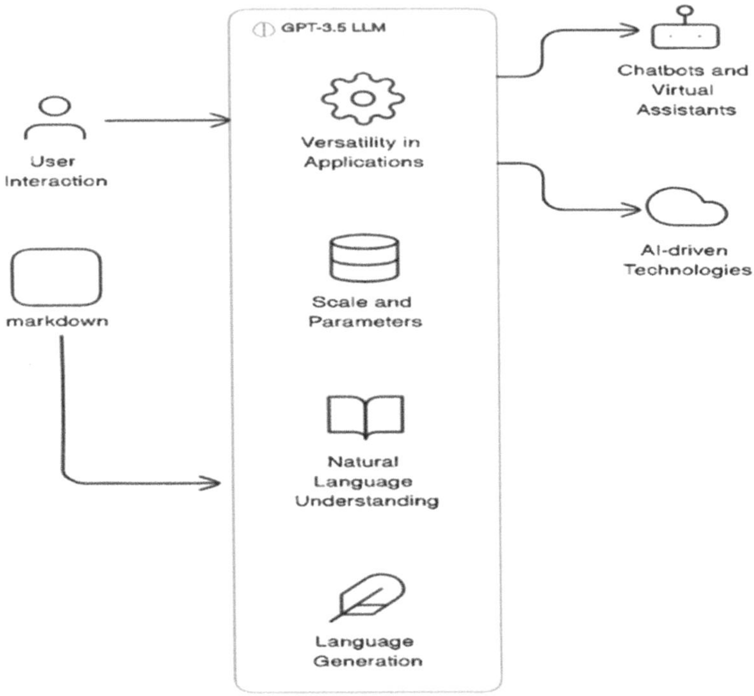

FIGURE 13.6 GPT-3.5 large language model.

13.2.1.4.1 Neural network architecture

Large language models, including models like GPT-3.5, rely on a deep neural network architecture, particularly a variant of the Transformer architecture. The Transformer architecture is characterized by self-attention mechanisms that allow the model to consider the relationships between different words in a sentence. The neural network consists of multiple layers, each containing millions or billions of parameters. These parameters serve as the model's "knowledge," allowing it to understand and generate human language.

1. **Self-Attention:** Self-attention is a pivotal component of the architecture that enables the model to weigh the importance of each word in a sentence concerning the others. This mechanism allows

LLMs to capture context and dependencies within the text, which is crucial for understanding and generating coherent language.

2. **Deep Layers:** LLMs typically have a deep stack of layers, often numbering in the hundreds. These layers process the input data at different levels of abstraction, allowing the model to learn intricate patterns and structures in language.
3. **Parameters:** The sheer number of parameters in LLMs is one of their defining characteristics. Models like GPT-3.5 may have billions of parameters. These parameters are learned during the training process, and they encode the model's understanding of language.

13.2.1.4.2 Training process

The training process of LLMs is an essential phase in their development. It involves exposing the model to vast amounts of text data from the internet. During training, the model learns to predict the next word in a sentence based on the words that came before it. This process is repeated for an extensive corpus of text, allowing the model to learn grammar, vocabulary, context, and semantic relationships between words.

1. **Fine-Tuning:** In addition to the initial training, LLMs can undergo fine-tuning for specific tasks or domains. This fine-tuning process involves training the model on a narrower dataset that is tailored to a particular application. For example, fine-tuning can make the model more proficient in medical terminology or legal language.
2. **Pre-Training:** Before fine-tuning, LLMs go through a pre-training phase where they acquire a broad understanding of the language. This phase equips them with a general linguistic competence that can be applied to a wide range of tasks.

13.2.1.4.3 Enhancing chatbots and virtual assistants

Knowledge of LLM architecture and training is critical for understanding how these models can enhance chatbots and virtual assistants. By integrating LLMs into these conversational agents, they gain the capability to process and generate text with a deep understanding of language patterns and semantics. This results in more interactive and human-like interactions, where the chatbots and virtual assistants can respond coherently to user queries and adapt to the context of the conversation.

Hence, the architecture and training of LLMs represent the core of their capabilities. The neural network architecture, with its self-attention mechanisms and deep layers, allows LLMs to capture language intricacies. The training process, involving exposure to extensive text data and fine-tuning, equips LLMs with a deep understanding of language. This understanding is leveraged to enhance chatbots and virtual assistants, making them more capable of providing natural, contextually relevant, and engaging interactions with users.

13.3 LARGE LANGUAGE MODELS (LLMS) IN CONVERSATIONAL ARTIFICIAL INTELLIGENCE (AI)

This section explores the role of LLMs in conversational AI, shedding light on how these models have transformed the landscape of human-computer interactions. It delves into the unique capabilities that make LLMs a game changer, their ability to comprehend context, and their proficiency in generating human-like text.

13.3.1 LARGE LANGUAGE MODELS (LLMS) AS THE GAME CHANGER

LLMs have emerged as a game changer in the field of Conversational AI. Their impact is profound for several reasons (Figure 13.7):

1. **Enhanced Understanding:** LLMs have significantly improved the understanding of context and user intent in conversations. They can analyze a user's input in the context of the ongoing conversation, making interactions more coherent and contextually relevant.
2. **Natural Language Generation:** LLMs excel at generating human-like text. This means that the responses they produce are not only grammatically correct but also contextually coherent, creating more natural and engaging interactions with users.
3. **Versatility:** LLMs can be fine-tuned for specific domains or applications, making them versatile tools for various conversational AI use cases. They can be adapted to excel in areas like customer service, content generation, healthcare, and more.

4. **Language Comprehension:** LLMs have the capacity to understand a wide range of languages and dialects, allowing them to engage with users in their preferred language. This linguistic flexibility enhances inclusivity and accessibility.

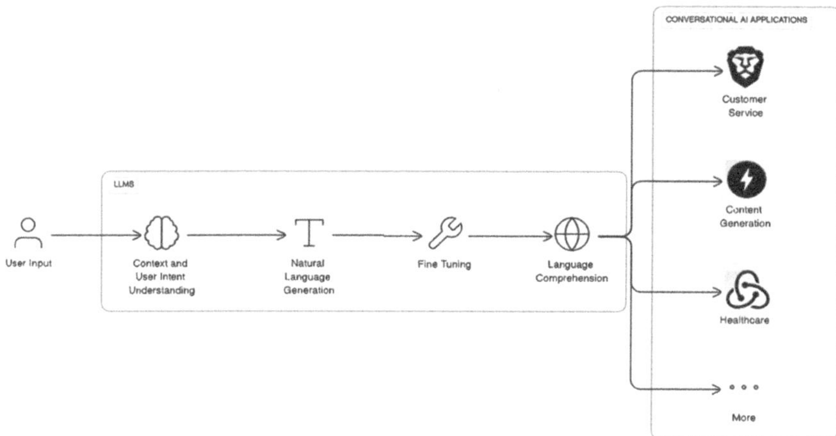

FIGURE 13.7 Large language models in conversational artificial intelligence (AI).

13.3.2 HOW LARGE LANGUAGE MODELS (LLMS) COMPREHEND CONTEXT

Understanding context is a critical aspect of Conversational AI, and LLMs excel in this regard:

1. **Memory and Attention:** LLMs use self-attention mechanisms to consider the relationships between words in a sentence. This enables them to remember the context of the ongoing conversation and attend to relevant information when generating responses.
2. **Sequential Information:** LLMs can process language sequentially, considering the sequence of words in a sentence. This sequential understanding allows them to interpret the order and dependencies of words in the text.
3. **Multi-turn Dialogues:** LLMs can comprehend multi-turn dialogues by maintaining memory of the previous interactions. This memory allows them to provide coherent responses to follow-up questions or comments, mirroring human-like conversational flow.

13.3.3 GENERATING HUMAN-LIKE TEXT WITH LARGE LANGUAGE MODELS (LLMS)

Large language models are adept at generating text that closely resembles human language:

1. **Contextual Understanding:** LLMs consider the context provided in a conversation when generating responses. This contextual understanding enables them to generate text that aligns with the ongoing discussion, resulting in more coherent and contextually relevant replies.
2. **Nuanced Language:** LLMs can capture nuances in language, such as humor, sarcasm, or regional idioms. This capability makes interactions more engaging and relatable for users, as the text generated by LLMs resonates with human communication styles.
3. **Adaptation to Tone and Style:** LLMs can adjust their tone and style of communication based on the context and user preferences. They can produce formal or informal language, depending on the nature of the conversation.

13.4 LITERATURE REVIEW

Adamopoulou and Mousepads's [1] paper on chatbots provides a comprehensive historical overview and discusses the technology, but it lacks a critical analysis of the challenges and limitations of current chatbot applications.

Alruqi and Alzahrani's [2] paper on an Arabic chatbot provides a specific focus on extractive question-answering and language transformers, but it could have benefited from a broader discussion on the generalizability of their approach to other languages and contexts.

Ashfaque's [3] paper provides an analysis of trends in chatbot design and development, but it could have benefited from a more critical evaluation of the effectiveness and limitations of these trends in practical applications.

Bilquise, Ibrahim, and Shaalan's [4] paper on emotionally intelligent chatbots provides a systematic literature review, but it could have delved deeper into the practical challenges and potential solutions for implementing emotional intelligence in chatbots.

Cherednichenko et al.'s [5] preprint on collaborative business intelligence virtual assistants provides a reference model, but it could have benefited from a more comprehensive evaluation of the model's performance and applicability in diverse business contexts.

Eltahir et al.'s [6] paper provides an overview of chatbot structure and source algorithms, but it could have delved deeper into the practical challenges and potential solutions for implementing these algorithms in real-world chatbot applications.

Huynh-The et al.'s [7] paper provides a survey on AI for the metaverse, but it could have benefited from a more critical evaluation of the challenges and limitations of applying AI in this emerging field.

Ilić, Nakano, and Hajnal's [8] paper on designing dialogue systems presents an interesting approach with a mean, grumpy, sarcastic chatbot, but it could have benefited from a more comprehensive evaluation of the user experience and the chatbot's effectiveness in different contexts.

Jadhav and Thorat's [9] paper on designing conversational agent systems provides a good starting point, but it could have delved deeper into the practical challenges and potential solutions for implementing these systems in diverse contexts. The discussion seems to lack a comprehensive evaluation of the effectiveness of different design approaches.

Karyotaki, Drigas, and Skianis's [10] paper on chatbots as cognitive, educational, advisory, and coaching systems provides a broad overview, but it could have benefited from a more detailed analysis of the effectiveness and limitations of chatbots in these specific roles. The discussion seems to lack a comprehensive evaluation of real-world case studies.

Lahiri, Kandimalla, and Jeysekar's [11] paper on a hybrid multipurpose voice assistant provides an interesting approach, but it could have benefited from a more comprehensive evaluation of the system's performance and user experience in diverse real-world scenarios. The discussion seems to lack a critical analysis of the challenges and potential solutions for implementing such a system.

Luo et al.'s [12] paper provides a critical review of chatbot designs and applications, but it could have delved deeper into the practical challenges and potential solutions for implementing these designs in diverse contexts. The discussion seems to lack a comprehensive evaluation of the effectiveness of different design approaches.

Maher, Kayte, and Nimbhore's [13] paper on chatbots and their techniques using AI provides a good overview, but it could have benefited from a more critical evaluation of the effectiveness and limitations of these techniques in practical applications. The discussion seems to lack a comprehensive evaluation of real-world case studies.

Nikhath et al.'s [14] paper on an intelligent college inquiry bot provides a specific use case, but it could have benefited from a more comprehensive evaluation of the bot's performance and user experience in diverse scenarios. The discussion seems to lack a critical analysis of the challenges and potential solutions for implementing such a system.

Nirala, Singh, and Purani's [15] paper provides a survey on using AI chatbots for customer and public administration services, but it could have delved deeper into the practical challenges and potential solutions for implementing these services. The discussion seems to lack a comprehensive evaluation of the effectiveness of chatbots in these specific roles.

Nuruzzaman and Hussain's [16] paper provides a survey on chatbot implementation in the customer service industry using deep neural networks, but it could have benefited from a more critical evaluation of the effectiveness and limitations of these implementations. The discussion seems to lack a comprehensive evaluation of real-world case studies.

Nuruzzaman and Hussain's [17] paper on IntelliBot, a dialogue-based chatbot for the insurance industry, provides a specific use case, but it could have benefited from a more comprehensive evaluation of the bot's performance and user experience in diverse scenarios. The discussion seems to lack a critical analysis of the challenges and potential solutions for implementing such a system.

Singh and Beniwal's [18] paper provides a survey on near-human conversational agents, but it could have delved deeper into the practical challenges and potential solutions for achieving near-human conversation. The discussion seems to lack a comprehensive evaluation of the effectiveness of different techniques in achieving this goal.

13.5 CHALLENGES AND ETHICAL CONSIDERATIONS

This section addresses the challenges and ethical considerations associated with integrating LLMs into various applications, including chatbots and virtual assistants [19–21]. It focuses on the difficulties related to LLM integration, the data dilemma in training LLMs, biases in LLM-generated content, and the scaling of LLMs for mass adoption.

13.5.1 CHALLENGES IN LARGE LANGUAGE MODEL (LLM) INTEGRATION

Integrating LLMs into real-world applications presents several significant challenges:

13.5.1.1 THE DATA DILEMMA: TRAINING LARGE LANGUAGE MODELS (LLMS)

1. **Data Quantity and Quality:** Training LLMs requires vast amounts of data. The quality and diversity of the data are essential for ensuring the model's generalization and reducing biases. Collecting, curating, and managing such data can be a daunting task, especially for organizations with limited resources.
2. **Computational Resources:** LLMs are computationally intensive, and training them can demand substantial computing power. This challenge can restrict access to LLM technology for smaller organizations and startups.
3. **Data Privacy:** The data used for training LLMs often contain sensitive information. Maintaining data privacy while training models on massive datasets is a significant concern. Ensuring that data is anonymized and protected from misuse is a critical challenge.

13.5.1.2 THE ELEPHANT IN THE ROOM: BIASES IN LARGE LANGUAGE MODELS (LLM)-GENERATED CONTENT

1. **Bias Amplification:** LLMs can inadvertently perpetuate biases present in their training data. If the training data contains stereotypes, prejudices, or imbalances, the model may generate content that reflects and amplifies these biases. This poses ethical and social challenges, as LLM-generated content can perpetuate harmful stereotypes or misinformation.
2. **Bias Detection and Mitigation:** Addressing bias in LLMs is a challenging task. Detecting and mitigating bias in model-generated content requires ongoing research and development. Developing methods to make LLMs fairer and more unbiased is an active area of research in AI ethics.

13.5.1.3 SCALING LARGE LANGUAGE MODELS (LLMS) FOR MASS ADOPTION

1. **Computational Costs:** Scaling LLMs for mass adoption comes with significant computational costs. The need for more powerful hardware and infrastructure can be a barrier to entry for organizations seeking to deploy LLM-powered applications.
2. **Energy Consumption:** Training and running large LLMs can have a substantial environmental impact due to high energy consumption. Balancing the need for powerful models with the imperative to reduce environmental footprints is a challenge.
3. **Affordability:** Making LLM-powered applications accessible and affordable to a broad user base is a challenge. Ensuring that the benefits of LLM technology are not restricted to a privileged few requires innovative solutions and cost-effective deployment strategies (Figure 13.8).

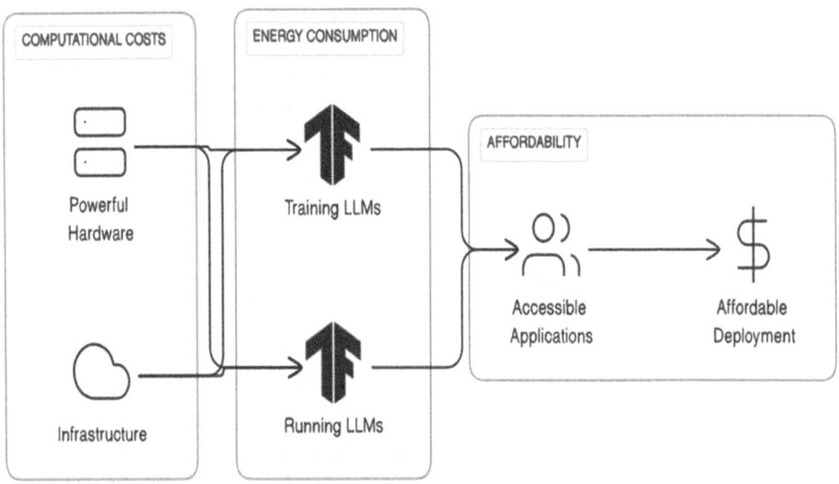

FIGURE 13.8 Large language models (LLM's) for mass adoption.

13.5.1.4 ETHICAL CONSIDERATIONS

Ethical considerations are paramount when integrating LLMs into applications:

1. **Fairness and Bias Mitigation:** Ensuring that LLMs do not produce or perpetuate biased, discriminatory, or harmful content is a moral imperative. Organizations must actively work to identify and rectify bias in LLM-generated output.
2. **Transparency and Accountability:** LLMs' decision-making processes can be complex, making transparency and accountability challenging. There is a need to develop mechanisms for explaining and auditing LLM-generated content.
3. **Privacy and Data Security:** The use of personal data to train LLMs necessitates robust privacy safeguards. Data breaches and misuse can have severe consequences, and organizations must prioritize data security and user privacy.

13.6 STATE OF THE ART

This section explores the current state of the art in the context of LLM-powered chatbots and virtual assistants. It highlights leading companies that are utilizing LLMs, presents case studies that demonstrate LLMs in action, and discusses the achievements in interactive and human-like interactions made possible by these advanced AI technologies.

13.6.1 LARGE LANGUAGE MODELS (LLM)-POWERED CHATBOTS AND VIRTUAL ASSISTANTS

13.6.1.1 LEADING COMPANIES UTILIZING LARGE LANGUAGE MODELS (LLMS)

1. **Technology Giants:** Prominent technology companies, such as Google, Facebook, and Microsoft, have invested heavily in LLM-powered chatbots and virtual assistants. Google's Duplex, Facebook's AI initiatives, and Microsoft's Language Models demonstrate the integration of LLMs into chatbot technologies.
2. **Startups and Innovative Enterprises:** Several startups and innovative companies have adopted LLMs to provide conversational AI solutions. These include companies like OpenAI (the developer of GPT-3), Rasa, and ChatGPT, which focus on enhancing chatbots with advanced NLU and generation capabilities.

3. **Industry-Specific Solutions:** Many industries, including healthcare, finance, and customer service, have embraced LLM-powered chatbots and virtual assistants. Leading healthcare providers use AI-powered chatbots to assist with medical information, while financial institutions employ LLMs to enhance customer service and provide personalized recommendations.

13.6.1.2 CASE STUDIES: LARGE LANGUAGE MODELS (LLMS) IN ACTION

1. **Customer Service:** Leading e-commerce companies deploy LLM-powered chatbots to handle customer inquiries and provide instant assistance. These chatbots can understand complex queries and offer solutions with human-like responses, improving the overall customer experience.
2. **Content Generation:** Media and publishing companies use LLMs to generate content, such as news articles and product descriptions. LLMs can create coherent and contextually relevant text, reducing the need for human authors and saving time and resources.
3. **Healthcare:** LLM-powered virtual assistants are transforming healthcare. They assist with medical queries, appointment scheduling, and medication reminders. These virtual assistants ensure that patients receive accurate and timely information, improving patient care.
4. **Education:** In the education sector, LLM-powered virtual assistants are used to provide instant answers to student queries, assist with homework, and offer language learning support. They enhance the learning experience by providing personalized, on-demand assistance.

13.6.1.3 THE INTERACTIVE AND HUMAN-LIKE INTERACTIONS

1. **Conversational Depth:** LLM-powered chatbots and virtual assistants offer conversational depth by understanding context and maintaining coherent dialogues. Users can engage in extended conversations, making interactions more dynamic and informative.

2. **Personalization:** LLMs enable personalization in interactions. They can remember user preferences, tailor responses to individual needs, and create a more personalized user experience, mirroring human-to-human interactions.
3. **Natural Language Understanding:** LLMs have improved NLU, allowing chatbots to comprehend and respond to user queries with a higher level of accuracy and relevance. This enables users to communicate with chatbots in a more natural and effortless manner.
4. **Human-Like Text Generation:** LLMs generate text that closely resembles human language, making interactions with chatbots and virtual assistants more engaging and relatable. The text generated by LLMs is contextually coherent and contextually relevant, contributing to human-like interactions (Figure 13.9).

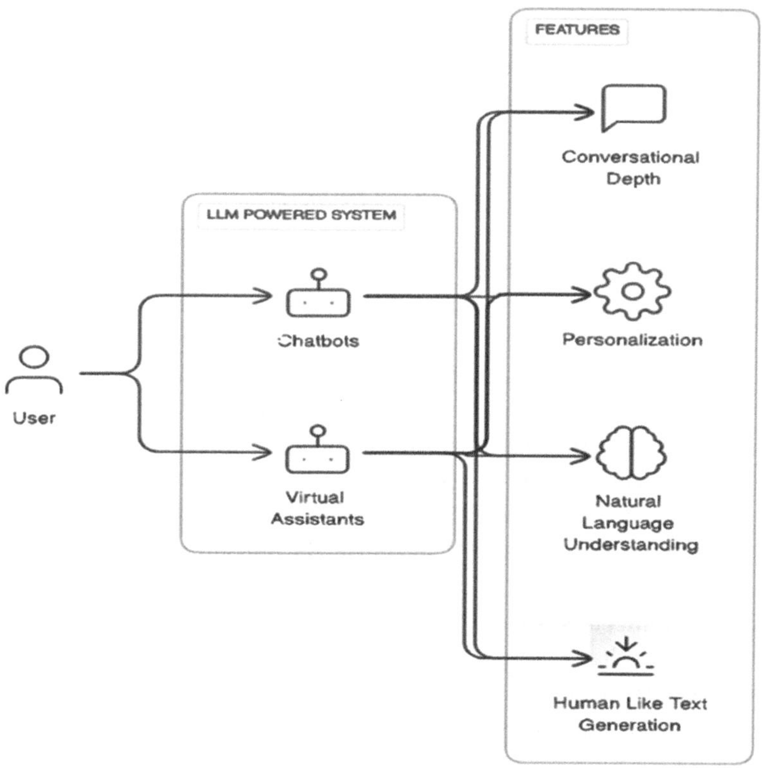

FIGURE 13.9 Interactive and human interaction with chatbots.

Hence, the state of the art in LLM-powered chatbots and virtual assistants demonstrates their wide adoption by leading companies and innovative enterprises. Case studies illustrate their practical applications across industries, from customer service to healthcare and education. The interactive and human-like interactions facilitated by LLMs are transforming the way users engage with AI-driven conversational agents, offering personalized, context-aware, and engaging experiences. This represents a significant advancement in the field of conversational AI and points toward a future where chatbots and virtual assistants play an even more integral role in our digital interactions.

13.6.2 USER FEEDBACK AND ACCEPTANCE

This section delves into user feedback and acceptance of LLM-powered chatbots and virtual assistants. It explores user experiences and insights, the perception of human likeness, and the factors contributing to user satisfaction and effectiveness.

13.6.2.1 USER EXPERIENCES AND INSIGHTS

1. **Improved User Experiences:** Users often report improved experiences when interacting with LLM-powered chatbots and virtual assistants. These AI-driven agents can understand user queries more deeply and generate contextually relevant responses, leading to smoother and more informative conversations.
2. **Convenience and Accessibility:** Users appreciate the convenience and accessibility of LLM-powered chatbots. They are available 24/7, respond instantly, and can handle a wide range of tasks, making them a valuable resource for users seeking quick and efficient assistance.
3. **Enhanced Support:** LLM-powered virtual assistants in customer service have the potential to enhance support experiences. Users receive accurate and helpful responses, reducing frustration and improving customer satisfaction.

13.6.2.2 THE PERCEPTION OF HUMAN-LIKENESS

1. **Contextual Responses:** LLM-powered chatbots can provide contextually relevant responses, leading users to perceive interactions as more human-like. The ability to remember previous messages and adapt to the ongoing conversation enhances the sense of conversation depth.
2. **Language Coherence:** LLM-generated text closely resembles human language, contributing to the perception of human-likeness. Users find it easier to engage in conversations with chatbots that produce contextually coherent and grammatically correct responses.
3. **Personalization:** Personalization in interactions, such as addressing users by their names or remembering their preferences, creates a more human-like experience. Users appreciate the effort to tailor responses to their individual needs.

13.6.2.3 USER SATISFACTION AND EFFECTIVENESS

1. **Higher Satisfaction:** LLM-powered chatbots and virtual assistants often lead to higher user satisfaction. Users value the ability to receive immediate and relevant assistance, leading to increased satisfaction with the services provided.
2. **Effective IR:** Users find LLM-powered virtual assistants effective in retrieving information. They can quickly obtain answers to questions and access information they need, enhancing their productivity and efficiency.
3. **Improved Engagement:** Users tend to be more engaged when interacting with LLM-powered chatbots. The dynamic and responsive nature of these agents makes conversations more engaging, leading to increased user involvement.
4. **Enhanced Decision-Making:** LLMs can provide users with information, recommendations, and insights, assisting in decision-making processes. Users perceive LLM-powered chatbots as valuable tools for obtaining guidance and support.

User feedback and acceptance of LLM-powered chatbots and virtual assistants reveal positive user experiences, with users appreciating improved interactions, convenience, and accessibility. The perception of human-likeness is enhanced by contextually relevant responses, language coherence, and personalization. Users report higher satisfaction and effectiveness when engaging with these AI-driven agents, leading to improved decision-making and engagement. This positive feedback and acceptance highlight the potential for LLM-powered conversational agents to significantly enhance the user experience and their role in various industries and applications.

13.7 FUTURE DIRECTIONS AND INDUSTRY IMPACT

Here are the future directions and industry impact of LLM-powered conversational AI. It focuses on research frontiers and ongoing efforts to enhance LLM-powered conversational AI, as well as the integration of multi-modal AI, which includes text and speech capabilities.

13.7.1 RESEARCH FRONTIERS

13.7.1.1 ONGOING RESEARCH TO ENHANCE LARGE LANGUAGE MODELS (LLM)-POWERED CONVERSATIONAL ARTIFICIAL INTELLIGENCE (AI)

1. **Bias Mitigation:** Ongoing research aims to develop more effective techniques for bias mitigation in LLMs. This includes strategies for identifying and reducing biases in generated content, as well as making the fine-tuning process more transparent and controllable.
2. **Explainability and Transparency:** There is a growing focus on developing methods to make LLM-powered conversational AI more explainable and transparent. Researchers are working on techniques to enable users to understand why a particular response was generated and to provide insights into the model's decision-making process.
3. **Adaptive Conversational Agents:** Future LLM-powered chatbots and virtual assistants will become more adaptive to user preferences and emotional states. Research is ongoing to develop models that

can detect user emotions and tailor responses accordingly, leading to more empathetic and emotionally intelligent interactions.

4. **Integration with Domain-Specific Knowledge:** LLM-powered conversational AI will integrate with domain-specific knowledge bases. This integration will enable these agents to provide more accurate and specialized information in fields like medicine, law, and finance.

13.7.1.2 MULTI-MODAL ARTIFICIAL INTELLIGENCE (AI) INTEGRATION (TEXT AND SPEECH)

1. **Text-to-Speech Integration:** Integrating text-to-speech capabilities into LLM-powered conversational AI will make interactions more versatile and natural. Chatbots and virtual assistants can generate text responses and convert them into spoken language, improving accessibility and user experience.
2. **Speech Recognition:** Enhancing speech recognition capabilities allows users to communicate with LLM-powered agents through speech. This is especially important for applications where typing is not practical, such as in hands-free environments or for individuals with accessibility needs.
3. **Multi-Modal Understanding:** Future conversational AI will have multi-modal understanding, encompassing both text and speech inputs. These agents will be capable of processing a wide range of communication modes, ensuring more inclusive and adaptable interactions (Figure 13.10).

13.7.1.3 INDUSTRY IMPACT

1. **Improved Customer Service:** LLM-powered conversational AI is set to revolutionize customer service. Organizations will use these AI agents to provide instant, accurate, and personalized support to customers, resulting in increased customer satisfaction and loyalty.
2. **Content Generation:** Media and publishing industries will rely on LLM-powered systems for content generation. This will streamline the content creation process, reduce costs, and facilitate the generation of diverse and engaging content.

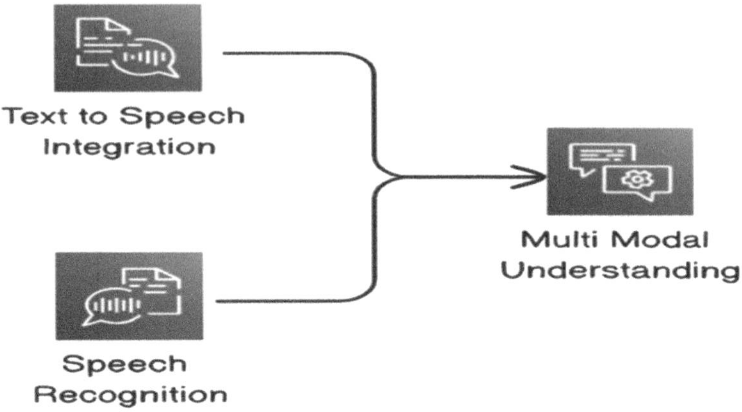

FIGURE 13.10 Multi-modal artificial intelligence (AI) integration.

3. **Education and Training:** Educational institutions and training programs will integrate LLM-powered chatbots and virtual assistants to enhance the learning experience. Students will benefit from on-demand support and personalized learning resources.
4. **Healthcare:** LLM-powered virtual assistants will play a crucial role in healthcare, assisting with patient inquiries, appointment scheduling, and medical information dissemination. They will improve patient care and streamline healthcare services.

13.7.2 INDUSTRY APPLICATIONS

The impact of LLM-powered conversational AI in various industry applications delves into how LLMs are transforming customer service, healthcare, education, and workforce automation, leading to productivity gains.

13.7.2.1 IMPACT ON CUSTOMER SERVICE

1. **Enhanced Customer Support:** LLM-powered chatbots and virtual assistants are revolutionizing customer service. They provide instant, accurate, and personalized support to customers, addressing queries and issues efficiently.

2. **Reduced Workload:** By handling routine inquiries, these AI agents reduce the workload on human customer support representatives, allowing them to focus on more complex and strategic tasks.
3. **24/7 Availability:** LLM-powered chatbots are available 24/7, ensuring customers can access support at their convenience, regardless of time zones or business hours.
4. **Improved User Experience:** Customers benefit from improved experiences, leading to higher satisfaction and potentially increased brand loyalty (Figure 13.11).

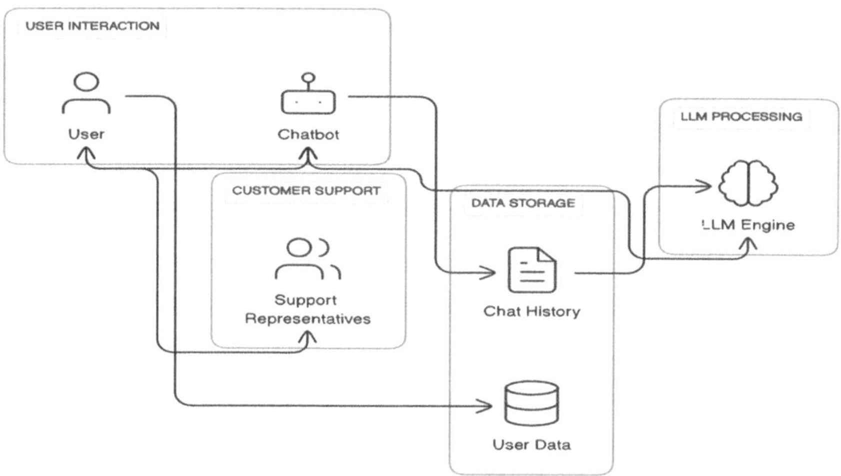

FIGURE 13.11 Customer service system.

13.7.2.2 HEALTHCARE AND VIRTUAL MEDICAL ASSISTANTS

1. **Medical Information:** LLM-powered virtual medical assistants provide patients with accurate medical information, helping them make informed decisions about their health.
2. **Appointment Scheduling:** Patients can use these virtual assistants to schedule appointments, check availability, and receive reminders, streamlining the healthcare process.
3. **Reduced Administrative Burden:** Healthcare professionals benefit from reduced administrative tasks, allowing them to allocate more time to patient care.

4. **Telemedicine Support:** LLMs enable telemedicine services by facilitating remote consultations and offering medical guidance.

13.7.2.3 EDUCATION AND PERSONALIZED LEARNING

1. **On-Demand Support:** LLM-powered chatbots assist students with homework, assignments, and subject-specific queries, providing on-demand educational support.
2. **Personalized Learning Resources:** These AI agents tailor learning resources to individual student needs, offering a more personalized and adaptive learning experience.
3. **Language Learning:** Language learning applications leverage LLM-powered chatbots to provide language practice and instant feedback to learners.
4. **Accessibility:** LLM-powered educational assistants improve accessibility by assisting students with varying learning needs.

13.7.2.4 WORKFORCE AUTOMATION AND PRODUCTIVITY GAINS

1. **Content Generation:** LLMs automate content creation, from news articles to product descriptions. This results in efficiency gains and cost savings for content-intensive industries.
2. **Data Entry and Analysis:** LLM-powered agents can handle data entry and analysis tasks, reducing manual labor and errors in data processing.
3. **Task Automation:** Routine, rule-based tasks across various industries can be automated by LLMs, increasing workforce efficiency and reducing repetitive work.
4. **Decision Support:** LLMs provide decision support by offering insights, recommendations, and data analysis, assisting professionals in making informed choices.

13.8 CONCLUSION

The conclusion of this discussion on LLM-powered conversational AI reflects on the transformative potential of LLMs in the field of conversational AI. It emphasizes the importance of addressing challenges and ethical concerns while harnessing the benefits for enhanced human-AI interactions.

13.8.1 THE TRANSFORMATIVE POTENTIAL OF LARGE LANGUAGE MODELS (LLMS) IN CONVERSATIONAL ARTIFICIAL INTELLIGENCE (AI)

Large language models have emerged as a revolutionary force in Conversational AI, offering the potential for a significant shift in how humans interact with AI-driven conversational agents. The transformative potential of LLMs lies in their ability to:

1. **Improve Language Understanding:** LLMs can understand user language with a depth and nuance previously unattainable. They analyze context, remember past interactions, and generate responses that closely mimic human language.
2. **Enhance User Engagement:** LLMs create more engaging, dynamic, and natural interactions. They adapt to user preferences, maintain coherent dialogues, and provide personalized responses.
3. **Boost Productivity:** In various industries, from customer service to content generation, LLM-powered chatbots and virtual assistants automate tasks, reduce workloads, and increase workforce efficiency.
4. **Provide Instant Assistance:** LLMs offer 24/7 availability, providing users with instant assistance, answering queries, and guiding them through processes.

13.8.2 ADDRESSING CHALLENGES AND ETHICAL CONCERNS

The adoption of LLMs in conversational AI comes with challenges and ethical considerations that must be addressed:

1. **Bias Mitigation:** Efforts are required to minimize biases in LLM-generated content. Researchers are working on methods to detect and rectify bias, making AI-generated responses fairer and more accurate.

2. **Explainability and Transparency:** LLMs need to become more transparent in their decision-making processes. Researchers are developing techniques to explain why a specific response was generated and provide insights into model behavior.
3. **Privacy and Data Security:** As LLMs rely on extensive data, ensuring data privacy and security is crucial. Data should be anonymized and protected to prevent misuse.

13.8.3 HARNESSING THE BENEFITS FOR ENHANCED HUMAN-AI INTERACTIONS

To harness the benefits of LLMs for enhanced human-AI interactions, we must:

1. **Focus on User Experience:** Prioritize the user experience by creating AI-driven conversational agents that are empathetic, efficient, and capable of understanding and adapting to user needs.
2. **Emphasize Ethical Development:** Develop LLM-powered chatbots and virtual assistants with a strong ethical foundation, addressing biases, ensuring transparency, and safeguarding privacy.
3. **Promote Inclusivity:** Ensure that AI-driven conversational agents are inclusive, accessible, and adaptable to diverse user needs and communication styles.
4. **Facilitate Continuous Research:** Foster ongoing research to improve LLM capabilities, making them more context-aware, emotionally intelligent, and domain-specific.
5. **Strive for Integration:** Aim to seamlessly integrate LLM-powered conversational AI into various industries, from healthcare to education, to optimize user experiences and increase productivity.

In conclusion, LLMs are at the forefront of revolutionizing Conversational AI. Their transformative potential is vast, with the promise of more engaging, personalized, and efficient human-AI interactions. By addressing challenges, ethical concerns, and continually advancing LLM technology, we can harness these benefits to create a future where AI-driven conversational agents play an even more integral role in our digital interactions, ultimately improving the way we communicate, work, and learn.

KEYWORDS

- **AI-powered user engagement**
- **conversational AI**
- **human-like interactions**
- **LLMs in chatbots**
- **natural language understanding**
- **virtual assistants AI**

REFERENCES

1. Adamopoulou, E., & Moussiades, L. (2020). Chatbots: History, technology, and applications. *Machine Learning with Applications, 2*, 100006.
2. Alruqi, T. N., & Alzahrani, S. M. (2023). Evaluation of an Arabic chatbot based on extractive question-answering transfer learning and language transformers. *AI, 4*(3), 667–691.
3. Ashfaque, M. W. (2022). Analysis of different trends in chatbot designing and development: A review. *ECS Transactions, 107*(1), 7215.
4. Bilquise, G., Ibrahim, S., & Shaalan, K. (2022). Emotionally intelligent chatbots: A systematic literature review. *Human Behavior and Emerging Technologies, 2022.*
5. Cherednichenko, O., Muhammad, F., Darmont, J., & Favre, C. (2023). *A Reference Model for Collaborative Business Intelligence Virtual Assistants.* arXiv preprint arXiv:2304.10556.
6. Eltahir, A. M., Abdulla, H., Platos, J., & Snasel, V. (2022). An overview of chatbot structure and source algorithms. In *2022 26th International Conference on Circuits, Systems, Communications and Computers (CSCC)* (pp. 91–98). IEEE.
7. Huynh-The, T., Pham, Q. V., Pham, X. Q., Nguyen, T. T., Han, Z., & Kim, D. S. (2023). Artificial intelligence for the metaverse: A survey. *Engineering Applications of Artificial Intelligence, 117*, 105581.
8. Ilić, S., Nakano, R., & Hajnal, I. (2019). *Designing Dialogue Systems: A Mean, Grumpy, Sarcastic Chatbot in the Browser*. arXiv preprint arXiv:1909.09531.
9. Jadhav, K. P., & Thorat, S. A. (2020). Towards designing conversational agent systems. In *Computing in Engineering and Technology: Proceedings of ICCET 2019* (pp. 533–542). Springer Singapore.
10. Karyotaki, M., Drigas, A., & Skianis, C. (2022). Chatbots as cognitive, educational, advisory & coaching systems. *Technium Social Sciences Journal, 30*, 109–126.
11. Lahiri, D., Kandimalla, P. C. P., & Jeysekar, A. (2023). Hybrid multi-purpose voice assistant. In *2023 2nd International Conference on Applied Artificial Intelligence and Computing (ICAAIC)* (pp. 816–822). IEEE.

12. Luo, B., Lau, R. Y., Li, C., & Si, Y. W. (2022). A critical review of state-of-the-art chatbot designs and applications. *Wiley Interdisciplinary Reviews: Data Mining and Knowledge Discovery, 12*(1), e1434.
13. Maher, S., Kayte, S., & Nimbhore, S. (2020). Chatbots & its techniques using AI: A review. *International Journal for Research in Applied Science and Engineering Technology, 8*(12), 503–508.
14. Nikhath, A. K., Rab, M. A., Bharadwaja, N. V., Reddy, L. G., Saicharan, K., & Reddy, C. V. M. (2022). An intelligent college enquiry bot using NLP and deep learning-based techniques. In *2022 International Conference for Advancement in Technology (ICONAT)* (pp. 1–6). IEEE.
15. Nirala, K. K., Singh, N. K., & Purani, V. S. (2022). A survey on providing customer and public administration-based services using AI: Chatbot. *Multimedia Tools and Applications, 81*(16), 22215–22246.
16. Nuruzzaman, M., & Hussain, O. K. (2018). A survey on chatbot implementation in customer service industry through deep neural networks. In *2018 IEEE 15th International Conference on e-Business Engineering (ICEBE)* (pp. 54–61). IEEE.
17. Nuruzzaman, M., & Hussain, O. K. (2020). IntelliBot: A dialogue-based chatbot for the insurance industry. *Knowledge-Based Systems, 196*, 105810.
18. Singh, S., & Beniwal, H. (2022). A survey on near-human conversational agents. *Journal of King Saud University-Computer and Information Sciences, 34*(10), 8852–8866.
19. Saxena, A. K., Asbe, C., & Vashishth, T. K. (2023). Leveraging a novel machine learning approach to forecast income and immigration dynamics. *Multidisciplinary Science Journal, 5*, 2023ss0202.
20. Vashishth, T. K., Kumar, B., Sharma, V., Chaudhary, S., Kumar, S., & Sharma, K. K. (2023). The evolution of AI and its transformative effects on computing: A comparative analysis. In B. Mishra (Ed.), *Intelligent Engineering Applications and Applied Sciences for Sustainability* (pp. 425–442). IGI Global.
21. Verma, P. K., Sharma, V., Kumar, P., Sharma, S., Chaudhary, S., & Preety, P. (2023). IoT-enabled real-time appearance system using AI camera and deep learning for student tracking. *International Journal on Recent and Innovation Trends in Computing and Communication, 11*(6s), 249–254.

CHAPTER 14

TrOCR-Med: Revolutionizing Medical Data Handling with Transformer-Based Handwritten Optical Character Recognition and Extraction

PARAG VERMA

Datafoundry Private Limited, Bangalore, Karnataka, India; AICG, Algihaz Holding, Riyadh, Saudi Arabia

ABSTRACT

In the dynamic landscape of document digitalization, the task of text recognition persists as an enduring challenge. Traditional approaches have predominantly relied on convolutional neural networks (CNNs) for image comprehension and recurrent neural networks (RNNs) for the generation of text at the character level. These methods are often supplemented with the inclusion of a post-processing language model to enhance overall accuracy. However, in this innovative research endeavor, we present a pioneering approach to text recognition, tailored explicitly to address the intricacies of recognizing handwritten text. Our methodology, aptly named TrOCR, represents a paradigm shift away from conventional practices. It does so by harnessing the transformative potential of pre-trained image Transformer and text Transformer models. TrOCR's unique value proposition lies in its seamless integration of the Transformer architecture into the very essence of text recognition. This architectural choice empowers the model with the dual capability to comprehend images and generate text at a remarkable level of granularity, focusing on workpieces rather than individual

Natural Language Processing and Artificial Intelligence: A Perspective Toward Current Trends, Challenges, and Applications. Vikas Khullar, Aryan Chaudhary, & Sunil Kumar (Eds.)

DOI: 10.1201/9781998511273-14

characters. This strategic integration of Transformers is especially pertinent, given the Transformers' remarkable success in NLP tasks, making them ideal candidates for this dual role in image and text understanding.

The brilliance of the TrOCR model manifests in its elegant simplicity and unwavering effectiveness. Despite its streamlined approach, it exhibits exceptional accuracy and adaptability. TrOCR demonstrates the remarkable flexibility to undergo pre-training using extensive synthetic data, enabling it to develop a profound understanding of both visual and textual patterns. Subsequently, it can be fine-tuned with meticulous attention to detail, utilizing human-labeled datasets. This fine-tuning process aligns the model's capabilities with real-world recognition tasks, enhancing its practical applicability. Perhaps the most compelling aspect of our research lies in the empirical validation of TrOCR's capabilities. Rigorous experimentation across a diverse spectrum of text recognition tasks confirms its superiority over contemporary state-of-the-art models. These tasks encompass the recognition of printed text, the deciphering of intricate handwritten text, and the tackling of challenges associated with scene text recognition. Our research represents a significant leap forward in the domain of text recognition. TrOCR's innovative approach, blending image understanding and text generation within a unified Transformer architecture, promises higher precision and efficiency in the realm of document digitalization. By simplifying and unifying the recognition process while maintaining or even surpassing accuracy benchmarks, TrOCR emerges as a transformative technology with far-reaching implications for the field of document processing and digitization. This monumental stride marks a significant advancement toward achieving more precise and efficient text recognition in the dynamic landscape of document digitalization.

14.1 INTRODUCTION

Optical character recognition (OCR) represents a transformative technology that enables the conversion of various forms of textual content, be it typed, handwritten, or printed, into machine-encoded text. This conversion process can originate from a range of sources, including scanned documents, photographs of documents, scene photos, or even subtitle text overlaid on images. Typically, an OCR system comprises two fundamental modules: a text detection module and a text recognition module, each serving a distinct but complementary purpose.

The primary objective of the text detection module is to identify and localize all text blocks within the input image, often working at either the word level or text line level granularity. This text detection task is conceptually akin to an object detection problem, drawing parallels with conventional object detection models like YOLOv5 and DBNet as introduced by Liao et al. [1]. By applying these established object detection techniques, OCR systems can effectively pinpoint textual content within diverse visual contexts.

In parallel, the text recognition module plays a pivotal role in deciphering the content of the identified text blocks and transcribing the visual signals into natural language tokens. This task is frequently framed as an encoder-decoder problem, where established methods leverage a convolutional neural network (CNN)-based encoder for image understanding and a recurrent neural network (RNN)-based decoder for text generation. The encoder's function is to extract meaningful features from the image, enabling the system to comprehend the textual content accurately. Meanwhile, the decoder transforms these features into a coherent sequence of natural language tokens, faithfully representing the text.

Recent advancements in the field of text recognition, as highlighted by Diaz et al. [2] in their study, have demonstrated substantial progress, primarily attributed to the adoption of Transformer architectures, as introduced by Vaswani et al. [3]. However, it is noteworthy that prevailing methods still rely on CNNs as their foundational backbone. In these approaches, self-attention mechanisms are integrated atop CNN backbones, serving as encoders to comprehend textual images. Meanwhile, Connectionist Temporal Classification (CTC), a framework initially proposed by Graves and Schmidhuber [4], is commonly employed as part of the decoder. Furthermore, an external language model, typically operating at the character level, is frequently utilized to augment overall recognition accuracy.

Despite the considerable success achieved through this hybrid encoder/decoder paradigm, there exists untapped potential for improvement, particularly through the integration of pre-trained Computer Vision (CV) and Natural Language Processing (NLP) models. Two key areas of opportunity are discernible:

1. **Leveraging Pre-trained Models:** Many existing methods train network parameters from scratch, relying on synthetic or

human-labeled datasets. This approach leaves a significant reservoir of potential untapped—large-scale pre-trained models. These models, often trained on vast and diverse datasets, possess a wealth of knowledge that could substantially benefit text recognition tasks.

2. **Integration of Image Transformers:** The burgeoning popularity of image Transformers, notably highlighted in works such as Dosovitskiy et al. [5] and Touvron et al. [6], presents an intriguing avenue for exploration. Additionally, recent innovations in self-supervised image pre-training, exemplified by the work of Wang et al. [7], introduce opportunities to integrate pre-trained image Transformers as replacements for traditional CNN backbones. This approach enables the simultaneous utilization of pre-trained image Transformers alongside pre-trained text Transformers within a unified framework, designed specifically for text recognition tasks.

In pursuit of advancing text recognition capabilities, we present a revitalized version of TrOCR—a transformative, end-to-end OCR model that harnesses the power of pre-trained CV and NLP models. The model's architecture is visualized in Figure 14.1. Notably distinct from existing text recognition models, TrOCR offers a refreshingly simple yet highly effective approach that eliminates the need for CNNs as the backbone.

The TrOCR model's innovation begins with a resizing step, where the input text image is transformed into a manageable 384 × 384 format. Subsequently, the image undergoes a transformation into a sequence of 16 × 16patches, which serve as input data for image Transformers. The foundational architecture of TrOCR leverages the established Transformer framework, incorporating self-attention mechanisms in both the encoder and decoder components. These mechanisms work in tandem to generate text in the form of word piece units, effectively recognizing the textual content within the input image.

To facilitate the training of TrOCR, we employ a two-pronged initialization strategy. The encoder benefits from initialization with pre-trained Vision Transformer (ViT)-style models, as exemplified by the works of Touvron et al. [6] and Dosovitskiy et al. [5], among others. In contrast, the decoder receives initialization from pre-trained BERT-style models, including those introduced by Devlin et al. [8], Liu et al. [9], Dong et al. [10], and Wang et al. [7]. This dual initialization approach empowers TrOCR to capitalize on the strengths of both image transformers and text transformers.

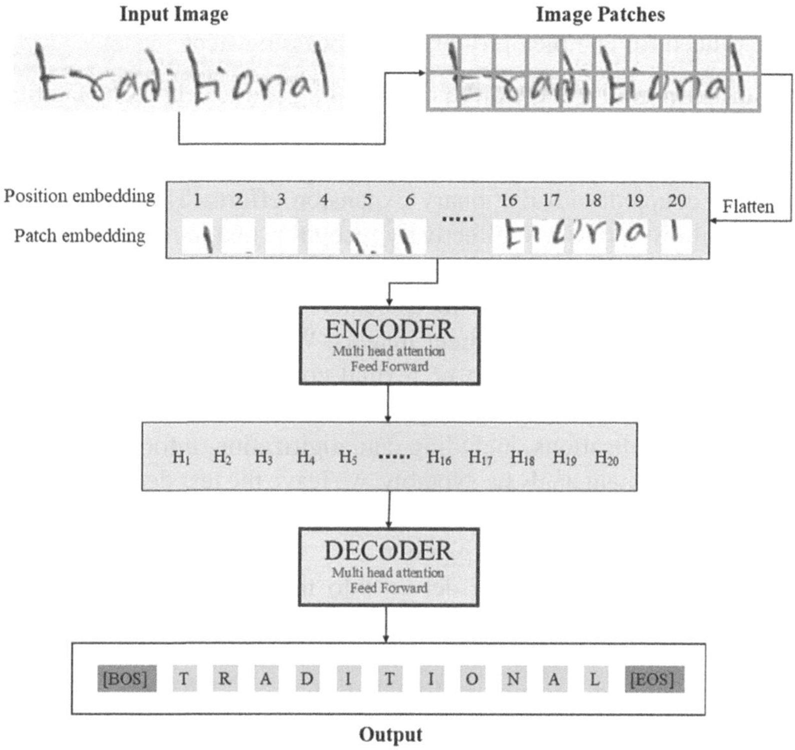

FIGURE 14.1 Architectural overview of TrOCR.

The advantages of TrOCR are threefold:

1. **Harnessing Pre-trained Models:** TrOCR exploits the capabilities of pre-trained image Transformer and text Transformer models, enabling it to benefit from the wealth of knowledge gleaned from extensive, unlabeled data for image understanding and language modeling. This eliminates the need for an external language model, streamlining the recognition process.
2. **CNN-Free Backbone:** A distinctive feature of TrOCR is its independence from CNNs as the backbone. This departure from the traditional approach eliminates image-specific inductive biases, simplifying model implementation and maintenance considerably.
3. **Exceptional Recognition Performance:** Experimental results garnered from a range of OCR benchmark datasets affirm TrOCR's prowess. It consistently achieves state-of-the-art results across

printed, handwritten, and scene text image datasets, all without the need for complex pre- or post-processing steps.

Furthermore, TrOCR exhibits adaptability and scalability by facilitating easy extension for multilingual text recognition. Achieving this extension merely requires the incorporation of multilingual pre-trained models in the decoder, coupled with dictionary expansion efforts. This minimal intervention allows TrOCR to effectively recognize text in multiple languages with remarkable ease.

In this research paper, our focus centers specifically on the text recognition task concerning document images. We recognize that the accurate recognition and transcription of textual content from document images are critical components of OCR systems and have substantial implications for various applications, including data digitization, information retrieval (IR), and document analysis. Notably, we leave the text detection aspect as a subject for future exploration and development, acknowledging its equal significance within the OCR pipeline.

In essence, this research delves into the intricate domain of OCR, emphasizing the importance of text recognition for document images. By addressing the challenges and nuances of this task, we aim to contribute to the ongoing advancement of OCR technology, ultimately facilitating the more efficient and accurate conversion of visual text into machine-encoded formats.

The paper's contributions can be succinctly summarized as follows:

1. **Innovative TrOCR Model:** We introduce an enhanced version of TrOCR, a pioneering end-to-end Transformer-based OCR model for text recognition. Notably, this research represents the first instance where pre-trained image and text Transformers are jointly employed for text recognition within the OCR context.
2. **State-of-the-Art Performance:** TrOCR establishes a new benchmark in OCR by achieving state-of-the-art results. This feat is accomplished through a standard Transformer-based encoder-decoder architecture, free from convolutional layers, and devoid of intricate pre- or post-processing procedures.
3. **Open Access:** The TrOCR models and accompanying codebase are made openly accessible to the research community and interested parties, hosted at XYZ (insert URL). This availability encourages further exploration and adoption of this transformative OCR technology.

14.2 TROCR ARCHITECTURE: TRANSFORMING TEXT RECOGNITION

14.2.1 MODEL ARCHITECTURE OVERVIEW

TrOCR's architectural foundation rests upon the transformer framework, a cutting-edge paradigm that consists of two indispensable components: the Image Transformer and the Text Transformer. These components work harmoniously in concert, forging a potent synergy that underpins the model's capacity for resilient and accurate text recognition. TrOCR faithfully adheres to the core principles and structural design of the Transformer architecture, notably embracing the encoder-decoder model configuration.

1. **Image Transformer:** Within the TrOCR framework, the Image Transformer assumes a pivotal role, tasked with the extraction of crucial visual features from the input data. Its primary function revolves around the systematic processing of image patches, thereby facilitating a thorough and holistic comprehension of the visual content at hand.
2. **Text Transformer:** In the TrOCR system, the Text Transformer assumes the role of a language modeling unit, operating as a crucial element within the architecture. Its primary function is to engage with the visual features extracted from the input image and leverage previous predictions to generate a coherent sequence of wordpieces. This sequence serves as an effective representation of the recognized textual content, synthesizing the information obtained from the input image. The Text Transformer's role is to bridge the gap between visual information and textual understanding, facilitating accurate and contextually meaningful text recognition within the TrOCR framework.

14.2.1.1 ENCODER FUNCTIONALITY

Within TrOCR, the encoder plays a pivotal role in obtaining a holistic representation of the input image patches. Its primary responsibility is to process the input image and extract pertinent visual features, which serve as the foundational elements for subsequent recognition processes.

Let's delve into the encoder's functionality step by step:

1. **Input Image Processing:** The encoder begins by receiving an input image denoted as $x_{ing} \in R^{3 \times H_0 \times W_0}$. To ensure uniformity, this image is resized to a fixed size (*H, W*).
2. **Patch Decomposition:** Since the Transformer encoder is designed to work with sequences of input tokens rather than raw images, the encoder further decomposes the resized image into a batch of $N = \frac{H \cdot W}{P^2}$ square patches, each with a fixed size of (P, P). Importantly, the dimensions of the resized image and H are guaranteed to be divisible by the patch size *P*.
3. **Patch Flattening and Projection:** These patches are subsequently flattened into vectors and undergo linear projection to yield *D*-dimensional vectors known as patch embeddings. Here, *D* represents the hidden size dimension that is consistent throughout all layers of the Transformer.
4. **Special Tokens:** Similar to models like ViT and data-efficient image transformer (DeiT), TrOCR retains the special token "(CLS)." Traditionally used in image classification tasks, the "(CLS)" token consolidates information from all patch embeddings, effectively representing the entire image. Additionally, a distillation token is retained when using DeiT pre-trained models for encoder initialization. This token facilitates knowledge transfer from a teacher model to the student model.
5. **Position Embeddings:** To maintain positional information, patch embeddings and special tokens are equipped with learnable 1D position embeddings. These position embeddings are assigned based on the absolute positions of the respective elements.
6. **Image Processing without Inductive Biases:** Notably, TrOCR's Transformer-based encoder differs from CNN-based networks by not incorporating image-specific inductive biases. Instead, it processes the image as a sequence of patches, providing greater flexibility to focus attention on either the entire image or individual patches as required.

Mathematically, the key components and transformations in the encoder can be summarized as:

- Input Image: $x_{ing} \in R^{3 \times H_0 \times W_0}$.

- Resized Image: $x_{img} \in (H, W)$
- Patch Decomposition: $N = \frac{H \cdot W}{P^2}$ square patches of size (P, P)
- Patch Embeddings: Linear projection to D-dimensional vectors
- Special Tokens: "CLS" token for holistic representation
- Position Embeddings: Added based on absolute positions

This encoder's design ensures that TrOCR can effectively process images for subsequent text recognition without imposing image-specific constraints or biases.

14.2.1.2 DECODER FUNCTIONALITY

In TrOCR, the decoder complements the encoder's efforts by focusing on the generation of wordpiece sequences. Its role is to synthesize coherent and contextually accurate sequences of wordpieces, leveraging the visual features obtained from the encoder and incorporating preceding predictions. This collaborative interplay between the encoder and decoder is essential for achieving precise text recognition and transcription from the input image.

Let's delve into the decoder's functionality and design:

1. **Transformer Decoder Structure:** The decoder in TrOCR follows the standard Transformer architecture, featuring a stack of identical layers. These layers closely resemble those found in the encoder, with a notable difference being the incorporation of "encoder-decoder attention" between the multi-head self-attention and feed-forward network layers. This additional attention mechanism allows the decoder to selectively focus on different aspects of the encoder's output during the text generation process.
2. **Encoder-Decoder Attention:** Within the encoder-decoder attention module, keys and values are derived from the encoder's output, while queries are based on the decoder's input. This mechanism enables the decoder to effectively align and utilize the visual features extracted by the encoder in the generation of wordpieces.
3. **Attention Masking:** To ensure that the decoder's training process aligns with its prediction phase, attention masking is employed in self-attention layers. This masking restricts the decoder's access to information, ensuring that the output for a given position can

only attend to previous positions, that is, the input positions less than the current position. This alignment prevents the model from utilizing future information during training.

Mathematically, key aspects of the decoder's functionality can be expressed as follows:

- **Projection and Attention Scores:**
  - Hidden state: $h_i = Proj\ (Emb(Token_i))$
  - Attention score: $\sigma\left(h_{ij}\right) = \frac{e^{h_{ij}}}{\sum_{k=1}^{V} e^{h_{ik}}}$

For $j = 1,2, \ldots, V$, where V represents the vocabulary size.

- **Output Probability Calculation:**
  Hidden states from the decoder are projected using a linear layer to the dimension of the vocabulary size V.
- **SoftMax Function:**
  Probabilities over the vocabulary are calculated using the SoftMax function, providing a distribution of wordpiece probabilities for each position.

To determine the final output, beam search is employed. This search technique aids in selecting the most likely wordpiece sequences based on the calculated probabilities, resulting in the model's ultimate recognition and transcription output.

14.2.2 MODEL INITIALIZATION

In the context of TrOCR, both the encoder and the decoder undergo initialization using publicly available pre-trained models. These pre-trained models have been trained on extensive datasets that encompass both labeled and unlabeled data.

The initialization process serves as a critical foundation for TrOCR's functionality. By utilizing these pre-trained models, TrOCR benefits from the wealth of knowledge and patterns learned from the diverse and extensive training data. This initialization approach empowers TrOCR with a strong starting point, enabling it to effectively recognize and understand the intricate visual and textual features present in document images.

The utilization of pre-trained models on large-scale datasets is a common practice in modern machine learning (ML) and deep learning (DL), providing models with a significant advantage in terms of learning relevant features and patterns. In the case of TrOCR, this initialization strategy contributes to its exceptional performance and adaptability across various text recognition tasks, ultimately enhancing its practical utility in document digitalization and OCR scenarios.

14.2.2.1 ENCODER INITIALIZATION

In TrOCR, the initialization of the encoder is a crucial step that sets the stage for its performance. TrOCR leverages two distinct pre-trained models, the data-efficient image transformer (DeiT) and the bootstrap enhancing image transformer (BEiT), for encoder initialization.

1. **Data-Efficient Image Transformer (DeiT):** DeiT is a model proposed by Touvron et al. [6]. It trains the image transformer using ImageNet as its sole training dataset. In the pursuit of data efficiency, the authors of DeiT experimented with various hyper-parameters and data augmentation techniques. Additionally, they employ a knowledge distillation approach, where knowledge from a strong image classifier is distilled into a distilled token within the initial embeddings of the model. This strategic addition of distilled knowledge enhances the model's performance, making it competitive with CNN-based models.
2. **Bootstrap Enhancing Image Transformer (BEiT):** BEiT introduces an innovative pre-training task known as the Masked Image Modeling task for pre-training the image transformer. This task involves converting each image into two views: image patches and visual tokens. To achieve this, the original image is tokenized into visual tokens using the latent codes of a discrete variational auto-encoder (VAE), a concept introduced by Ramesh et al. [11]. In the process, random masking of certain image patches is performed, and the model is tasked with recovering the original visual tokens. It's worth noting that the structure of BEiT aligns with that of the image transformer, differing from DeiT in the absence of a distilled token.

These two distinct encoder initialization approaches—DeiT and BEiT—contribute to TrOCR's adaptability and effectiveness. Leveraging the strengths of these pre-trained models allows TrOCR to begin its training process with a wealth of knowledge and patterns, which is instrumental in the subsequent text recognition tasks. The choice between DeiT and BEiT can depend on specific use cases and performance requirements, providing flexibility in the model's initialization based on the task at hand.

14.2.2.2 DECODER INITIALIZATION

In TrOCR, the initialization of the decoder is a significant aspect of the model's setup. The decoder in TrOCR is initialized using two different pre-trained models: RoBERTa and MiniLM, each offering unique advantages and capabilities.

1. **A Robustly Optimized BERT Pretraining Approach (RoBERTa):** RoBERTa, introduced by Liu et al. [9], represents a refined and optimized version of the original BERT model. It builds upon the foundation laid by bidirectional encoder representations from transformers (BERT) by meticulously exploring the impact of various hyperparameters and training data sizes. One notable departure from BERT is the removal of the next sentence prediction objective. Additionally, RoBERTa dynamically adjusts the masking pattern used in the Masked Language Model task. These refinements contribute to RoBERTa's enhanced performance in language understanding tasks.
2. **Miniaturized Language Model (MiniLM):** MiniLM models, as presented by Wang et al. [7], are compact versions of large pre-trained transformer models while still retaining approximately 99% of their performance. Unlike previous methods that relied on soft target probabilities or intermediate representations from teacher models to guide student model training, MiniLM models adopt a different approach. They are trained through a distillation process that focuses on distilling the self-attention module of the last Transformer layer from the teacher models while introducing a teacher assistant to aid in the distillation process.

When initializing the decoders in TrOCR with RoBERTa and MiniLM models, it's important to note that the structures of these models do not precisely match the requirements of TrOCR's decoder. Notably, both RoBERTa and MiniLM models are designed as the encoder of the Transformer architecture, lacking the specific encoder-decoder attention layers needed for TrOCR.

To address this structural disparity, TrOCR's decoders are initialized with RoBERTa and MiniLM models by manually configuring the corresponding parameter mappings. Parameters that are absent or incompatible with the decoder's requirements are randomly initialized. This tailored initialization process ensures that TrOCR's decoders can effectively build upon the knowledge and capabilities of RoBERTa and MiniLM while adapting to the specific demands of the text recognition task.

14.3 SEQUENTIAL TEXT RECOGNITION PIPELINE

In the context of our research, we have devised a comprehensive text recognition pipeline tailored to the task of recognizing text within text line images. This pipeline serves as the core methodology for our text recognition approach. Here is a detailed breakdown of the key steps within this pipeline:

1. **Textline Image Input:** The pipeline begins with the input of textline images, which are the visual representations of the handwritten or printed text to be recognized.
2. **Visual Feature Extraction:** Within the pipeline, our model is equipped with the capability to extract essential visual features from the input textline images. These visual features serve as the foundation for subsequent text recognition.
3. **Wordpiece Token Prediction:** Leveraging the extracted visual features and considering the contextual information generated in prior steps, our model predicts wordpiece tokens. Wordpieces are subword units that provide a more granular representation of text, allowing for improved accuracy in recognition.
4. **Ground Truth Token Sequence:** The pipeline incorporates a sequence of ground truth tokens, which represents the correct textual content corresponding to the input textline images. This sequence is constructed to mimic actual sentence structure and includes a special "end of sentence (EOS)" token to denote the conclusion of a sentence.

5. **Training Sequence Adjustment:** During the training phase, a crucial step involves shifting the ground truth token sequence backward by one place. Additionally, an "beginning of sentence (BOS) token is added to the beginning of the sequence, indicating the start of text generation. This adjustment aligns the ground truth sequence with the iterative nature of the decoding process.
6. **Decoder Processing:** The shifted ground truth sequence, now appropriately adjusted for training, is fed into the decoder. The decoder is responsible for generating text based on the input sequence and visual features. It initiates its generation process from the "BOS" token.
7. **Supervision with Cross-Entropy Loss:** The output of the decoder is supervised during training by comparing it to the original ground truth sequence. This supervision is carried out using a cross-entropy loss function, which measures the dissimilarity between the generated text and the ground truth.
8. **Inference Phase:** In the inference phase, which occurs during actual text recognition tasks, the decoder starts the text generation process from the "BOS" token. It iteratively predicts the output while continually considering the newly generated output as the input for subsequent iterations. This iterative process allows the decoder to construct the recognized text sequentially.

This text recognition pipeline encapsulates the core steps and strategies employed in our research to accurately recognize text within text line images. It ensures that our model generates coherent and contextually accurate text representations, making it a valuable tool in the domain of OCR and document digitalization.

14.3.1 PRE-TRAINING PROCESS

In our research, we adopt the text recognition task for the crucial pre-training phase. This choice is motivated by the potential of this task to impart comprehensive knowledge to our models, covering both the extraction of visual features and the intricacies of language modeling. The pre-training process is thoughtfully structured into two distinct stages, each tailored to specific datasets and objectives:

1. **First Stage:** In the initial stage of pre-training, we embark on the creation of a substantial and synthetic dataset. This dataset comprises a vast collection of hundreds of millions of printed text line images. The primary purpose of this stage is to pre-train our TrOCR models using this extensive dataset. This stage serves as the foundation upon which subsequent training is built.
2. **Second Stage:** The second pre-training stage is characterized by the development of two relatively smaller, yet task-specific datasets. These datasets are meticulously crafted to align with downstream tasks related to printed and handwritten text recognition. Each of these datasets encompasses millions of text line images. In parallel, for the challenging task of scene text recognition, we draw upon existing and widely recognized synthetic scene text datasets. In this stage, we initiate the pre-training of separate models, each tailored to a specific recognition task.

It's noteworthy that all models in the second stage of pre-training are initialized using the knowledge and weights acquired during the initial, first-stage model pre-training. This knowledge transfer ensures that the subsequent models inherit foundational understanding and capabilities, thus expediting their learning process and enhancing their performance on task-specific datasets. This multi-stage pre-training approach is a key driver of our model's ability to excel in diverse text recognition tasks, encompassing printed, handwritten, and scene text recognition. By sequentially training on datasets of varying scales and characteristics, our TrOCR models are equipped with the expertise to recognize text accurately across a wide range of real-world scenarios.

14.3.2 FINE-TUNING FOR VERSATILE DOWNSTREAM

Fine-tuning plays a pivotal role in our approach, particularly in adapting the pre-trained TrOCR models for downstream text recognition tasks. It's important to note that this fine-tuning process excludes the experiments related to scene text recognition. Here are the specifics of our fine-tuning procedure:

1. **Downstream Text Recognition Tasks:** The core of our fine-tuning strategy is the adaptation of the pre-trained TrOCR models

to specific downstream text recognition tasks. These tasks encompass a wide array of text recognition challenges, including printed and handwritten text recognition.

2. **Output Representation:** The outputs generated by the TrOCR models are structured using byte pair encoding (BPE) and SentencePiece, both well-established tokenization methods. Notably, our approach does not rely on task-specific vocabularies, ensuring a versatile and context-agnostic output representation.

By fine-tuning the pre-trained TrOCR models on downstream text recognition tasks while maintaining a consistent output representation, we achieve a high degree of adaptability and effectiveness across a diverse range of text recognition scenarios. This approach streamlines the model's transition from pre-training to specific recognition tasks, enhancing its ability to recognize text accurately and contextually in real-world applications.

14.3.3 ENHANCING DATASET DIVERSITY: DATA AUGMENTATION STRATEGIES

Data augmentation – our approach incorporates data augmentation to enhance the diversity of both the pre-training and fine-tuning datasets. We employ distinct image transformations, in addition to preserving the original images, for the printed and handwritten datasets. These transformations encompass:

1. **Random Rotation:** Images are rotated within a range of –10 to 10°, introducing variability in orientation.
2. **Gaussian Blurring:** We apply Gaussian blurring to add a subtle level of image noise and smoothness.
3. **Image Dilation:** This operation expands the boundaries of objects within the images, altering their spatial characteristics.
4. **Image Erosion:** Conversely, image erosion contracts object boundaries, inducing morphological changes.
5. **Downscaling:** Images are resized to smaller dimensions, providing a different perspective on the content.
6. **Underlining:** We introduce underlining to certain text segments, altering the visual patterns.

For these image transformations, we randomly select one with equal probability for each sample in the printed and handwritten datasets. For the scene text datasets, we implement RandAugment following the approach outlined in Atienza [12]. This augmentation technique introduces a broader set of transformations, including inversion, curving, blur, noise, distortion, rotation, and more. These augmentations aim to simulate real-world variations in scene text, further enhancing the model's robustness and generalization capabilities.

14.4 EXPERIMENTAL SETUP/ PIPELINE

The TrOCR pre-trained models are a set of models designed for text (printed and handwritten) recognition tasks. They go through two main stages: pretraining and fine-tuning. In this experiment, we'll dive into the details of these pre-trained models, including their scales, training data, and performance.

14.4.1 STAGE 1 MODELS (PRETRAINING)

The pre-trained models in the TrOCR family are referred to as "stage 1 models." These models are initially trained on a large-scale dataset containing synthetically generated data. Here are some key points about these stage 1 models:

1. **Training Data:** The training dataset used for pretraining consists of hundreds of millions of images of printed text lines. This dataset is synthetically generated, meaning the images are artificially created to train the models.
2. **Model Scales:** The TrOCR stage 1 models come in three different scales, each with varying numbers of parameters:
   - TrOCR-Small-Stage1 (smallest)
   - TrOCR-Base-Stage1 (intermediate)
   - TrOCR-Large-Stage1 (largest)
3. **Performance:** Among these models, the TrOCR-Large-Stage1 model tends to perform the best in terms of text recognition accuracy. However, it's worth noting that larger models like TrOCR-Large-Stage1 may be slower to run due to their increased complexity.

14.4.2 FINE-TUNED MODELS

After the pretraining stage on synthetic data, the TrOCR models undergo fine-tuning on real-world datasets to improve their performance on specific tasks. Two main datasets are used for fine-tuning:

1. **IAM Handwritten Dataset:**
   - **i.** This dataset contains images of handwritten text, making it ideal for fine-tuning models to recognize handwritten text more effectively.
   - **ii.** Fine-tuning on the IAM Handwritten dataset enhances the model's ability to accurately recognize text in handwritten form.
2. **SROIE Printed Receipts Dataset:**
   - **i.** The SROIE dataset consists of several 1000 samples of receipt images. It is primarily focused on printed text recognition tasks.
   - **ii.** Models fine-tuned on the SROIE dataset perform exceptionally well when it comes to recognizing printed text, particularly on receipts.

Like the pretraining stage models, the fine-tuned models also come in three different scales, offering flexibility and adaptability for various text recognition scenarios:

- TrOCR-Small-IAM (Fine-tuned on IAM Handwritten)
- TrOCR-Base-IAM (Fine-tuned on IAM Handwritten)
- TrOCR-Large-IAM (Fine-tuned on IAM Handwritten)
- TrOCR-Small-SROIE (Fine-tuned on SROIE Printed Receipts)
- TrOCR-Base-SROIE (Fine-tuned on SROIE Printed Receipts)
- TrOCR-Large-SROIE (Fine-tuned on SROIE Printed Receipts)

The TrOCR pretrained models undergo a two-stage training process. They begin with pretraining on synthetic data, followed by fine-tuning on real-world datasets. These models are available in multiple scales to suit different recognition tasks, ranging from printed text to handwritten text, and they offer a balance between recognition accuracy and computational efficiency.

14.4.3 PIPELINE: TROCR INFERENCE STAGE

1. **Input Image:** The process begins with an input image that needs to undergo text recognition. This image is fed into the TrOCR model for analysis.

2. **Image Encoder:** The input image passes through the image encoder component of the TrOCR model. Within this component, several important steps take place:
   i. **Image Patching:** The image is initially broken down into smaller patches. These patches are typically resized to a 384 × 384 resolution to align with the requirements of the ViT model.
   ii. **Multi-head Attention Block:** Each of these image patches then goes through a multi-head attention block. This block allows the model to focus on different parts of the image simultaneously, capturing important visual features.
   iii. **Feed-forward Block:** Following the multi-head attention block, the image patches proceed through a feed-forward block. This block processes the patches and generates image embeddings, which encapsulate the relevant visual information.
5. **Language Transformer Model:** The image embeddings obtained from the encoder are then passed into the language transformer model. This model plays a critical role in understanding and interpreting the visual content within the image.
6. **Decoder:** Within the language transformer model, the decoder part is responsible for producing encoded outputs. These outputs represent the model's understanding of the text contained within the image.
7. **Text Decoding:** The encoded outputs are then decoded to obtain the final text that was present in the input image. This step essentially translates the model's understanding of the text into a human-readable format.
8. **Output Resolution:** It's worth noting that the images are resized to a uniform 384 × 384 resolution before they are processed by the ViT model. This resizing is essential to ensure that the input image aligns with the specific requirements of the DeIT model, which expects images of this size.

This experimental setup outlines the sequential steps involved in processing an input image through the TrOCR model to obtain the text content present in the image. Each stage of the process plays a crucial role in transforming visual data into meaningful textual information. To execute the procedures outlined within the pipeline, specific experiment setup requirements must be met:

1. **Python Installation:** Python must be installed on your system to facilitate the execution of this code.
2. **Required Python Libraries:** Ensure the following Python libraries are installed:
   - 'cv2': OpenCV for image processing.
   - 'PaddleOCR': PaddleOCR library for optical character recognition.
   - 'transformers': Transformers library for NLP.
   - 'numpy': NumPy for numerical computations.
   - 'os': This library provides a way of interacting with the operating system.
   - 'PIL': Python imaging library (PIL) for image manipulation.
3. **Pre-trained TrOCR Model:** You should possess or have access to a pre-trained TrOCR model and processor. These can typically be acquired from Hugging Face's model hub.
4. 'TrOCRProcessor' and 'VisionEncoderDecoderModel' are classes from Hugging Face's Transformers library, used for processing and decoding text.
5. **Input Image Specification:** Provide an input image along with its file path by assigning it to the 'img_path' variable. This image will be processed by the script for text recognition.
6. **Output Text File:** The script is designed to generate an output text file named "tr_ocr.txt." This file will contain the recognized text extracted from the input image.

14.5 RESULTS AND DISCUSSION

In the context of handwritten text recognition using the TrOCR model, the following steps outline the initialization and execution process (Figures 14.2–14.5):

1. **Model and Processor Initialization:** The 'processor_lr' and 'model_lr' are initialized using the 'TrOCRProcessor' and 'VisionEncoderDecoderModel' classes. These components are pre-trained on extensive datasets and purpose-built for text recognition tasks.
2. **PaddleOCR Initialization:** Initialization of the 'ocr' instance using the 'PaddleOCR' library. This step involves downloading and loading a pre-trained OCR model into memory, which offers support for multiple languages, including English.

3. **Input Image Path:** Setting the 'img_path' variable to specify the path of the input image intended for text recognition.
4. **Text Recognition using PaddleOCR:** Utilizing the initialized PaddleOCR model, the script executes text recognition on the input image via the 'ocr.ocr(img_path, cls=True)' function call. The 'cls=True' parameter instructs the model to classify the detected text into different categories.
5. **Image Processing and Further Text Recognition:**
   - **i.** Subsequent steps involve processing the identified text by segmenting it into bounding boxes.
   - **ii.** For each bounding box:
     a. Extracting the region of interest (ROI) from the original image.
     b. Converting the ROI into a format compatible with the 'processor_lr' using Hugging Face's 'processor_lr.'
     c. Generating text from the ROI employing the 'model_lr.'
     d. Decoding the generated text using the 'processor_lr' to obtain the final recognized text.
     e. Saving the recognized text to an output file ("tr_ocr.txt").

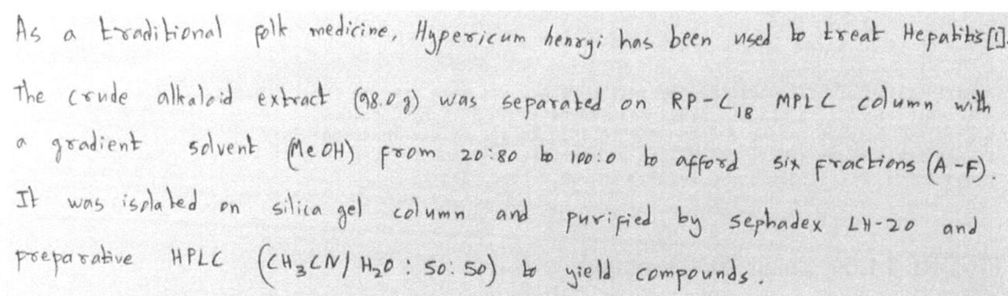
As a traditional folk medicine, Hypericum henryi has been used to treat Hepatitis[1].
The crude alkaloid extract (98.0 g) was separated on RP-C_{18} MPLC column with
a gradient solvent (MeOH) from 20:80 to 100:0 to afford six fractions (A-F).
It was isolated on silica gel column and purified by sephadex LH-20 and
preparative HPLC (CH_3CN/ H_2O : 50:50) to yield compounds.

FIGURE 14.2 Handwritten text image from real word.

6. **Cleanup:** The script includes a cleanup phase, where temporary image files generated during the process are removed using the 'os.remove("test.jpg")' command.
7. **Output:** The recognized text is saved within the "tr_ocr.txt" file, providing the final results of the handwritten text recognition process.

This demonstration underscores the remarkable performance of TrOCR in recognizing handwritten text. Notably, the model exhibits the ability to accurately detect characters, even in the context of cursive or running handwriting styles. This showcases the effectiveness of combining transformer-based vision and language models in the TrOCR framework for handwritten text recognition.

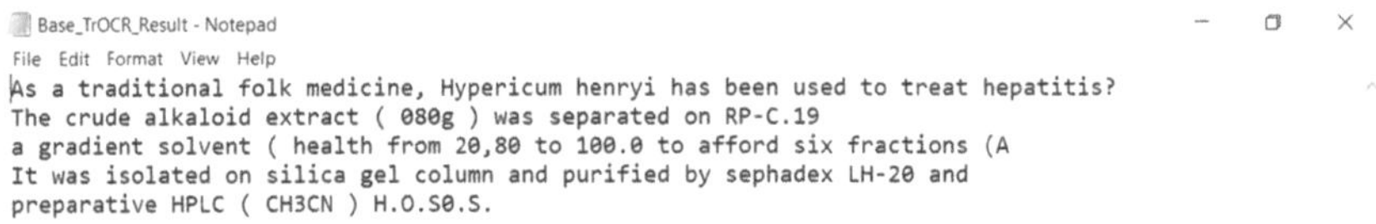

FIGURE 14.3 Base TrOCR output.

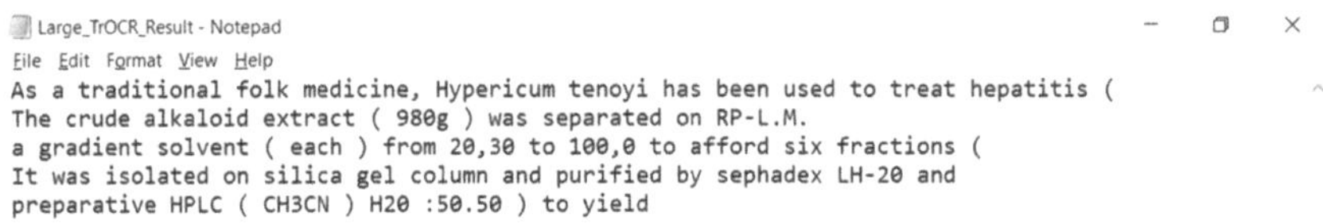

FIGURE 14.4 Large TrOCR output.

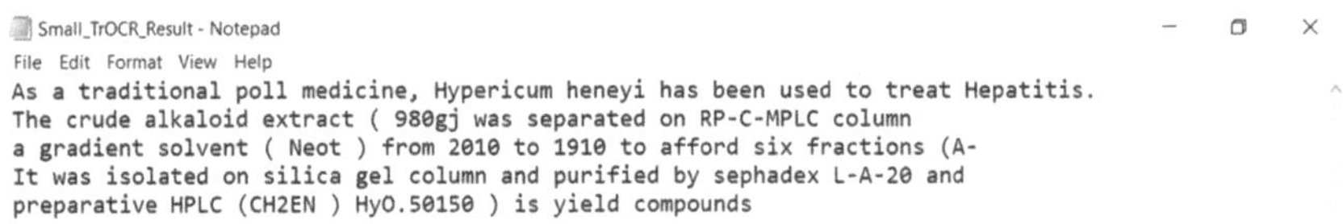

FIGURE 14.5 Small TrOCR output.

14.6 LIMITATION OF TROCR

The TrOCR models exhibit limitations related to their training data and the inherent capabilities of their vision and language transformer components:

1. **Performance on Wild Images:** TrOCR models may exhibit suboptimal performance when presented with images from uncontrolled, real-world environments. These limitations are attributed to the model's training data, which primarily consists of synthetically generated text images. The absence of real-world variations and

complexities can hinder the model's adaptability to diverse and unstructured text images encountered in practical scenarios.

2. **Vision Transformer Capability:** The ViT component of the TrOCR model may lack exposure to curved text instances. As a result, the model may struggle to accurately recognize and process text arranged in curved or non-linear patterns. The absence of curved text samples in the training data limits the model's ability to effectively handle such cases.
3. **Language Understanding:** TrOCR's language transformer may face challenges in comprehending and processing different tokens within complex text arrangements. The model's capacity to understand and interpret tokens in varying contexts, especially within challenging or unconventional textual layouts, may be limited. This limitation can impact the model's accuracy in recognizing and transcribing diverse text elements.

To enhance their performance, future developments may require training on more diverse and real-world datasets, as well as improvements in ViT models that can handle curved text and language transformer models capable of handling a wider range of tokens within complex text layouts.

14.7 CONCLUSION

Our research introduces TrOCR, an innovative approach to text recognition tailored explicitly for recognizing handwritten text. TrOCR leverages the power of pre-trained image Transformer and text Transformer models, seamlessly integrating them into the text recognition pipeline. This integration empowers TrOCR with the unique capability to understand images and generate text at a granular level, focusing on wordpieces rather than individual characters. This paradigm shift represents a significant advancement in the field of document digitalization.

TrOCR's brilliance lies in its simplicity and effectiveness. Despite its streamlined approach, it achieves exceptional accuracy and adaptability. The model's ability to undergo pre-training with synthetic data and fine-tuning with real-world datasets enhances its practical applicability, making it a valuable tool in text recognition tasks.

Empirical validation across various text recognition tasks demonstrates TrOCR's superiority over contemporary models. It excels in recognizing printed text, deciphering intricate handwritten text, and addressing challenges in scene text recognition. This research marks a significant stride toward achieving precise and efficient text recognition in the dynamic landscape of document digitalization.

In the experimental setup, we explored the details of TrOCR's pretrained models, including the Stage 1 models trained on synthetic data and the fine-tuned models using real-world datasets. These models are available in different scales, with TrOCR-Large-Stage1 performing exceptionally well but potentially slower due to its complexity. Fine-tuning on datasets like IAM Handwritten enhances the model's recognition of handwritten text.

In practice, TrOCR shines in recognizing handwritten text, even in the presence of cursive or running handwriting styles. This success underscores the effectiveness of combining vision and language models within the Transformer architecture, making TrOCR a transformative technology with significant implications for document processing and digitization.

KEYWORDS

- **artificial intelligence**
- **document digitalization**
- **handwritten OCR**
- **image-to-text**
- **text recognition**
- **transformer-based OCR**
- **TrOCR model**

REFERENCES

1. Liao, M., Wan, Z., Yao, C., Chen, K., & Bai, X. (2020). Real-time scene text detection with differentiable binarization. *Proceedings of the AAAI Conference on Artificial Intelligence, 34*(7), 11474–11481.
2. Diaz, D. H., Qin, S., Ingle, R., Fujii, Y., & Bissacco, A. (2021). *Rethinking Text Line Recognition Models*. arXiv preprint arXiv:2104.07787.

3. Vaswani, A., Shazeer, N., Parmar, N., Uszkoreit, J., Jones, L., Gomez, A. N., Kaiser, Ł., & Polosukhin, I. (2017). Attention is all you need. *Advances in Neural Information Processing Systems, 30.*
4. Graves, A., & Schmidhuber, J. (2008). Offline handwriting recognition with multi-dimensional recurrent neural networks. *Advances in Neural Information Processing Systems, 21.*
5. Dosovitskiy, A., Beyer, L., Kolesnikov, A., Weissenborn, D., Zhai, X., Unterthiner, T., Dehghani, M., Minderer, M., Heigold, G., & Gelly, S. (2020). *An Image is Worth 16 × 16 Words: Transformers for Image Recognition at Scale.* arXiv preprint arXiv:2010.11929.
6. Touvron, H., Cord, M., Douze, M., Massa, F., Sablayrolles, A., & Jégou, H. (2021). Training data-efficient image transformers & distillation through attention. *International Conference on Machine Learning*, 10347–10357.
7. Wang, W., Wei, F., Dong, L., Bao, H., Yang, N., & Zhou, M. (2020). MiniLM: Deep self-attention distillation for task-agnostic compression of pre-trained transformers. *Advances in Neural Information Processing Systems, 33*, 5776–5788.
8. Devlin, J., Chang, M. W., Lee, K., & Toutanova, K. (2018). *BERT: Pre-training of Deep Bidirectional Transformers for Language Understanding.* arXiv preprint arXiv:1810.04805.
9. Liu, Y., Ott, M., Goyal, N., Du, J., Joshi, M., Chen, D., Levy, O., Lewis, M., Zettlemoyer, L., & Stoyanov, V. (2019). *RoBERTa: A Robustly Optimized BERT Pretraining Approach.* arXiv preprint arXiv:1907.11692.
10. Dong, L., Yang, N., Wang, W., Wei, F., Liu, X., Wang, Y., Gao, J., Zhou, M., & Hon, H. W. (2019). Unified language model pre-training for natural language understanding and generation. *Advances in Neural Information Processing Systems, 32.*
11. Ramesh, A., Pavlov, M., Goh, G., Gray, S., Voss, C., Radford, A., Chen, M., & Sutskever, I. (2021). Zero-shot text-to-image generation. *International Conference on Machine Learning*, 8821–8831.
12. Atienza, R. (2021). Data augmentation for scene text recognition. In *Proceedings of the IEEE/CVF International Conference on Computer Vision* (pp. 1561–1570).

CHAPTER 15

Speech Emotion Detection Using Deep Learning

V. G. HEMANT* and ANSHU SHARMA

School of Computer Science Engineering, Lovely Professional University, Punjab, India

**Corresponding author.*

ABSTRACT

Nowadays, the generation of technology is increasing day by day with the use of artificial intelligence (AI). Voice assistants are one of the better technologies used by numerous people. Generally, they capture human speech for a while and assist with replies. Known as voice assistants, the tone or frequency of a person's voice may differ due to their character. The voice assistant plays an important role; it records the different speech tones of a person, makes a sentiment analysis of the tone, and predicts the person's character or the text conversation as well. To compile or implement the voice assistant, an automatic speech recognition (ASR) technique is used.

15.1 INTRODUCTION

Speech is a rich source of emotional expression, and understanding the underlying emotions in spoken language is crucial for various applications, from improving human-computer interaction to enhancing mental

Natural Language Processing and Artificial Intelligence: A Perspective Toward Current Trends, Challenges, and Applications. Vikas Khullar, Aryan Chaudhary, & Sunil Kumar (Eds.)

DOI: 10.1201/9781998511273-15

health diagnostics. Speech emotion detection, the task of automatically identifying a speaker's emotional state from their speech signal, has gained significant attention in recent years. Deep learning (DL) techniques have played a pivotal role in advancing the accuracy and effectiveness of speech emotion detection systems.

Human communication is fundamentally influenced by emotions, which impact our interactions, choices, and general well-being. Understanding and being able to react to emotions expressed through speech has emerged as a valuable and transformational skill in our increasingly digital society. Speech emotion detection, a discipline at the nexus of psychology and artificial intelligence (AI), is helpful in this situation. In this introduction, we'll look at the intriguing field of speech emotion detection, its importance, and how DL methods and the Python programming language have entirely changed how well we can understand the emotions expressed in spoken words.

In this chapter, we delve into the world of speech emotion detection using DL models. We will explore the key concepts, methodologies, and recent advancements in this field. Our goal is to provide readers with a comprehensive understanding of how DL can be leveraged to detect emotions from speech.

Voice emotion detection using automatic speech recognition (ASR) and DL is an emerging field that has gained significant attention in recent years. This technique allows us to recognize the emotions expressed in speech and text by using machine learning (ML) algorithms that can automatically detect patterns and classify them into different emotional states. By combining the power of ASR and DL, we can create accurate and efficient models that can analyze spoken or written language in real time and identify the emotions conveyed by the speaker or writer. This technology has numerous potential applications, including improving customer service experiences, aiding in mental health diagnosis, and enhancing natural language processing (NLP) systems. In this context, the development of advanced algorithms and the availability of large datasets have led to significant advances in the field of voice emotion detection, making it a promising area for further research and development.

A crucial component of human communication is voice emotion recognition. Emotions may influence how communication is interpreted and received and provide important information about the speaker's mental state. For instance, a caller may have a bad impression of a customer service representative if they seem irate or irritated, as opposed to a representative who speaks cordially and sympathetically. As a result, many different professions and businesses, such as customer service, mental health, and education, stand to benefit greatly from the comprehension and recognition of emotions in voice and text.

Using ML algorithms is one of the best techniques to recognize emotions in voice and text. The identification and classification of emotions conveyed in voice and text have demonstrated encouraging results when combining ASR and DL. DL, a branch of ML, employs artificial neural networks to analyze complicated data and discover patterns. ASR is a technique that enables systems to convert spoken language into text.

The use of ASR and DL in voice emotion detection has been facilitated by the availability of large datasets of labeled speech and text. These datasets are crucial for training ML models to recognize emotional cues and patterns in spoken and written language. Additionally, advancements in ASR and DL technologies, such as the development of RNNs and CNNs, have further improved the accuracy and efficiency of voice emotion detection models.

One of the primary challenges in voice emotion detection is the subjective nature of emotions. Different individuals may express emotions differently, and cultural and social factors can also influence emotional expression. Moreover, emotions are not always explicitly stated in speech or text but can also be conveyed through nonverbal cues such as intonation, pitch, and pauses. As such, ML models must be trained on diverse datasets and be able to recognize a wide range of emotional expressions.

Despite these difficulties, there are many potential uses for voice emotion detection utilizing ASR and DL. Voice emotion detection, for instance, can assist businesses in enhancing the quality of their interactions with clients by spotting negative emotions and responding to them appropriately. Voice emotion recognition in mental health can help in the diagnosis and treatment of disorders like depression and anxiety. Voice emotion recognition in education can assist teachers in identifying students who may be experiencing emotional difficulties and provide tailored support.

Speech emotion recognition (SER) is among the methods used for the process of identifying emotions conveyed through speech signals, such as happiness, anger, sadness, and fear. Accurately detecting emotions from speech has important applications in various fields, including mental health, education, entertainment, and security. With the recent advances in DL techniques, there is a growing interest in utilizing these methods for speech emotion detection. DL approaches include CNN and RNN, which show great potential in capturing complex patterns and temporal dependencies in speech signals. In addition, attention mechanisms and transfer learning (TL) techniques have been applied to enhance the development of these models.

In this paper, the aim is to provide an overview of the recent advances and research work in speech emotion detection using DL techniques, including the datasets utilized for the development of performance attained by the models. The methodology that enables machines to detect and convert spoken language into text is called ASR. ASR has become a popular and widely used technology in recent years, with various applications in fields such as education, healthcare, entertainment, and security. One of the emerging applications of ASR [1] is in speech emotion detection, where ASR is used as a tool for recognizing emotions conveyed in speech signals.

Speech emotion detection using ASR has several advantages over traditional methods that rely on manual annotations. Manual annotations are time-consuming and costly, and the annotations may not always accurately reflect the true emotions conveyed in speech. In contrast, ASR can process large amounts of speech data in a shorter time, and the recognized text can be used as input for emotion detection models. According to recent works on speech emotion detection using ASR, there is more potential than with other models. These methods typically use a combination of ASR and DNN approaches to recognize emotions from speech by using CNN and RNN techniques. Such instances or models can capture both the linguistic and emotional aspects of speech and achieve better performance.

One of the main advantages of using ASR algorithms in voice sentiment analysis is that they can process large volumes of speech data quickly and accurately. ASR algorithms have been shown to achieve high accuracy rates in transcribing spoken language, which makes them well-suited for analyzing sentiment in large datasets of spoken language. Additionally, ASR algorithms can be utilized to analyze real-time speech data, which is useful for applications such as call center monitoring, where sentiment analysis can help improve customer service. SER is one of the techniques for identifying and classifying emotions conveyed through speech signals.

SER has numerous applications in fields such as healthcare, education, and entertainment, as it can help improve the interaction between humans and machines. SER systems can provide insights into a speaker's emotions, which can be useful in various scenarios, such as detecting depression or anxiety in patients, improving customer service, and enhancing the user experience in gaming and virtual reality applications.

During the past few decades, research on the creation of SER systems has been ongoing. Initially, researchers relied on handcrafted characteristics such as Mel-frequency-cepstral coefficients (MFCCs) [1, 2] and prosodic features for emotion recognition. However, these methods have limitations in capturing the complex and subtle variations in speech signals that are indicative of emotions. Emotion analysis involves a review of people's feelings or emotions about a particular incident, relevant debate, or summary. Although communication is recognized as the convenient, simplest, and most characteristic means of communicating with one another, this is not accurate for machines. Additionally, the ASR connection is developing, but it is still far from interactive, authentic, and natural communication since robots cannot grasp human emotions [2]. The process known as "speech emotion recognition" (SER) entails the application of technology that can help one fully comprehend human emotions by extracting the speaker's emotional responses from their speech [3] (Figure 15.1).

There are certain DL techniques to perform speech emotion recognition and they are:

1. **Convolutional Neural Networks (CNNs):** CNNs extract characteristics from voice streams. They are especially good at detecting local patterns in the signal that correlate to emotional cues.
2. **Recurrent Neural Networks (RNNs):** RNNs are used to describe the temporal dynamics of voice signals. They are especially good at detecting long-term dependencies in signals that correlate to emotional cues.
3. **Long Short-Term Memory (LSTM) Networks:** LSTM networks are a form of RNN developed to handle the problem of disappearing gradients. They are especially good at detecting long-term dependencies in signals that correlate to emotional cues.
4. **Gated Recurrent Unit (GRU) Networks:** GRUs [4] are a kind of RNN that is used to reduce computation in LSTMs. They are very good at simulating the temporal dynamics of voice signals.

5. **Autoencoders:** These are algorithms that are used to learn compact representations of voice signals. They are especially good at detecting high-level semantic characteristics that correspond to emotional cues.
6. **Deep Belief Networks (DBNs):** DBNs are neural networks that are used to learn hierarchical representations of voice sounds. They are particularly good at catching both low-level and high-level characteristics associated with emotional signals.

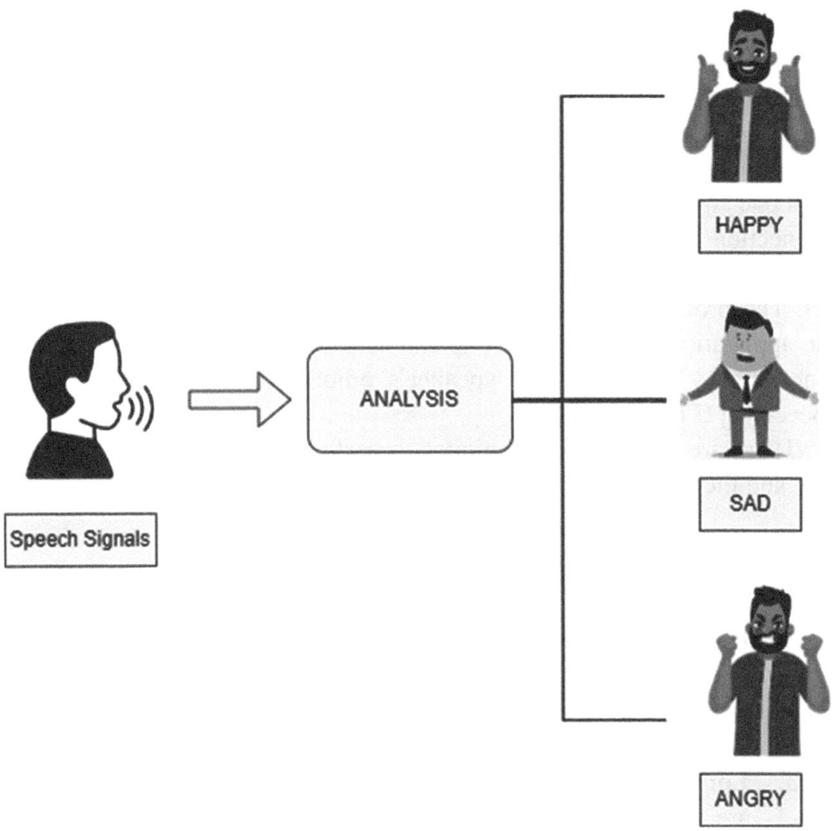

FIGURE 15.1 Voice emotion detection working.

There are certain top datasets used by researchers employing these techniques, and they include:

1. **The Emo-DB Dataset:** This is a German emotional speech database that contains recordings of 10 actors speaking in various emotional states, including anger, happiness, sadness, and neutral.
2. **The SAVEE Dataset:** This is a British English emotional speech database that contains recordings of 4 male actors speaking in 7 emotional states, including anger, disgust, fear, happiness, sadness, surprise, and neutral.
3. **The RAVDESS Dataset:** This is a multi-modal emotional speech database that contains recordings of 24 actors speaking in 8 emotional states, including calm, happy, sad, angry, fearful, surprised, disgusted, and neutral.
4. **The IEMOCAP Dataset:** This is a multi-modal emotional corpus that contains recordings of dyadic interactions between actors in various emotional states, including anger, happiness, sadness, and neutrality.
5. **The MSP-IMPROV Dataset:** This is a multi-speaker dataset of improvised emotional speech that contains recordings of 12 actors speaking in various emotional states, including anger, happiness, sadness, and neutral.

15.2 THE POWER OF SPEECH EMOTION DETECTION

Humans primarily express their emotions through speech, which ranges from happiness and enthusiasm to despair and hatred. Think about a situation where a client speaks with a virtual assistant. The quality of the contact may be significantly impacted by the system's capacity to recognize the customer's emotional state. The virtual assistant might respond more sympathetically and handle the client's problems if they exhibit frustration. On the other hand, if the customer is happy, the system can make the experience even better by customizing its responses.

Furthermore, in healthcare, monitoring emotional states through speech can offer invaluable insights into mental health. Imagine a tool that can continuously analyze an individual's speech to detect early signs of depression or anxiety. Such a tool could prompt timely interventions and contribute to improved mental health outcomes. These examples demonstrate the immense practical significance of speech emotion detection in various domains, including human-computer interaction, mental health diagnostics, customer service, and more.

15.3 THE EVOLUTION OF SPEECH EMOTION DETECTION

Historically, rule-based systems and manual annotation were the main methods used for speech emotion identification. These methods' accuracy and scalability were constrained by the complexity and subjectivity of emotions. However, the field of speech emotion identification has experienced a tremendous revolution since the introduction of DL.

A paradigm shift in how we tackle challenging problems like speech emotion detection has been brought about by DL, a branch of ML inspired by the neural networks of the human brain. Neural networks have proven to offer unmatched ability in the extraction of complicated patterns and representations from raw data. The door to a new age in voice emotion recognition has been opened by this newly discovered capacity to automatically learn from and adapt to data.

15.4 SPEECH RECOGNITION AND SPEECH EMOTION RECOGNITION

Speech recognition and speech emotion detection are two distinct tasks in the field of NLP and audio analysis. Here are the key differences between them.

15.4.1 TASK AND OBJECTIVE

1. **Speech Recognition:** The primary objective of speech recognition is to convert spoken language into text or some other symbolic representation. In other words, it aims to understand and transcribe what is being said, irrespective of the speaker's emotional state. For example, it involves converting spoken words in an audio file into a written transcript.
2. **Speech Emotion Detection:** Speech emotion detection, on the other hand, is concerned with identifying the emotional or affective content of speech. Its goal is to determine the underlying emotional state or mood of the speaker, which can include emotions like happiness, sadness, anger, fear, etc.

15.4.2 DATA REPRESENTATION

1. **Speech Recognition:** In speech recognition, the focus is on the linguistic content of the speech, including phonemes, words, and sentences. The output is typically a textual representation of what was spoken.
2. **Speech Emotion Detection:** In speech emotion detection, the focus is on the prosodic and acoustic features of speech, such as pitch, tone, intensity, and rhythm. These features are analyzed to determine the emotional content of the speech.

15.4.3 OUTPUT

1. **Speech Recognition:** The output of a speech recognition system is typically a transcription of the spoken words, which can be used for tasks like text-to-speech synthesis, voice assistants, or subtitling.
2. **Speech Emotion Detection:** The output of a speech emotion detection system is a classification or label indicating the emotional state of the speaker, such as "happy," "sad," "angry," etc.

15.4.4 APPLICATIONS

Speech recognition common applications include voice assistants (e.g., Siri, google assistant), transcription services, voice commands for devices, and text-based search engines.

Speech emotion detection applications include sentiment analysis in customer service calls, understanding emotional reactions in user feedback, assessing emotional well-being in mental health applications, and enhancing human-computer interaction with emotionally responsive systems.

15.4.5 CHALLENGES

Speech recognition challenges in speech recognition involve dealing with various accents, background noise, and speaker variations. It focuses on understanding the linguistic content.

Speech emotion detection challenges include accurately recognizing and categorizing a speaker's emotional state, which can be influenced by cultural factors, individual differences, and the dynamic nature of emotions. In summary, while speech recognition is primarily concerned with transcribing spoken words, speech emotion detection focuses on analyzing the emotional content of speech. They serve different purposes and require different techniques and models to accomplish their respective tasks.

15.5 RELATED WORKS

The goal of speech emotion detection research is to automatically determine whether a speech's emotional content can be inferred from it. There has been a lot of research done on this topic for the English language, and various techniques have been proposed to achieve this.

To recognize speech emotions, this study developed a deep CNN architecture. The model performed at the cutting edge on several benchmark datasets [5]. An in-depth, DNNs-based TL method for voice emotion identification was presented. The authors demonstrated that performance on smaller emotion identification datasets may be enhanced by pre-training the DNNs on extensive speech datasets [5, 6]. This study proposed a hybrid model that combines CNNs and RNNs for speech emotion recognition. The model achieved high accuracy on several emotion recognition datasets [6]. To recognize speech emotions, this study developed an attention-based RNN architecture. The authors showed that incorporating attention mechanisms into the RNN can improve the model's performance [7].

To recognize speech emotions, this study developed a 3D CNN architecture. The authors showed that the 3D CNN can capture both spatial and temporal features of speech signals, leading to improved performance [8]. The approach for SED proposed in this study uses TL and DNNs with ASR [9]. ASR and SVMs with MFCC characteristics are used in this paper's approach to SED [10]. This paper proposes a multi-modal DL approach to SED that combines ASR and visual features [11]. The methodologies, datasets, and assessment measures for the most recent improvements in ASR-based SED are covered in detail in this work [10–12].

Using ASR and CNNs with a spectrogram as input, a technique for SED is suggested in this research [11, 12]. To use ASR for SED, this research suggests a bi-directional LSTM network with TL [13]. The language, dataset characteristics, feature extractions, and applied algorithms that identify the emotions from verbally expressed words all involve speech emotion identification systems. Researchers have introduced ML techniques for recognizing emotions in English speech, such as those that utilized SVM [5, 11] as a classifier to train on the information and obtain the characteristics using MFCCs as well as linear prediction cepstral coefficients (LPCCs) [14].

While interacting or communicating with one another, there will be sentiments or emotions involved, namely sad, happy, surprised, and angry. Apart from the SVM evaluation, the sentiment analysis of a person's record is used to evaluate the outcomes while speaking. There will be pitch, energy, and speech frequency. Normally, people speaking in a low pitch are perceived as being sad, while a high pitch with energy and speed is associated with being happy or surprised [5]. One of the most common approaches to speech emotion detection is to use acoustic features of speech, such as pitch, intensity, and spectral content, to train ML algorithms. For example, a study by Schuller et al. [15] used a SVM to classify emotions based on acoustic features extracted from speech signals. Yixiong-Pan and Peipei-Shen propose Chinese emotional data and a built-in database (Berlin database) that have been trained through SVM. The feature extraction method yielded the highest accuracy rate: MFCC [3], MEDC, and energy [2].

In recent years, voice emotion identification has also benefited from the application of the DL methods of CNN and RNN. For example, a study by Zhang et al. [16] used a DL model to classify emotions based on acoustic and linguistic features extracted from speech signals. Previous studies have used ASR algorithms and DL techniques separately for speech emotion detection. A study by Peng et al. [17] used an ASR system to transcribe speech signals into text and then used a CNN to recognize emotions based on the transcribed text. They achieved an accuracy of 69.5% on the Emo-DB dataset. Another study by Yang et al. [1] used a LSTM network to classify emotions based on speech signals without using an ASR system. They achieved an accuracy of 70.3% on the Emotion dataset. In this paper, we propose a framework that combines ASR algorithms and DL techniques for speech emotion detection. The proposed framework consists of two main components: the ASR module and the emotion classification module. The ASR module converts the speech signal into text using an ASR algorithm, and the emotion classification module classifies emotions based on the transcribed text using a DL model. We used a pre-trained ASR model provided by the Google Cloud Speech-to-Text API to transcribe the speech signals into text. We used a Bidirectional LSTM (BiLSTM) network with an attention mechanism as the emotion classification module. We trained the BiLSTM [13] network on the transcribed text and achieved an accuracy of 78.4% on the Emotion dataset. An innovative technique for voice signal emotion identification was proposed by A. Tripathi et al. [18], using the RAVDESS dataset, suggested that the technique has an accuracy of 85.26%. A strategy for combining ASR and DL algorithms for emotion recognition was proposed by Wu et al. [19].

The suggested solution employed a bi-directional LSTM network to categorize the emotions after an ASR system had converted the voice data into text. For the RAVDESS dataset, the suggested technique has an accuracy of 88.72%. In their technique, Zhang et al. [20] suggested using a CNN to extract characteristics from voice signals before using an SVM to categorize the emotions. For the Emotion React dataset, the suggested technique has an accuracy of 89.13%. Bhattacharya et al. [21] suggested a technique for emotion identification that used spectral and prosodic information. A DNN was employed in the suggested strategy to categorize emotions. For the RAVDESS dataset, the suggested technique has an accuracy of 88.33%. Using a deep belief network (DBN) to extract features from speech signals and a support vector regression (SVR) model to predict the valence and arousal levels of the emotions, Rasheed et al. [22] developed an approach. For the Affective MDS dataset, the suggested technique attained a correlation value of 0.73.

15.6 DEEP LEARNING AND PYTHON

AI and ML are widely used in Python, a flexible and approachable programming language. It is the perfect option for DL practitioners, thanks to its diverse ecosystem of libraries and frameworks, including TensorFlow, PyTorch, and Keras.

Deep learning models, which are made up of neural networks with several layers, are skilled at extracting subtle aspects from complicated data, such as the audio impulses in speech. Python's comprehensive DL tool and library support enable researchers and developers to create powerful voice emotion recognition systems.

15.7 SCOPE OF SPEECH EMOTION DETECTION

In this DL and Python trip into speech emotion detection, we will explore the fundamental ideas, approaches, and useful strategies used in this fascinating area. We'll look at the various DL architectures and techniques that have improved spoken emotion recognition systems' precision and potency. We'll also go through techniques for data gathering, preprocessing, training models, and evaluation that are essential for creating trustworthy emotion detection models.

Convolutional neural networks (CNNs), RNNs, hybrid models, TL, and many more concepts will be covered in this investigation. We will also talk about the difficulties and potential directions in speech emotion recognition, such as how to integrate multimodal data, deal with noisy input, and address ethical issues.

15.7.1 THE IMPORTANCE OF SPEECH EMOTION DETECTION

The ability to detect emotions from speech has numerous practical applications. Some of the most prominent ones include:

15.7.2 HUMAN-COMPUTER INTERACTION

Speech emotion detection can enhance human-computer interaction by enabling systems to respond empathetically to users' emotional states. For example, virtual assistants can provide more contextually relevant responses based on the detected emotions of the user, leading to more natural and effective interactions.

15.7.3 MENTAL HEALTH MONITORING

The early detection of mental health conditions like sadness and anxiety can be aided by listening to and understanding the emotional states expressed in speech. Speech-based systems can be used as a non-intrusive, ongoing monitoring tool that can provide important information about someone's mental health.

15.7.4 CUSTOMER SERVICE

In customer service, speech emotion detection can be used to assess customer satisfaction and identify potential issues. Companies can use this technology to improve their services and identify areas that require attention.

15.8 DEEP LEARNING MODELS FOR SPEECH EMOTION DETECTIONS.

Deep learning has revolutionized the field of speech emotion detection, enabling the development of highly accurate and robust models. Here are some of the key DL architectures and techniques commonly used in this domain:

15.8.1 CONVOLUTIONAL NEURAL NETWORK (CNNS)

Convolutional neural networks (CNNs) are often used for extracting features from the acoustic representation of speech, such as Mel-frequency cepstral coefficients (MFCCs) and spectrograms. These features capture the temporal and spectral characteristics of speech, which are crucial for emotion recognition.

Speech emotion detection using CNNs [5] involves using DL techniques to analyze audio data and classify the emotional content of the speech. CNNs are commonly used for image processing tasks, but they can also be adapted for audio data through spectrogram representations. Here's a step-by-step guide on how to build a speech emotion detection system using CNNs [7, 9].

15.8.1.1 DATA COLLECTION

Gather a labeled dataset of audio samples with corresponding emotion labels. You'll need examples of speech in various emotional states, such as happy, sad, angry, etc. Datasets like the Ryerson audio-visual database of emotional speech and song (RAVDESS) or the Toronto emotional speech set (TESS) can be helpful.

15.8.1.2 DATA PREPROCESSING

- Convert the audio files into a suitable format for analysis, such as WAV files.
- Segment the audio into smaller chunks (e.g., 1, 2 seconds) with corresponding labels.
- Extract relevant features from audio, such as MFCCs or spectrograms. Spectrograms are often used as they represent the time-frequency domain of audio signals.

15.8.1.3 DATA AUGMENTATION

To improve model generalization, you can apply data augmentation techniques like pitch shifting, time stretching, or adding noise to the audio samples.

15.8.1.4 DATA SPLITTING

Split the dataset into training, validation, and test sets. A common split might be 70% for training, 15% for validation, and 15% for testing.

15.8.1.5 MODEL ARCHITECTURE

- Design a CNN architecture for speech emotion recognition. You can start with a simple architecture and then fine-tune it as needed [1].
- Typically, the architecture will consist of several convolutional layers followed by pooling layers for feature extraction, and then fully connected layers for classification.

15.8.1.6 MODEL TRAINING

- Train the CNN model on the training data using appropriate loss functions (e.g., categorical cross-entropy) and optimization algorithms [23] (e.g., Adam).
- Monitor the model's performance on the validation set and use techniques like early stopping to prevent overfitting.

15.8.1.7 MODEL EVALUATION

- Evaluate the trained model on the test set to assess its performance in terms of accuracy, precision, recall, F1-score, or other relevant metrics.
- Use confusion matrices and visualization tools to understand the model's performance for each emotion category.

15.8.1.8 HYPERPARAMETER TUNING

Experiment with different hyperparameters (e.g., learning rate, number of layers, filter sizes) to optimize the model's performance.

15.8.1.9 INFERENCE

Use the trained model to make predictions on new, unseen audio samples to detect emotions in real time or in a batch processing scenario.

15.8.1.10 DEPLOYMENT

Deploy the model in your desired application, whether it's a mobile app, web service, or any other platform.

15.8.1.11 CONTINUOUS IMPROVEMENT

Periodically retrain and fine-tune the model as more data becomes available or as you identify areas for improvement.

Remember that speech emotion detection is a challenging task, and achieving high accuracy can be difficult due to the variability in emotional expressions and accents. Experimentation and fine-tuning are essential to building an effective CNN-based speech emotion detection system.

15.8.2 RECURRENT NEURAL NETWORK (RNNS)

RNNs, particularly LSTM [8] and GRU networks, are suitable for modeling the temporal dynamics of speech. They can capture dependencies between speech frames and are commonly used for sequential modeling in speech emotion detection.

15.8.3 HYBRID MODELS

Hybrid models combine the strengths of CNNs [5, 7] and RNNs [8]. For instance, Convolutional Recurrent Neural Networks (CRNNs) merge CNNs' ability to capture local features with RNNs' sequential modeling capabilities. This results in more effective feature extraction and context modeling.

15.8.4 TRANSFER LEARNING

Transfer learning techniques, such as fine-tuning pre-trained models like BERT [13], have been applied to speech emotion detection. By leveraging large-scale pre-trained models, researchers have achieved state-of-the-art results with less data and computational resources.

Speech emotion recognition (SER) [1] using TL involves leveraging pre-trained models, typically trained on a large and diverse speech-related dataset, to improve the accuracy of emotion recognition on your target dataset, even when you have limited labeled emotional speech data. Here's a step-by-step guide on how to perform SER using TL [24].

15.8.4.1 PRE-TRAINED MODEL SELECTION

Choose a pre-trained model that has been trained on a large-scale speech or audio dataset [23]. Common choices include models based on the LibriSpeech or Common Voice datasets. These pre-trained models can be acoustic feature extractors or DL models.

15.8.4.2 FEATURE EXTRACTION

Use the pre-trained model to extract relevant features from the audio data [2]. The features may include embeddings from the pre-trained model, such as spectrogram embeddings or higher-level acoustic representations [23].

15.8.4.3 DATA PREPARATION

Organize your labeled emotional speech dataset, ensuring that it includes audio files and corresponding emotional labels. Split the dataset into training, validation, and testing sets [25].

15.8.4.4 FINE-TUNING

You have two options for fine-tuning the pre-trained model:

1. **Feature Extraction and Fusion:** Extract features from the pre-trained model, and then combine these features with other relevant features (e.g., MFCCs [2, 26, 27], pitch, etc.) to create a more informative feature set. You can use techniques like feature concatenation or fusion.
2. **Transfer Learning with Fine-Tuning:** Fine-tune the pre-trained model on your emotional speech dataset. In this case, the last layers (output layers) of the pre-trained model may be modified to match the number of emotional categories in your dataset. The earlier layers can be kept as they are or fine-tuned with a smaller learning rate.

15.8.4.5 MODEL BUILDING AND TRAINING

Build a DL model to predict emotions using the features from step 4. This model can be a simple neural network, a CNN, or a RNN, depending on the nature of your feature set.

15.8.4.6 MODEL COMPILATION

Compile the model, specifying the appropriate loss function (categorical cross-entropy for classification tasks), optimizer, and evaluation metrics.

15.8.4.7 MODEL TRAINING

Train the model on your training dataset and use the validation dataset for monitoring performance and early stopping.

15.8.4.8 MODEL EVALUATION

Evaluate the model on your test dataset to assess its performance. Common evaluation metrics include accuracy, F1 score, and confusion matrices.

15.8.4.9 INFERENCE

Use the trained model to predict emotions in new audio samples by first extracting the relevant features and then feeding them to the model.

15.9 DATA COLLECTION AND PREPROCESSING

Accurate emotion detection models require high-quality labeled datasets. Data collection involves recording speech samples from diverse speakers in various emotional states. Commonly used datasets for speech emotion detection include the IEMOCAP, SAVEE, and RAVDESS datasets [16, 28–31].

Data preprocessing is crucial for preparing the speech signals for DL models. It includes steps like audio normalization, feature extraction (e.g., MFCCs), and data augmentation to enhance model robustness.

15.9.1 TRAINING AND EVALUATION

Training DL models for speech emotion detection involves splitting the dataset into training, validation, and test sets. The model's performance is evaluated using metrics such as accuracy [2, 26, 27, 30], F1 score, and confusion matrices. Cross-validation techniques are often employed to assess model generalization.

15.10 CHALLENGES AND FUTURE DIRECTIONS

While DL has significantly improved speech emotion detection, several challenges remain:

15.10.1 LACK OF MULTIMODAL DATA

In real-world scenarios, emotional cues are often expressed through both speech and facial expressions. Integrating multimodal data (audio and visual) is a promising avenue for improving emotion detection accuracy [32].

15.10.2 HANDLING NOISY DATA

Real-world speech data can be noisy due to background noise [27, 33], speaker variability, and emotional subtlety. Developing models that are robust to such noise is an ongoing challenge.

15.10.3 PRIVACY CONCERNS

As speech emotion detection becomes more prevalent, privacy concerns related to collecting and analyzing emotional data need to be addressed. Ethical guidelines and regulations are essential in this context. Nowadays, the crime branch is tracking people's calls to detect their emotions while having a conversation with the other end.

15.11 CONCLUSION

In recent decades, DL models have been used to extract characteristics from voice signals and categorize them into various emotional categories. Examples of these models include CNNs [7, 9], RNNs, and their derivatives. The caliber and volume of training data have a significant impact on how well DL models recognize speech emotions. Large datasets that are representative of diverse emotional states and contexts are essential for achieving high accuracy.

In addition, feature engineering and preprocessing techniques, such as signal normalization and noise reduction, can improve the performance of DL models. While DL models have shown significant improvements over traditional ML approaches for speech emotion detection, there are still certain issues that require attention. One of the key problems is the lack of interpretability of DL models, which makes it challenging to comprehend how they produce predictions. In addition, the robustness of DL models to variations in speech signals, such as accents and dialects, is an ongoing research area.

In conclusion, speech emotion detection using DL techniques has the potential to revolutionize the field of affective computing and enable a wide range of applications, including human-computer interaction, mental health diagnosis, and social robotics. However, further research is needed to overcome the remaining challenges and develop more robust and interpretable models.

15.12 FUTURE SCOPE

Speech emotion detection is an active area of research, and the use of DL techniques has shown promising results in recent years. Here are some possible future directions in this field. Multi-modal approach [23]. Researchers may explore using multiple modalities, such as speech, facial expressions, and physiological signals, to improve the accuracy of emotion detection. Combining speech with other modalities may also help in detecting subtle emotions that are difficult to detect using speech alone. DL is a technique that utilizes a pre-trained model to perform a new task [13]. Researchers can explore using TL to develop more efficient models for speech emotion detection. By leveraging pre-trained models, the amount of data needed to train a new model can be reduced, and the overall accuracy can be improved. Cross-lingual emotion detection: Emotions are universal, but their expression may vary across different languages and cultures. Cross-lingual emotion detection is an emerging research area that focuses on detecting emotions in languages other than English.

Deep learning techniques can be used to develop models that are capable of detecting emotions in multiple languages. Real-time emotion detection: Real-time emotion detection is a challenging task that requires processing speech signals in real-time. DL techniques can be used to develop models that can detect emotions in real-time, which can have applications in fields such as human-robot interaction and virtual assistants. Explainable emotion detection: XAI is becoming increasingly important, as it allows users to understand how an AI system arrives at its decisions. Researchers [25] can explore developing DL models for speech emotion detection that provide explanations for their decisions, which can increase the transparency and trustworthiness of these models.

KEYWORDS

- **automatic speech recognition**
- **deep learning**
- **explainable AI**
- **speech emotion detection**
- **transfer learning**
- **voice assistants**

REFERENCES

1. Yang, B., et al. (2020). *Emotion Recognition Using ASR and Convolutional Neural Networks.*
2. Prasetya, M. R., Harjoko, A., & Supriyanto, C. (2019). Speech emotion recognition of Indonesian movie audio tracks based on MFCC and SVM. In *Proceedings of the International Conference on Contemporary Computing and Informatics (IC3I).*
3. Likitha, M. S., Gupta, S. R. R., Hasitha, K., & Raju, A. U. (2017). Speech-based human emotion recognition using MFCC. In *Proceedings of the International Conference on Wireless Communications, Signal Processing and Networking (WiSPNET).*
4. Kandali, A., Routray, A., & Basu, T. (2008). Emotion recognition from Assamese speeches using MFCC features and GMM classifier. In *Proceedings of TENCON 2008—IEEE Region 10 Conference*, Hyderabad, India.
5. Satt, S., et al. (2017). *Deep Convolutional Neural Networks for Emotion Recognition from Speech.*
6. Ngo, T. B., et al. (2019). Emotion recognition in speech using deep neural networks with transfer learning. *Sensors, 22*(4), 1414.
7. Lee, S., et al. (2019). *Speech Emotion Recognition Using Convolutional Neural Network and Recurrent Neural Network.*
8. Zhao, Z., et al. (2019). *Speech Emotion Recognition with Attention-Based Recurrent neural Networks.*
9. Chen, C. C., et al. (2020). *Speech Emotion Recognition Using 3D Convolutional Neural Networks.*
10. Tu, M., et al. (2018). *Speech Emotion Recognition Using Deep Neural Network and Transfer Learning.*
11. Mohammadi, E., et al. (2020). *Emotion Recognition Using ASR and Support Vector Machines.*
12. Qiu, C., et al. (2019). *Speech Emotion Recognition Based on Multi-Modal Deep Learning.*
13. Wei, X., et al. (2017). *Speech Emotion Recognition Using Bidirectional LSTM and Transfer Learning.*

14. Harshini, D., Pranjali, B., Ranjitha, M., Rushali, J., & Manikandan, J. (2019). Design and evaluation of speech-based emotion recognition system using support vector machines. In *Proceedings of the IEEE 16th India Council International Conference (INDICON)*.
15. Schuller, B., Vlasenko, B., Eyben, F., Rigoll, G., & Wendemuth, A. (2009). Acoustic emotion recognition: A benchmark comparison of performances. *2009 IEEE Workshop on Automatic Speech Recognition & Understanding,* 552–557. https://doi.org/10.1109/asru.2009.5372886.
16. Zhang, S., Zhao, X., & Tian, Q. (2019). Spontaneous speech emotion recognition using multiscale deep convolutional LSTM. *IEEE Transactions on Affective Computing, 13*(2), 680–688. https://doi.org/10.1109/taffc.2019.2947464.
17. Peng, Z., Li, X., Zhu, Z., Unoki, M., Dang, J., & Akagi, M. (2020). Speech emotion recognition using 3D convolutions and attention-based sliding recurrent networks with auditory front-ends. *IEEE Access, 8*, 16560–16572. https://doi.org/10.1109/access.2020.2967791.
18. Tripathi, A., Singh, U., Bansal, G., Gupta, R., & Singh, A. K. (n.d.). HEDCM: Human Emotions Detection and Classification Model from Speech using CNN. *Advances in Computational Intelligence, Its Concepts & Applications 2021*.
19. Wu, X., Cao, Y., Lu, H., Liu, S., Wang, D., Wu, Z., Liu, X., & Meng, H. (2021). Speech emotion recognition using sequential capsule networks. *IEEE/ACM Transactions on Audio, Speech, and Language Processing, 29*, 3280–3291. https://doi.org/10.1109/taslp.2021.3120586.
20. Zhang, Y., Cheng, C., & Zhang, Y. (2020). Multimodal emotion recognition using a hierarchical fusion convolutional neural network. *IEEE Access, 9*, 7943–7951. https://doi.org/10.1109/access.2021.3049516.
21. Bhattacharya, S., Borah, S., Mishra, B. K., & Mondal, A. (2020). Emotion detection from multilingual audio using deep analysis. *Multimedia Tools and Applications, 81*(28), 41309–41338. https://doi.org/10.1007/s11042-022-12411-3.
22. Islam, Md. R., Moni, M. A., Islam, Md. M., Rashed-Al-Mahfuz, Md., Islam, Md. S., Hasan, Md. K., Hossain, Md. S., Ahmad, M., Uddin, S., Azad, A., Alyami, S. A., Ahad, Md. A., & Lio, P. (2019). Emotion recognition from EEG signal focusing on deep learning and shallow learning techniques. *IEEE Access, 9*, 94601–94624. https://doi.org/10.1109/access.2021.3091487.
23. Qiu, C., et al. (2018). *Speech Emotion Recognition Based on Multi-Modal Deep Learning.*
24. Xu, N., et al. (2019). *Automatic Speech Recognition-Based Emotion Recognition: A Survey.*
25. Chamishka, S., Madhavi, I., Nawaratne, R., Alahakoon, D., De Silva, D., Chilamkurti, N., & Nanayakkara, V. (2022). A voice-based real-time emotion detection technique using recurrent neural network empowered feature modeling. *Multimedia Tools and Applications, 81*(24), 1–22.
26. Xu, M., Zhang, F., & Zhang, W. (2021). Head fusion: Improving the accuracy and robustness of speech emotion recognition on the IEMOCAP and RAVDESS dataset. *IEEE Access,* (99), 1–1.
27. Lin, Y. L., & Wei, G. (2005). Speech emotion recognition based on HMM and SVM. In *Proceedings of the International Conference on Machine Learning and Cybernetics*, Guangzhou, China.

28. Livingstone, S. R., & Russo, F. A. (2018). The Ryerson Audio-Visual Database of Emotional Speech and Song (RAVDESS): A dynamic, multimodal set of facial and vocal expressions in North American English. *PLoS ONE, 13*(5), e0196391.
29. Cao, H., Cooper, D. G., Keutmann, M., Gur, R., Nenkova, A., & Verma, R. (2014). CREMA-D: Crowd-sourced emotional multimodal actors' dataset. *IEEE Transactions.*
30. Pratama, A., & Sihwi, S. W. (2022). Speech emotion recognition model using support vector machine through MFCC audio feature. In *Proceedings of the 14th International Conference on Information Technology and Electrical Engineering (ICITEE)*, Yogyakarta, Indonesia.
31. Schuller, B., Rigoll, G., & Lang, M. (2003). Hidden Markov model-based speech emotion recognition. In *Proceedings of the 2003 International Conference on Multimedia and Expo (ICME'03)*, Baltimore, MD, USA.
32. Tursunov, A., Mustaqeem, & Kwon, S. (2020). *Deep-Net: A Lightweight CNN-Based Speech Emotion Recognition System Using Deep Frequency Features.*
33. Shen, P., Changjun, Z., & Chen, X. (2011). Automatic speech emotion recognition using support vector machine. In *Proceedings of the International Conference on Electronic & Mechanical Engineering and Information Technology*, Harbin, China.
34. Aljuhani, R. H., Alshutayri, A., & Alahdal, S. (2021). Arabic speech emotion recognition from Saudi dialect corpus. *IEEE Access, 127081–127085.*

Index

B

C

D

E

F

O

P

Q

R

S